EVOLUTION AND EXTINCTION RATE CONTROLS

FLUX, FLUX, ALL IS FLUX

Project:

ECOSTRATIGRAPHY

Developments in Palaeontology and Stratigraphy, 1

EVOLUTION AND EXTINCTION RATE CONTROLS

Arthur J. Boucot

Department of Geology, Oregon State University, Corvallis, Ore., U.S.A.

ELSEVIER SCIENTIFIC PUBLISHING COMPANY

Amsterdam, Oxford, New York, 1975

ELSEVIER SCIENTIFIC PUBLISHING COMPANY
335 Jan van Galenstraat
P.O. Box 211, Amsterdam, The Netherlands

AMERICAN ELSEVIER PUBLISHING COMPANY, INC.
52 Vanderbilt Avenue
New York, New York 10017

Library of Congress Card Number: 73-85219

ISBN 0-444-41182-8

Printed in The Netherlands

PREFACE

Traditionally the role of paleontology in the study of evolution has been confined almost entirely to the working out of phylogenetic relations based on fossils. These phylogenetic relations have been studied by means of both changes in time of morphology of closely related forms, and of changes in morphology evidenced in the ontogeny of individual specimens, where this is apparent, and changes during ontogeny shown by suites of specimens belonging to the same species, where changes occurring during growth are not evident on individual specimens. A few studies employing fossils have been directed at the problems of rates of evolution, employing progressive changes in morphology (morphologic evolution). A few studies employing fossils have also been directed at the problems of rates of evolution employing family trees (phyletic and cladogenetic evolution), and employing number of taxa per time interval (taxonomic evolution) for particular groups.

The purpose of this treatment is to take the information about the areas occupied by individual taxa during specific time intervals (on the assumption that the area occupied will be a direct function of population size[1]), and derive from these data conclusions about rate of evolution as a function of size of interbreeding population. Shallow-water marine benthos of the Lower Paleozoic are well suited to this type of investigation because of their widespread occurrence on the continents during the Lower Paleozoic as contrasted with later time intervals where restriction to far smaller areas on continental shelves for the most part make the sampling problem more difficult. Shallow-water marine benthos are better suited for such studies than non-marine vertebrates because of the far greater abundance and wider distribution of the former.

The following factors that influence the size of interbreeding populations for the Lower Paleozoic (in terms of area occupied by taxa) are considered: location and size of biogeographic units, location and size of animal communities, areas of the continents subject to shallow-water marine sedimentation (regression and transgression), climatic factors ("warm"-water and

---

[1]Population size is defined as the worldwide number of individuals belonging to a taxon.

"cold"-water regions), areas of hypersaline conditions, and areas occupied by reef-like environments. Populations have also been considered in the time dimension, as there is a strong correlation between taxa with widespread geographic distribution, numerical abundance of specimens and time duration (cosmopolitan taxa persist longer!). The interplay of these factors helps to determine the size of interbreeding populations.

The overall conclusion arrived at in this treatment is that the first-order factor governing rate of evolution (taxonomic, cladogenetic and phyletic evolution) is size of interbreeding population. Small populations are concluded to contain taxa that evolve more rapidly than do large populations. This conclusion has, of course, been first worked out long ago using living material but fossils have not been used for this purpose on a broad scale. Geneticists studying rate of change of gene frequencies and students of population genetics have contributed to the problem, following in the footsteps of the nineteenth-century naturalists who first recognized the relation.

The companion question of the controls governing rate of extinction is also considered. It is concluded from studying Late Ordovician through Late Devonian data that there is strong evidence for differing rates of extinction correlating with marked changes in provincialism (the change from high to low may coincide with increased extinction rate), between marked changes in climate (strong climatic gradients as opposed to weaker climatic gradients), development of fewer communities through the evolution of taxa with wider niche breadth, elimination of reef environments, and also of profound physical events whose nature is not understood at present.

The information from the geologic record is vast in its geographic and time extent. It offers biologists interested in problems of evolution and extinction a large sample unlike those of the present. Is it fair to compare the test tube full of fruit flies   and the worldwide fossil distribution pattern to the microscope and the telescope: can differing conclusions bearing on the same basic problems be drawn from them — I have a hunch that they can. The statistical averaging of these samples is analogous in some ways to those of the thermodynamicist. Many questions difficult or impossible to study with living material, may be considered with the aid of the fossil sample. The geologist may greatly improve both the quality of his stratigraphic correlations, on which so much of geologic interpretation rests, and his understanding of them by paying more attention to the type of information dealt with here.

The emphasis in much of paleontology during the past has been on faunal similarities. Similarities have been emphasized in order to provide the geologist with a more reliable time-stratigraphic framework, as well as to provide the paleontologist with a better framework against which to unravel the progress of organic evolution. Today it is becoming ever more important to also emphasize faunal differences. The emphasizing of faunal differences permits us to study contemporary questions of biogeography and community ecology. These two approaches reinforce each other if done concurrently; they should not be considered as opposing or contradictory in tendency. Any effort to extract all the data provided for us by the fossil record in terms biologic and geologic must first weigh similarities and then differences for the information that they convey. Failure to consider both similarities and differences leaves one open to making errors in time-stratigraphic, biogeographic and community-ecology studies.

The thrust of my treatment is to employ as much information from the geologic record as is available in attempting to understand these basic questions of evolutionary rates and extinction rates. We have available to us data about the size of biogeographic entities, the size of animal community entities, numbers of taxa, the abundance of reefs, the area of the evaporitic facies, the area occupied by marine sedimentation, and the times of occurrence of marked physical events such as glaciation, changes in climate and the regions occupied by animal communities as well as their total numbers in time. Information about the size of potentially interbreeding population is available for all time intervals of the Phanerozoic. Why not use it?

The evidence developed in this treatment strongly supports Mayr's (1970, p.345) statement: "... it is becoming increasingly clear that population size is by far the most important determinant of genetic change. The smaller the population, other things being equal, the greater the probability of rapid genetic change. The fossil record, however, favors widespread species rich in individuals. These, we now suspect, are an exceptional group of species with by far the slowest rates of evolution and speciation. The fossil record, thus, represents a badly skewed sampling of species and the chances are all against the fossil preservation of remnants of a rapidly speciating, small geographic isolate." Mayr's point is well taken for the great bulk of the specimens available to us, but there are enough specimens representing a middle ground to strongly support his concept of the importance of small population size for rapid speciation.

No mention is made in this treatment of continental drift, plate tectonics and like questions because it is not felt that the biologic evidence presented here has much direct bearing on the problems involved. This view will be taken amiss by the inflamed advocates on both sides of the many interesting questions, but after taking what I consider to be a dispassionate view of the subject I see nothing gained by claiming that the permissive data collected here greatly strengthens (or weakens) the stands taken by either side. Animals have the capability of moving about in a variety of ways that may not be directly related to either the stability or instability of portions of the earth's crust; many of these ways are still poorly understood. Biogeographic boundaries common to more than one continent may be interpreted in more than one manner (i.e., consistent with various plate-tectonic models or inconsistent with them using the same data). Our knowledge of worldwide taxonomic diversity is not yet complete enough to afford unique solutions to these questions. Future accretion of new information may enable the biologic data to be decisive in resolving some of the questions of this type, but the information presented in this volume is still ambiguous in my opinion except as regards a lowered probability for extensive north—south continental movements since the Silurian.

I hope that the treatment presented here will encourage students of animal communities and their functioning to work more closely with those concerned with problems of biogeography. In fact, it is my hope that these groups will cease thinking of the problems of community ecology as something apart from problems of biogeography; the two questions are inextricably welded together.

The biogeographer's task has been made far easier in the last three years by the appearance of *Geol. J., Spec. Issue*, 4 (Middlemiss et al., 1971) entitled *Faunal Provinces in Space and Time*, Elsevier's *Atlas of Palaeobiogeography* (Hallam, 1973) and the *Palaeontol. Assoc. Spec. Pap.*, 12 (Hughes, 1973) entitled *Organisms and Continents Through Time*. All three publications have extensive coverage devoted to the biogeography of shallow-water marine invertebrates in addition to related topics.

I have been heavily influenced in my thinking, during the course of preparing the manuscript, by the work of E. Mayr and G.G. Simpson. Their comments about how evolution appears to have proceded under natural conditions have been of great value to me, although they should not be held to blame for any misunderstanding or errors I have committed.

The book is organized into a part I, containing four Chapters that deal primarily with the problems of evolution and extinction rates, followed by a Part II, containing three Chapters that deal primarily with supporting paleoecologic and biogeographic data taken from the Silurian-Devonian record, terminated by a Part III, Summary.

I am greatly indebted for assistance in this work to a large number of people. My studies began in 1948 and I have received help and advice from a very large number of colleagues and friends. The following list only conveys thanks to those who specifically assisted with the present treatment; it by no means conveys my debt to the far larger number who have helped over the years.

I am very grateful to Edwin J. Anderson, Temple University, Philadelphia; W.B.N. Berry, University of California, Berkeley; P. André Bourque, Dept.

of Natural Resources, Quebec;Peter Bretsky, State University of New York, Stony Brook; Preston E. Cloud, Jr., University of California, Santa Barbara; John M. Dennison, University of North Carolina, Chapel Hill; Richard Flory, Peter Isaacson and David Rohr, Oregon State University, Corvallis; Günter Fuchs, Landessammlungen für Naturkunde Karlsruhe, Karlsruhe; Edmund Gill, National Museum of Victoria, Melbourne; Jane Gray, William Holser and Norman Savage, University of Oregon, Eugene; Charles W. Harper, Jr., University of Oklahoma, Norman; Erle Kauffman, U.S. National Museum, Washington; Helen Tappan Loeblich, University of California, Los Angeles; Anders Martinsson, Department of Palaeobiology, Uppsala University, Uppsala; Ernst Mayr, Museum of Comparative Zoology, Harvard University, Cambridge, Massachusetts; Stuart McKerrow, Oxford University, England; Arthur A. Meyerhoff, A.A.P.G.,Tulsa; William A. Oliver, Jr., U.S. Geological Survey, Washington, D.C.; Allen Ormiston, Amoco Production Company, Tulsa, Oklahoma; Allison R. Palmer, State University of New York at Stony Brook; Alfred Potter, Oregon State University, Corvallis; Alan Shaw, Amoco Production Company, Denver; Lawrence Rickard, New York State Museum and Science Service, Albany; Peter Sheehan, University of Western Ontario, London; George Gaylord Simpson, The Simroe Foundation, Tucson; John A. Talent, Macquarie University, North Ryde, N.S.W.; James Valentine and Jere Lipps, University of California at Davis; Peigi Wallace, Imperial College, London; Victor Walmsley, University of Wales, Swansea; Rodney Watkins, Oxford University, England; Anthony Wright, Wollongong University College, Wollongong, N.S.W.; Dan H. Yaalon, Hebrew University of Jerusalem, Jerusalem; Alfred M. Ziegler, University of Chicago, Chicago; and an anonymous critic whose acid-tipped pen I am very grateful to and have tried to benefit from; for carefully reviewing and criticizing the manuscript, and helping me to understand better the problems dealt with, though these persons should in no way be considered responsible for any mistakes or confusion that may be present.

Messrs. Grover Murray, Lubbock, Texas and Ing. J. Carillo B., Petroleos Mexicanos, Tampico, kindly provided specimens of *Baturria mexicana* from Tamaulipas. Peter Carls, Wurzburg, permitted me to use advance information dealing with his new genus *Baturria*. I am indebted to William A. Oliver, Jr., U.S.G.S., Washington, for identifying a Schoharie coral from northern Alabama kindly provided by Thomas J. Carrington, Auburn University, Auburn. A.W. Norris, Geological Survey of Canada, Calgary, kindly provided information about the Hudson Platform and the Williston Basin. In addition, Norris critically read the manuscript, as did Hans Trettin, Helen Belyea, and Brian Norford, Geological Survey of Canada, Calgary. F.G. Poole, U.S. Geological Survey, Denver, kindly reviewed and criticized the paleogeographic information regarding the western U.S., as did John Dennison, University of North Carolina, Chapel Hill, for the Central and Southern Appalachians. I am indebted to Leif Størmer, Oslo University, Oslo for his well based opinion that the Kjellesvig-Waering eurypterid communities conform closely to the data available to Størmer. Robert M. Garrels, University of Hawaii, Honolulu, generously made available some concepts dealing with the process of dolomitization in bottom sediment.

I am very indebted to the following paleontologists for their assistance in checking the information presented in Fig.27: John Temple, Birkbeck College, London, U.K.; Jan Bergstrom, Lund University, Lund, Sweden; J.G. Johnson, Oregon State University, Corvallis; John Carter, Carnegie Museum, Pittsburgh, Pennsylvania; Thomas W. Amsden, Oklahoma Geological Survey, Norman; J.T. Dutro, U.S. Geological Survey, Washington; V.G. Walmsley, University of Wales, Swansea; L.R.M. Cocks, British Museum (Natural History), London; Paul Sartenaer, Institut Royal des Sciences Naturelles de Belgique, Brussels; Wolfgang Struve, Natur-Museum Senckenberg, Frankfurt am Main; Valdar Jaanusson, Riksmuseet, Stockholm; and A.D. Wright, Queen's University of Belfast, Belfast, Northern Ireland.

I am indebted to Professor T.H. van Andel, School of Oceanography, Oregon State University, for providing me with the results of his rich experience regarding the distribution of bivalves on the modern deep-sea floor as well as with an appreciation of the virtual absence of bivalves from Cenozoic-age deep-sea cores; additionally he has encouraged me in a willingness to consider that the calcium-carbonate compensation depth of the past may have varied considerably from its present position with a position near the shelf margin not necessarily being absurd during some time intervals.

Last, but by no means least, I am indebted to the patience of my students with whom I have exhaustively discussed piecemeal the questions treated here over the years. Their questions and even "lack of understanding" have greatly assisted me in trying to formulate the thoughts presented here. I hope I have puzzled and confused them enough to have them go ahead and straighten things out.

CONTENTS

Part III. Summary

PART I.
EVOLUTION AND
EXTINCTION RATE CONTROLS

CHAPTER 1

Introduction

Since Darwin's time, biologists have been aware that speciation can proceed more rapidly in small, isolated populations than in large ones as exemplified by the contrast between populations on oceanic islands and nearby continents (e.g., Darwin's finches). The concept that evolution may proceed most rapidly in small populations has been developed over the years almost entirely from study of living examples taken from such specialties as population genetics and island biogeography. This book presents evidence from the fossil record in support of this concept and demonstrates that what we can conclude to be happening today has happened repeatedly in the past. Incorporated here is the necessary third dimension, the distribution of populations not only in a single time plane, the present, but in geologic time.

Experience with Silurian-Devonian brachiopods since 1948 on a world-wide basis has gradually brought me to an awareness of the relation between area occupied by populations and rate of evolution. In this treatment "rate of evolution" refers chiefly to production rate of taxa, mostly genera, both related and unrelated. This production rate of taxa is Simpson's (1953) "taxonomic evolution". In a few instances we have information about the phyletic relations between taxa in time that enables one to assess rate of evolution using Simpson's (1953) term "phyletic evolution". In addition, consideration is given to the rate of "branching" of closely related taxa (cladogenetic evolution, following Mayr, 1965) independent of the time duration of the individual taxa. Simpson's (1953) term "morphologic evolution" for rate of change of morphology in a phyletically related group has not been employed, nor has attention been paid to mosaic evolution within the various groups. The outlining of biogeographic entities makes it clear that production rate of new taxa (as distinguished from appearance rate of taxa migrating from unknown sources) is greatest during times of greatest provincialism and consequent greater abundance of isolated communities than during times of cosmopolitanism.

Fig.1 diagrams the factors which strongly correlate with population size, that may be determined by recourse to information present in the

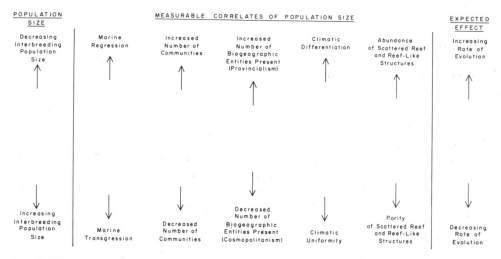

Fig. 1. Diagram showing effects on rate of evolution and size of interbreeding population by a number of relatively independent factors. When all of these factors point in one direction for a time interval they strongly reinforce each other, but when they point in differing directions, they tend to result in intermediate rates of evolution and sizes of interbreeding population.

geologic record. These factors, whose relative importance to each other changes from time interval to time interval, are marine regression and transgression; the number of animal communities present; the number of biogeographic entities present (related, of course, to the number and position of barriers to faunal migration present at any one time); the presence of a uniform or a highly differentiated climatic regime; and the rarity or abundance of reefs and reef-like structures. All of these factors may correlate strongly with the size of interbreeding populations, as well as the actual number of interbreeding populations. During some time intervals these factors are disposed so as to reinforce each other, whereas in other time intervals they oppose each other to a greater or lesser degree.[1] The resulting rates of evolution and extinction for any one time interval as a whole, as well as for individual population units, are the resultant derived from the interplay of these factors. It appears that the intensity and disposition of these factors at any one time interval has never been quite the same as for any other time interval. This conclusion is also apparent in that during times of both cosmopolitanism and provincialism production rate of new taxa appears to have been greatest within communities occupying small rather than large areas ("small" and "large" are used here as the total, worldwide area, including all patches). If one correlates area occupied by an interbreeding population with population size then the

_____

[1] For list of footnotes, see pp. 377—394.

conclusion is obvious that rate of evolution has varied inversely with population size as has been concluded for living organisms. There is a strong tendency towards higher rates of evolution among shallow-water benthic organisms that thrived in the epicontinental seas and their geosynclinal margins during times of regression with consequent smaller-size populations than during times of wide-spread marine transgression *if* the other factors governing the size of interbreeding populations remain constant. Strongly differentiated climatic regimes act to divide up the available shallow-water regions in much the same manner as barriers to migration and, in fact, strongly differentiated climatic regimes may be considered to provide a special type of barrier. The presence or absence of scattered reefs and reef-like bodies affect the total rates of evolution by providing specialized, peripheral environments where small, endemic, reef and reef-related populations may develop. These reef-type environments may be thought of as a special type of peripheral environment. Valentine (1968) has been thinking along similar lines in explaining evidence derived in large part from the Cenozoic of western North America. This work is a summary of the biogeographic, animal community, reef, climatic, regression-transgression and faunal-barrier distribution data on which my conclusions concerning the first-order control over rate of evolution exercized by population size are based.

Most of the material concerning brachiopods employed in this treatment discusses genera rather than species. This is the case because our knowledge of brachiopod genera is reliable on a worldwide basis, whereas the species belonging to the various genera are not well enough known over the world to form an adequate basis for discussion. It might be argued that many of the "genera" considered in this book correspond to what a neontologist would call a species, but the general experience of taxonomists dealing with fossils tends to confirm that the species of the paleontologist approximate those of the neontologist.

It is also necessary in arriving at these conclusions to consider the worldwide paleogeographic and lithofacies reconstruction developed for the Silurian during preparation of correlation charts for the period, and similar information for the Devonian. Moreover, it is necessary to consider some of the possible barriers to faunal migration potentially responsible for permitting the development of provincial conditions. Large masses of land and of hypersaline water on the platforms are two such potential faunal barriers. The presence and position of currents of sea water having distinct temperature characteristics and relations to hypersaline water masses and land masses is also considered to have great potential in explaining the occurrence of barriers to faunal migration.

Since our (Boucot et al., 1969) summary of Early Devonian brachiopod

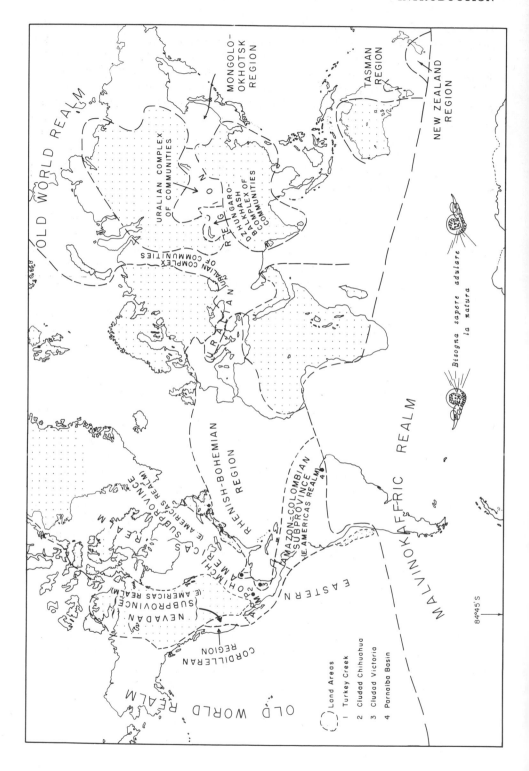

zoogeography, Johnson's (1971a) Middle Devonian brachiopod zoogeographic summary, and my Siluro-Devonian community occurrence and distribution summary (Boucot, 1970) additional information has become available for the Devonian (Fig.2).

The size of interbreeding populations is viewed with these data as far more important in determining rates of evolution than either taxonomic diversity within a community or the factors that determine it. This conclusion, based on Paleozoic worldwide samples agrees well with conclusions based on living material (Mayr, 1963).

The question of food supply [83] is certainly important as a control for size of interbreeding populations. But, it is a difficult variable for the geologist to evaluate. The best that could be done is to consider, on the assumption that most limestone is of biogenic origin, that the original total volume of limestone (and replacement dolomite) per time interval (prior to later erosion) is a direct function of biomass and ultimately of food supply for benthic marine organisms. If one admits the logic of this conclusion then the Cambro-Silurian would have been a time of greater food supply than the post-Silurian. Within the Devonian it would be clear from the relative limestone abundances that the Early Devonian was a time of lower food supply than the Middle Devonian than the Late Devonian. It is also clear, following this logic, that higher trophic-level animals should generally evolve more rapidly than lower trophic-level organisms. If, however, total amount of limestone is not a valid measure of total biomass and food supply then these conclusions do not follow. For example, we have no good measure of the admittedly large volume of organic-walled microfossil plankton (acritarchs, chitinozoans, etc.) that may have been an important component of the Lower Paleozoic marine food supply. An even more serious objection is the probability that the bulk of the organogenic calcium carbonate deposited on the platforms will have been dissolved penecontemporaneously thus rendering the remaining sample meaningless as an estimate of original biomass. This being the case, no serious effort has been directed toward measuring the volume of limestone per time interval.

Finally, it must be emphasized that although the conclusions presented here strongly indicate that size of interbreeding populations is the first-

---

Fig. 2. Generalized Early Devonian Biogeographic Realm and Region. (Boundaries modified after Boucot et al., 1969, with additions.)

The shoreline in British Columbia and adjacent areas of the Yukon may now be moved further to the east into a position at least fifty miles east of the Rocky Mountain Trench owing to the find of Siegen-Ems age brachiopods (including well developed *Cortezorthis*) in the Lloyd George Icefield region and of Siegen-Ems monograptids further south.

order control on rate of evolution for animals it by no means indicates that it is the only control. There are a host of biological and physical factors that influence rate of evolution very significantly at a lower level. After normalizing for the effects of differing population sizes we may look forward to the time when the relative effects of these many other factors may be evaluated more effectively.

The theme of this treatment is that the first-order control over rate of evolution is size of interbreeding population. The size of interbreeding populations is governed by a number of factors (Fig.1) that vary in time. Therefore, our ability to estimate the absolute time duration of various intervals is critical to the argument, and is dealt with subsequently.

Estimating Size of Interbreeding Population for a Species

Throughout this treatment I estimate the size of an interbreeding population by reference to the area in which the species occurs. Such an estimate is critical to the argument about the relation between rate of evolution and size of interbreeding population. Just how are these areas estimated, and just how are the population sizes related to the areas?

Several types of data are employed. Firstly, it is necessary to estimate the area occupied by the biogeographic unit, or units, in which the taxon is known to occur. It is clear that a taxon associated with a small-area biogeographic unit will include a smaller population of that taxon than one which is associated with a large-area biogeographic unit. The assumption is made here that the population densities found in both cases will be comparable.

Secondly, a taxon within a biogeographic unit associated with a specific community, or communities, will occupy a smaller area than a taxon associated with all the communities within that biogeographic unit. If the community is known to occupy only a small part of the biogeographic unit the population will be correspondingly small. Again, the assumption is made here that the population density for the various taxa will be comparable.

Thirdly, it is known that reef-type environments during the Silurian-Devonian occupy only a small fraction of the shallow-water continental regions. Therefore, taxa associated with the reef-type environment will occupy a small part of the area of any biogeographic unit. It is unlikely that reef-type environments during any interval of the Silurian-Devonian occupied as much as 10% of the available shallow-water environment.

Obviously, it is necessary to arrive at an estimate of the area occupied by the community in which the taxon of interest is known to occur. For some time intervals this may be done with ease (viz., the *Virgiana*,

Pentamerus and *Eocoelia* Communities of the Llandovery for the North Silurian Realm). For other time intervals this is not possible, but an upper limit may be obtained by estimating the area of the biogeographic unit in which the taxon occurs (viz., the *Karpinskia* Community of the Uralian Region during the Early Devonian).

The assumption that taxa of articulate brachiopods will have similar population densities serves as a first approximation, although in detail it certainly is true that different species will have different population densities (from packing considerations alone!). It is also true that a few taxa do not appear to ever occur except as very rare scattered specimens. However, the great bulk of the articulate brachiopods considered here occur as aggregations which are consistent with a fairly dense spatfall in any one area. The occasional rare specimen of such taxa is easily considered either as having occurred near the boundary of a community for which it was very abundant, or as having been transported (normally only for a short distance) away from its area of growth. Actual counts of specimens in individual collections, although few, are consistent with these assumptions and the conclusions drawn from them.

Stratigraphic Framework

The relative time framework for the Silurian-Devonian is based on faunal successions from stratigraphic sections, and their worldwide correlation. The reliability of this framework is partly a function of the degree to which the evolution of the different groups of related taxa is presently understood, the degree of provincialism during each time interval, the difficulty or ease with which precise correlation between differing biogeographic units has been made, the degree of community differentiation present during each time interval, and the difficulty or ease with which precise correlation has been made between differing communities (both benthic and planktic).

Berry and Boucot (1970; 1972c; 1973b—g) have prepared summaries of the lithofacies, paleogeography, correlation (about 800 columns), and fossil bases for correlation for the Silurian formations of the world.

Berdan et al. (1969) discussed Early Devonian correlation; Boucot and Johnson (1967a) discussed eastern North American Early Devonian correlation; Johnson et al. (1968) have discussed western North American Early Devonian correlation.

CHAPTER 2

Silurian-Devonian Community Framework

In order to assess the size of interbreeding populations in terms of the areas they occupied it is necessary to measure the area of the communities in which they occur. This section presents the bases with which such communities are defined and the newly defined Benthic Assemblage scheme employed to classify these communities systematically. Adoption of the Benthic Assemblage scheme makes possible a hierarchic subdivision of community information (Benthic Assemblages, Community Groups, Complex of Communities, and Communities of this book). Further conceptual discussion of this kind of community classification is given by Watkins et al. (1973).

In simplified form I conceive of a Benthic Assemblage as a group of communities that occur repeatedly in different parts of a region (during some times even worldwide) in the same position relative to shoreline. These Benthic Assemblages are probably temperature-controlled as well as highly correlated with depth. The correlation with depth very likely follows from temperature varying consistently in most places with depth. The different communities in a Benthic Assemblage are present or absent depending on various other controls, many of which are considered later. The Benthic Assemblages are arrayed in a systematic manner away from the shoreline as shown in Table I and Figs. 3—10. Although the Benthic Assemblage boundaries are arrayed systematically away from shoreline it must be emphasized that they do not occur systematically at the same distance away from shoreline. The actual distance from shoreline is controlled by other factors of which bottom slope is one of the most important.

Table I outlines the marine benthic animal community framework used in this book. Ziegler's (1965) "communities", as is made clear in the following sections of this treatment, do not appear to be homologs of the communities employed by most marine ecologists studying the modern environment. Ziegler's (1965) scheme for the Late Llandovery was applied to the entire Silurian, into the Early Devonian, and partly reinterpreted by Boucot (1970). Stratigraphic correlation within each community belonging to a particular Benthic Assemblage from place to place is more reliable than between communities of the same age that share relatively few

TABLE I

Depth-related Benthic Animal Assemblage zonal scheme employed in this book for the manyfold communities

1	2	3
Lingula Community	*Lingula* Benthic Marine Life Zone	Benthic Assemblage 1
Eocoelia Community	*Eocoelia* Benthic Marine Life Zone	Benthic Assemblage 2
Pentamerus Community	*Pentamerus* Benthic Marine Life Zone	Benthic Assemblage 3
Stricklandia Community	*Stricklandia* Benthic Marine Life Zone	Benthic Assemblage 4
Clorinda Community	*Clorinda* Benthic Marine Life Zone	Benthic Assemblage 5
		Benthic Assemblage 6
Pelagic Community		
		Pelagic Community

1 = Ziegler and Boucot's (in: Berry and Boucot, 1970) and Boucot's (1970) modification of Ziegler (1965); 2 = modification of Ziegler and Boucot (in: Berry and Boucot, 1970) by Berry and Boucot (1971), Berry and Boucot (1972b), and Gray and Boucot (1972); 3 = usage followed in this book.

Elles (1939) clearly envisaged a benthic assemblage scheme very close to that subsequently devised by Ziegler (1965). Her "in shore region, Large Brachiopods" is equivalent to our Benthic Assemblage 2 (note the emphasis on large brachiopods!); her "off shore region, inner, Trilobites and Brachiopods" is equivalent to our Benthic Assemblages 3 through 5; her "off shore region, middle, Trilobites, and outer, Graptolites and Trilobites" are equivalent to our Benthic Assemblage 6; and her "plankton region, quiet waters, Graptolites" is equivalent to our Pelagic Community. Her summary for the Ordovician and Silurian also points out the increasing taxic diversity of the graptolites as one moves offshore from one benthic assemblage to another (interpreted by Berry and Boucot, 1972, as depth zonation). Her conclusions, unfortunately, were presented in such a low ley, modest manner as to have escaped attention from both geologists and paleontologists until interest in the problems, stimulated by Ziegler in largest part, had been aroused more than twenty-five years after her paper appeared.

common taxa. Berry and Boucot (1971, 1972b) proposed that Ziegler's term "community" be replaced by the term "benthic marine life zone", both being prefixed by a generic name, as many of the diverse fossil associations encountered within each of Ziegler's "communities" are more comparable to what the ecologist would term a community. Ziegler's term "community" is more closely related to, although not identical to the ecologist's biome, or possibly floral life zone. Because there are already so many kinds of "zones" in geology I propose the term Benthic Assemblage to be employed as shown in Table I. Adoption of the term Benthic

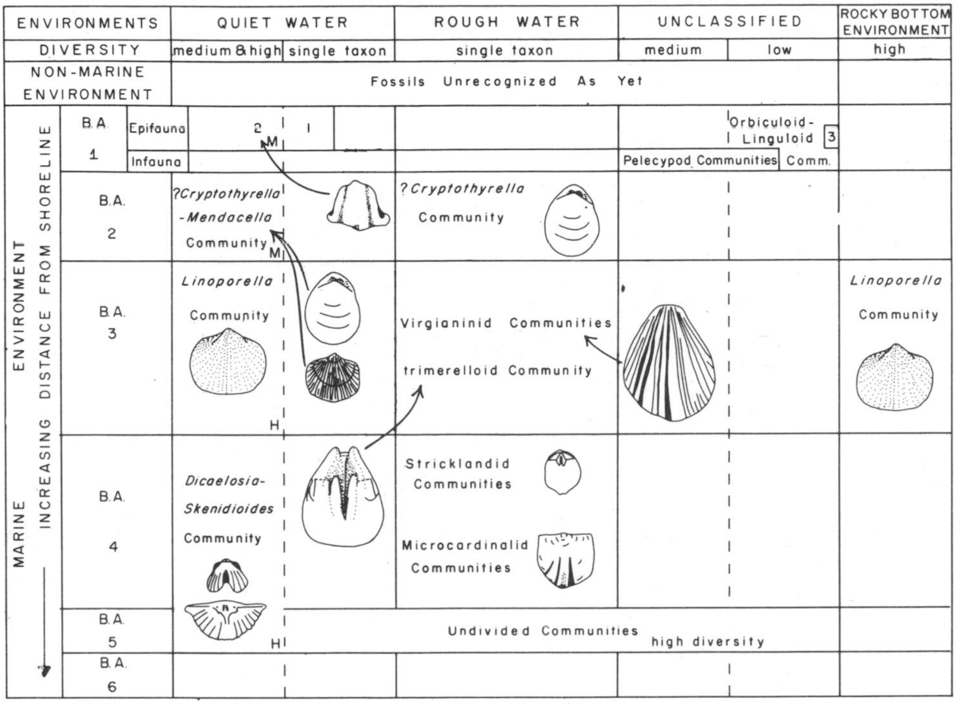

ENVIRONMENTS	QUIET WATER		ROUGH WATER	UNCLASSIFIED		ROCKY BOTTOM ENVIRONMENT
DIVERSITY	medium & high	single taxon	single taxon	medium	low	high
NON-MARINE ENVIRONMENT	Fossils Unrecognized As Yet					
B.A. 1 — Epifauna	2 1				Orbiculoid-Linguloid Comm. 3	
B.A. 1 — Infauna				Pelecypod Communities		
B.A. 2	?Cryptothyrella -Mendacella Community		?Cryptothyrella Community			
B.A. 3	Linoporella Community		Virgianinid Communities trimerelloid Community			Linoporella Community
B.A. 4	Dicaelosia- Skenidioides Community		Stricklandid Communities Microcardinalid Communities			
B.A. 5			Undivided Communities high diversity			
B.A. 6						

(Left axis: MARINE ENVIRONMENT — INCREASING DISTANCE FROM SHORELINE)

Fig. 3. Lower and Middle Llandovery level-bottom, non-reef communities (North Silurian Realm). (All communities listed within each Benthic Assemblage, B.A., possibly span the entire B.A.)

1 = Rhynchonellid Community; 2 = Homolanotid-*Plectonotus* Community; 3 = *Arthrophycus* Community; M = medium diversity; H = high diversity.

(Modified and amplified after Boucot, 1970.)

Assemblage will permit one to also use parallel sets of planktic assemblages numbered in the same or an independent manner. The use of the term Benthic Assemblage will not necessarily imply anything about the environment such as actual depth, roughness of the water, temperature, proximity to the shoreline, etc., although the disposition of the Benthic Assemblages in any place is undoubtedly strongly correlated with a variety of such factors. At the present time I do not propose that this terminology be adopted for brackish-water and non-marine communities owing to lack of experience with the problems they may present. (Calver, 1968a,b, presents data for the brackish and possibly nearshore non-marine Lower Carboniferous indicating that when adequate data are available there is no problem in arriving at a satisfactory ecologic zonation.) The adoption of a numerical listing of benthic assemblages rather than continuing to employ

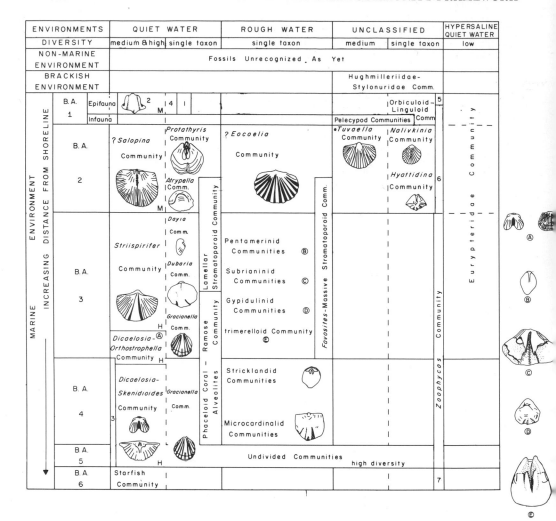

Fig. 4. Upper Llandovery—Ludlow level-bottom, non-reef communities (North Silurian Realm including Uralian-Cordilleran plus North Atlantic Regions of the Late Silurian). (All communities listed within each B.A. possibly span the entire B.A.) Reef complex of communities occurs in about B.A.-3 position.

• = Mongolo-Okhotsk Region; *1* = Rhynchonellid Community; *2* = Homolanotid-*Plectonotus* Community; *3* = Orthoceroid Limestone Community; *4* = Stromatolite Community; *5* = *Scolithus* Community; *6* = *Cruziana* Community; *7* = *Nereites* Community; *M* = medium diversity; *H* = high diversity.

(Modified and amplified after Boucot, 1970.)

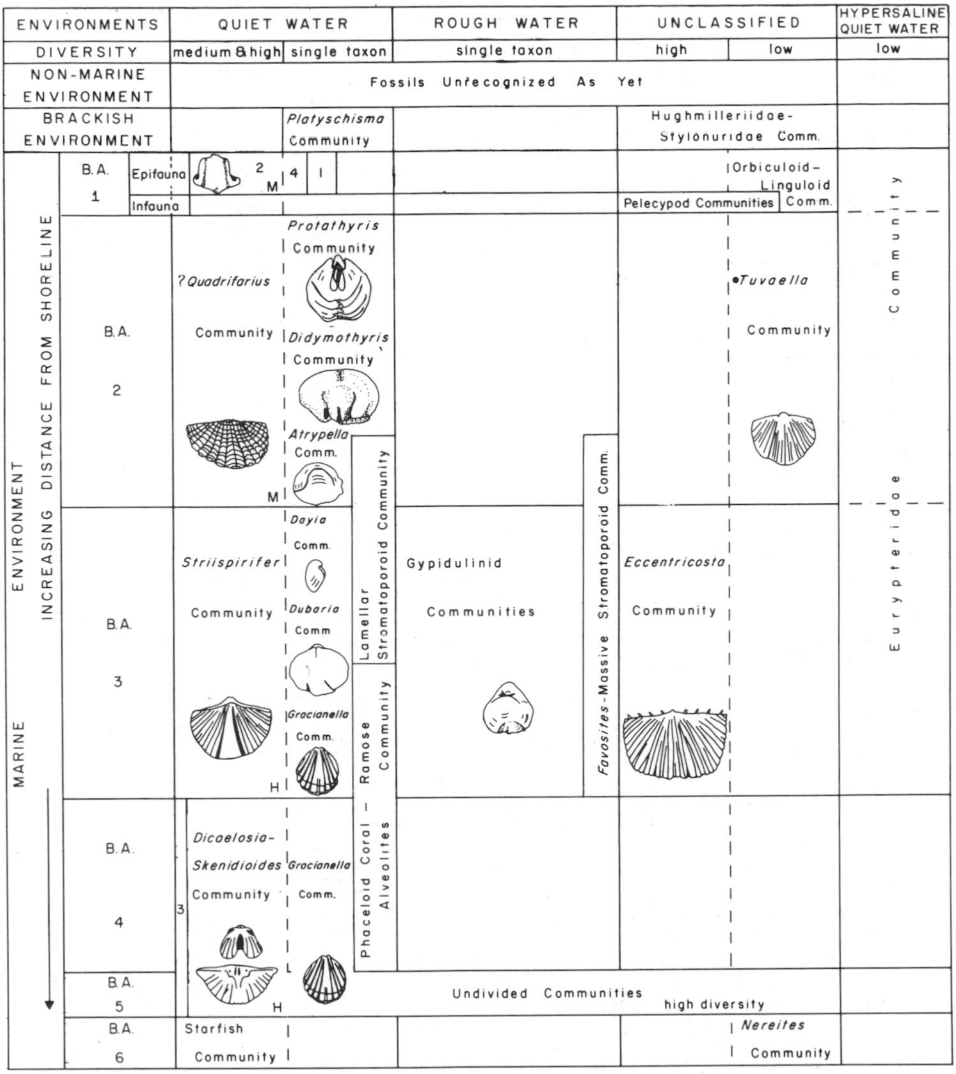

ENVIRONMENTS	QUIET WATER		ROUGH WATER	UNCLASSIFIED		HYPERSALINE QUIET WATER
DIVERSITY	medium & high	single taxon	single taxon	high	low	low
NON-MARINE ENVIRONMENT	Fossils Unrecognized As Yet					
BRACKISH ENVIRONMENT		Platyschisma Community		Hughmilleriidae-Stylonuridae Comm.		

Figure content (level-bottom communities arranged by B.A. 1–6, increasing distance from shoreline, marine environment):

B.A. 1 — Epifauna / Infauna (Quiet Water); | Orbiculoid–Linguloid Comm.; Pelecypod Communities

B.A. 2 — Protathyris Community; ?Quadrifarius Community; Didymothyris Community; Atrypella Comm. (M); •Tuvaella Community

B.A. 3 — Striispirifer Community; Dayia Comm.; Dubaria Comm.; Gracianella Comm. (H); Lamellar Stromatoporoid Community; Ramose Community; Gypidulinid Communities; Favosites-Massive Stromatoporoid Comm.; Eccentricosta Community

B.A. 4 — Dicaelosia-Skenidioides Community; Gracianella Comm.; Phaceloid Coral – Alveolites

B.A. 5 — Undivided Communities; high diversity

B.A. 6 — Starfish Community; Nereites Community

(right margin spanning: Community ... Eurypteridae)

Fig. 5. Pridoli level-bottom, non-reef communities (Uralian-Cordilleran and North Atlantic Regions). (All communities listed within each B.A. possibly span the entire B.A.) Reef complex of communities occurs in about B.A.-3 position.

• Mongolo-Okhotsk Region; *1* = Rhynchonellid Community; *2*= Homolanotid-*Plectonotus* Community; *3* = Orthoceroid Limestone Community; *4* = Stromatolite Community; *M* = medium diversity; *H* = high diversity.

(Modified and amplified after Boucot, 1970.)

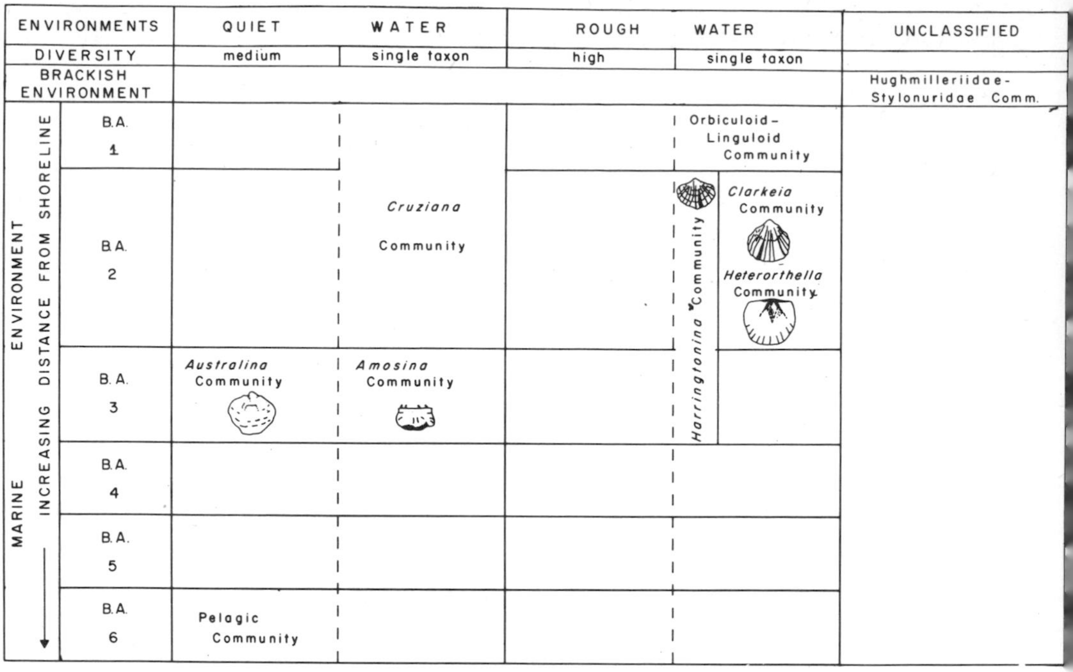

ENVIRONMENTS	QUIET	WATER	ROUGH	WATER	UNCLASSIFIED
DIVERSITY	medium	single taxon	high	single taxon	
BRACKISH ENVIRONMENT					Hughmilleriidae– Stylonuridae Comm.
B.A. 1				Orbiculoid– Linguloid Community	
B.A. 2		Cruziana Community		Clarkeia Community / Heterorthella Community	
B.A. 3	Australina Community	Amosina Community			
B.A. 4					
B.A. 5					
B.A. 6	Pelagic Community				

(MARINE — INCREASING DISTANCE FROM SHORELINE — ENVIRONMENT; Harringtonina Community spanning rough water column B.A. 2–3)

Fig. 6. Malvinokaffric Realm Silurian level-bottom, non-reef communities. (All communities listed within each B.A. possibly span the entire B.A.) Reef complex of communities absent in Malvinokaffric Realm.

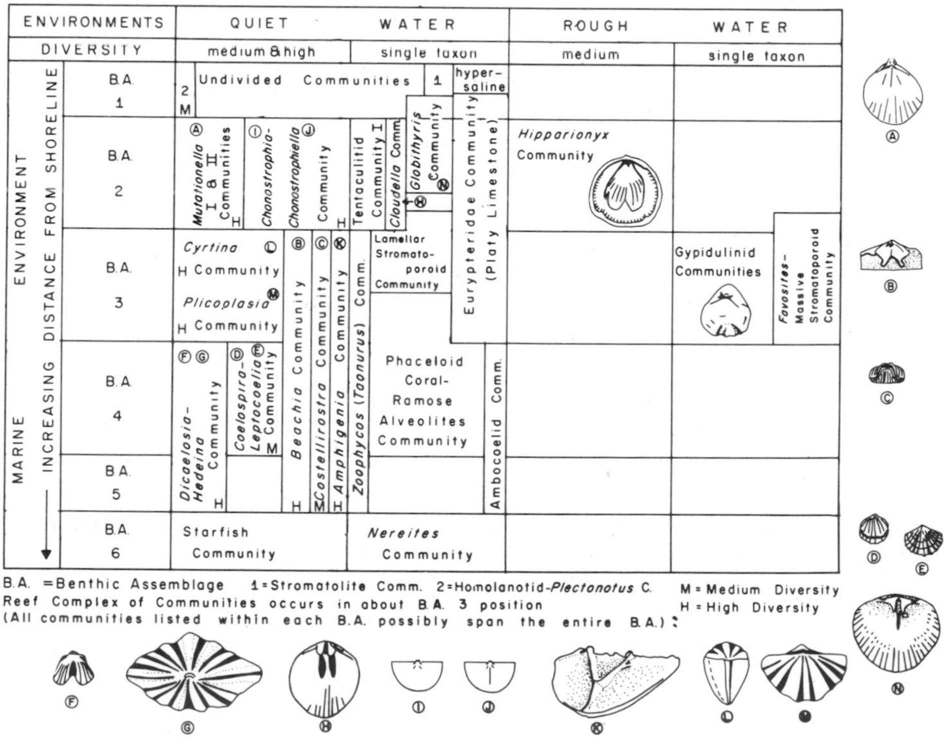

ENVIRONMENTS	QUIET	WATER	ROUGH	WATER
DIVERSITY	medium & high	single taxon	medium	single taxon
B.A. 1	2 M Undivided Communities 1	hyper-saline		
B.A. 2	Ⓐ Mutationella I & II Communities H / Chonostrophia– / Ⓘ Chonostrophella Community H / Tentaculitid Community I / Cloudella Comm. / Globithyris Community	Eurypteridae Community (Platy Limestone)	Hipparionyx Community	
B.A. 3	Cyrtina Ⓛ H Community / Plicoplasia Ⓜ H Community / Ⓑ Ⓒ Ⓧ / Beachia Community / Lamellar Stromatoporoid Community		Gypidulinid Communities	Favosites– Massive Stromatoporoid Community
B.A. 4	Ⓕ Ⓖ Coelospira– Leptocoelia Community M / Ⓓ Ⓔ Costellirostra Community / Amphigenia Community / Zoophycos (Taonurus) Comm.	Phaceloid Coral-Ramose Alveolites Community / Ambocoelid Comm.		
B.A. 5	Dicaelosia– Hedeina Community H			
B.A. 6	Starfish Community	Nereites Community		

B.A. = Benthic Assemblage   1 = Stromatolite Comm.   2 = Homolanotid-*Plectonotus* C.   M = Medium Diversity
Reef Complex of Communities occurs in about B.A. 3 position   H = High Diversity
(All communities listed within each B.A. possibly span the entire B.A.)

Fig. 7. Appohimchi Subprovince Early Devonian level-bottom communities.

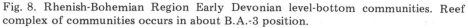

ENVIRONMENTS	QUIET WATER		ROUGH WATER	
DIVERSITY	medium & high	single taxon	high	single taxon
B.A. 1 — Epi-fauna	Homalonotid-*Plectonotus* Community Ⓐ	5 M		
B.A. 2	? *Quadrifarius* Community Ⓑ M — *Platyorthis-* Ⓒ *Proschizophoria* Comm. Ⓓ M	Tentaculitid Comm. M — *Mutationella* Comm. Ⓔ Ⅱ — Choretid Comm. M — *Rhenorensselaeria* Community Ⓕ		*Tropidoleptus* Community
B.A. 3	Rhenish Complex of Communities — Bohemian Complex of Communities — Rhenish Complex of Communities of H	4 — 3 H — 2 H — H (Orthoceroid Limestone C)	Lamellar Stromatoporoid C. — Phaceloid Coral-Ramose Alveolites Community	Gypidulinid Comm's Ⓖ — *Favosites-*Massive Stromatoporoid Community
B.A. 4				
B.A. 5				
B.A. 6	Starfish Community			
Pelagic Zone		Graptolitic Facies 1 6		

Fig. 8. Rhenish-Bohemian Region Early Devonian level-bottom communities. Reef complex of communities occurs in about B.A.-3 position.

1 = Graptolites found in B.A.-2 through Pelagic Zone; *2* = *Anathyris-Pradoia* Community; *3* = *Euryspirifer paradoxus* Community; *4* = "*Uncinulus*" *orbignyanus* Community; *5* = Stromatolite Community; *6* = *Nereites* Community; *M* = medium diversity; *H* = high diversity. *Cruziana* Community occurs in the Benthic Assemblage 1 position in this Region (Freulon, 1964).

(Modified and amplified after Boucot, 1970.)

generic names for them permits us to escape the absurdity of using a generic term in a time interval when the particular genus has been extinct for many hundreds of millions of years or conversely using it for a time interval hundreds of millions of years prior to the appearance of the genus. At the same time the adoption of such a nomenclature permits us to use a common reference system divorced from time; the spectre of a separate nomenclature for each small time interval is too awful to contemplate.

The practice of employing place names, formational names, or generalized faunal characters (Big Shell Community for example) as designations for communities is to be avoided in the interests of clarity. More than one community may commonly occur in a single formation or at a single locality and it is difficult to select unambiguous faunal characters (imagine the number of units in the Phanerozoic that are characterized by large shells!) Therefore, taxonomic designations have been employed wherever possible.

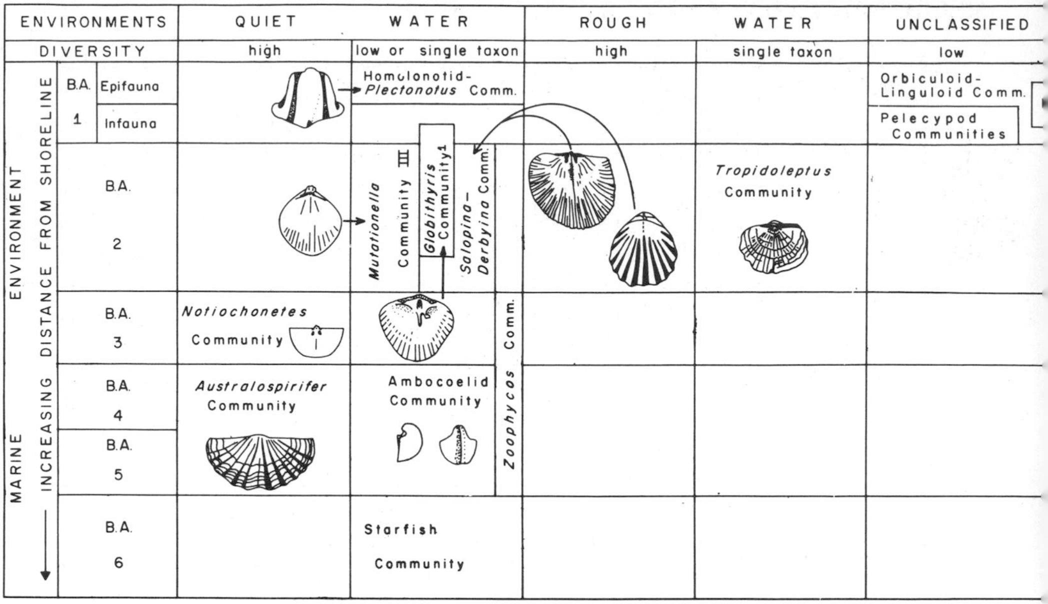

ENVIRONMENTS	QUIET	WATER	ROUGH	WATER	UNCLASSIFIED
DIVERSITY	high	low or single taxon	high	single taxon	low
B.A. 1 Epifauna		Homulonotid-Plectonotus Comm.			Orbiculoid-Linguloid Comm.
B.A. 1 Infauna					Pelecypod Communities
B.A. 2		Mutationella / Community III / Globithyris Community¹ / Salopina-Derbyina Comm.		Tropidoleptus Community	
B.A. 3	Notiochonetes Community		Zoophycos Comm.		
B.A. 4	Australospirifer Community	Ambocoelid Community			
B.A. 5					
B.A. 6		Starfish Community			

Fig. 9. Malvinokaffric Realm Early Devonian level-bottom communities. Reef complex of communities absent in Malvinokaffric Realm. (All communities listed within each B.A. possibly span the entire B.A.). *I* = *Cruziana* Community.
(Modified and amplified after Boucot, 1970.)
¹ Isaacson, in preparation.

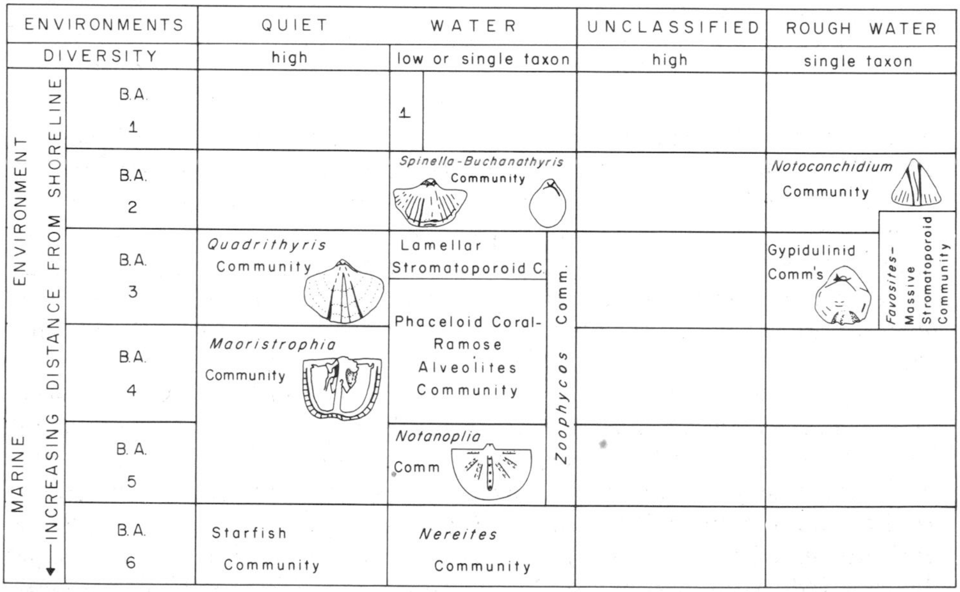

ENVIRONMENTS	QUIET	WATER	UNCLASSIFIED	ROUGH WATER
DIVERSITY	high	low or single taxon	high	single taxon
B.A. 1		¹		
B.A. 2		*Spinella-Buchanathyris* Community		*Notoconchidium* Community
B.A. 3	*Quadrithyris* Community	Lamellar Stromatoporoid C. / Phaceloid Coral-	Zoophycos Comm.	Gypidulinid Comm's / *Favosites-* Massive Stromatoporoid Community
B.A. 4	*Maoristrophia* Community	Ramose Alveolites Community		
B.A. 5		*Notanoplia* Comm		
B.A. 6	Starfish Community	*Nereites* Community		

Fig. 10. Tasman Region Early Devonian level-bottom communities. Reef complex of communities occurs in about B.A.-3 position. (All communities listed within each B.A. possibly span the entire B.A.)
I = Stromatolite Community.

Benthic Assemblage 6 occurs seaward of the "Communities" recognized by Ziegler (1965) in the Welsh Borderland. It is introduced to include less common shells, including abundant trilobites of certain types (see Sheehan, 1973b, for some Caradoc-Ashgill age examples in his "Mud-Clay Life Zone" from Sweden), Notanoplid Community and Starfish Community shells in certain regions. Shells are not commonly present in the Benthic Assemblage 6 position although plankton is abundant (graptolites and dacryoconarid tentaculitids in the Silurian and Devonian).[2]

Overlapping Ranges of Taxa; what is a Community?

Any compilation of the lateral distribution, during a single time interval, of a single taxon is apt in three cases out of four to show that the taxon does not invariably either occur by itself or with any fixed group of other taxa (Fig.11A, diagrammatic representation of this problem). It is common to find that the single taxon may occur in certain environments strictly by itself. It is also common to find that the single taxon will occur, within its total area of occurrence, with one or more other taxa or groups of taxa (see Fig.11A). Consideration of the implications and consequences of these associations suggest that the defining of communities as recurring associations of fossils[3] is not necessarily a simple matter. It is clear (using Fig.11A as a guide) that the recurring associations of our single taxon by itself, and in various combinations with other taxa tells us something about the environmental tolerances of all of the taxa. For part of their geographic range the individual taxa are exclusive because of limiting environmental factors, whereas in other portions of their ranges they share different environmental requirements in common with other taxa. In other words, an individual taxon dwells in an environmental range. Whittaker (1972) has treated the question of environmental gradients rigorously, but arrives at similar conclusions based on general considerations and also distributions of land plants.

Fig.11A diagrams the theoretical effects of two environmental variables in controlling the distribution of three taxa. In practice far more than two environmental variables may have to be considered, as well as far more than three taxa. This may be treated as a multidimensional hyperspace distribution, as used by Whittaker (1972), in which an extremely large number of taxic associations are possible. As few of these distributions will be of a "regular" nature (simple spherical or ellipsoidal distributions or their equivalents on a hyperspace basis), it can be seen that the reality might become very complex. However, it is also probable that many of the environmental variables may effect many groups in a similar enough manner to permit one not to consider them individually.[4]

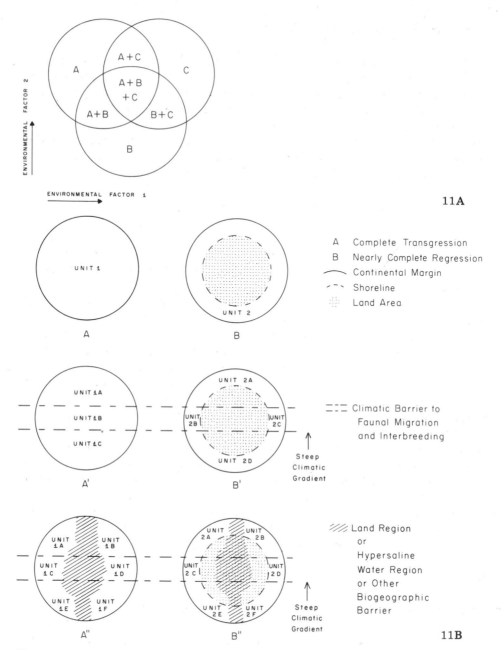

Fig. 11A. Diagram showing the theoretical ranges of taxa A, B, and C. They are arranged in such a way as to indicate the presence of seven communities, respectively (composed of A, B, C; A plus B; A plus C; C plus B; and A plus B plus C). It is important to note that each of the three taxa has an environmental range that permits it to overlap with the others to a certain extent. The presence of various taxonomic mixtures does not decrease the statistical validity of any of the seven communities. In other words, the fact that A may occur by itself, with C or with B or with both B and C does not detract from the multi-community concept.

Fig. 11B. Hypothetical continent subjected to conditions of high transgression and high regression (A, B); high transgression and high regression under conditions of a steep climatic gradient (A', B'); high transgression and high regression, steep climatic gradient plus a biogeographic barrier occurring at a high angle to the climatic gradient (A'', B''). Note the effects of these various conditions on area occupied.

In the classification of communities, following the above arguments, we must face the problem of erecting communities which will share taxa with other communities in many instances, although not invariably. The fact that some communities have taxa common to both should not be taken to mean that the communities are invalid concepts or identical to each other.

An additional complication is the potential patchiness on a very small scale (microenvironments) of associated communities in such a manner that a very minor amount of transportation, say no more than a few meters, may lead to a variety of post mortem mixtures. Kohn (1959) describes species of *Conus* occurring in a most complex distribution manner influenced heavily by a number of "minor" factors within a very small littoral area, that make this point clear.

This definition of the term community neither accepts or rejects the fact that a taxon or taxa will have had interdependent, dependent or no relation with associated taxa. In other words I am defining and using the term community without consideration of potential dependency relations (see Speden, 1966, p.411, for a review of the community definition question). It is very difficult to determine potential dependency relations with fossils (the absence of soft-bodied organisms seriously complicates the determination). Transport of fossils after death, as will be discussed shortly, is considered a relatively minor factor in most associations of fossils. It is considered conceptually unneccessary to have a separate term for the biocoenotic community complete with soft-bodied organisms as opposed to the remnant commonly preserved as fossils. We speak of fauna and flora from the past without feeling called on to state repeatedly that the fossil flora and fauna is no more than a small representation of the life of the past.[5]

The work summarized in this treatment shows the need for a classification of fossil communities suited to different situations and amounts of information. The term "Complex of Communities" is used for a situation in which we have inadequate data to diagnose the various communities present in an area but are forced to conclude that more than one community is present. The term "Community Group" is used to refer to a group of biotically closely related communities that commonly occur in the same Benthic Assemblage or Assemblages. We will eventually need a term like megacommunity to cover an intermediate category.

Benthic Assemblage and Community Continuity

Any attempt to map and classify both animal communities and Benthic Assemblages in time and space must cope with the question of their continuity, and how this continuity may be recognized in the face of continual change of environments and organic evolution. I have earlier (Boucot, 1970, pp.574—582) tried to indicate how study of the relative position and frequency of common taxa may aid in deciding whether two time-successive communities are related (one containing taxa derived in large part from the other) or unrelated. I also stressed the position of an individual community relative to others on either side of it laterally or vertically in time in trying to make this determination. Unless one develops the ability to reliably assign communities to Benthic Assemblages he will not be able to develop a method of analysis suited to a worldwide scheme. The following example of the logic involved in trying to assign the *Striispirifer* Community to a Benthic Assemblage position will serve to illustrate the approach used in this book for making these decisions. One must keep in mind that the *Striispirifer* Community has few, if any, taxa in common with the communities occurring to either side, above or below.

In his original concept of Upper Llandovery communities, Ziegler relied heavily on the presence of key genera (usually specified in community names, e.g., *Eocoelia* in the *Eocoelia* Community) to identify the community type to which a given assemblage of fossils should be assigned. Other taxa of megafossils, although not disregarded, were given little weight in making these assignments. Collections lacking the "key" genera proved difficult to assign to a "Community" unless they could be related to nearby collections that contained "key" genera. In the Welsh Borderland this problem did not prove serious, but on the platform Ziegler's "key" genera are absent from many widely distributed communities making assignment difficult at best. In the Welsh Borderland margining the Caledonian geosyncline the original distribution of various communities is patchy on a very small scale, which also affords more chance for short-distance post mortem mixing providing "key" genera in areas where they did not live. The overall steeper bottom slope in the Welsh Borderland helps to delineate the trends of shoreline-proximity-related Benthic Assemblages and their communities more easily than can be done on the excessively low-slope platforms. Currently the only recourse lies in noting both the lateral and vertical relations of communities lacking the "key" genera to communities possessing the "key" genera (Fig.12).

For example, the *Striispirifer* Community Rochester Shale (Fig.13) lacks any of the "key" genera. The Rochester Shale containing the *Striispirifer* Community grades eastward into the Herkimer Sandstone. The western member of the Herkimer also possesses a *Striispirifer* Community fauna, but the eastern member is characterized by a fauna with attributes of Benthic Assemblages 1 and 2. Westward the Rochester Shale grades into the Fossil Hill Formation that contains some beds with a

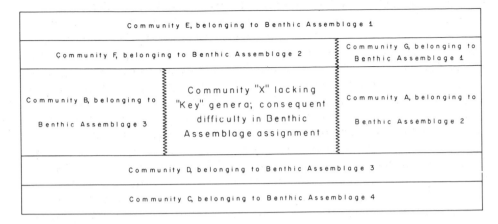

Fig. 12. Diagram indicating how Community "*X*", which lacks "key" taxa that permit it to be assigned to a Benthic Assemblage, may be bracketed into a Benthic Assemblage position by reference to adjacent communities whose Benthic Assemblage position has been previously determined.

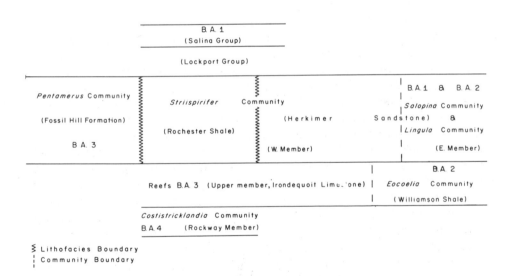

Fig. 13. Relations of the *Striispirifer* Community which lead to the conclusion that a Benthic Assemblage 3 position is reasonable. Other interpretations may be made from this data, but additional information from other regions supports a Benthic Assemblage 3 position.

Benthic Assemblage 3 Pentamerinid Community fauna. Vertically the Rochester Shale is underlain by the Irondequoit Limestone which has a Benthic Assemblage 4 Stricklandid Community fauna (*Costistricklandia*) in the lower, Rockway Member, and reef-like bodies that suggest Benthic Assemblage 3 (all of the Silurian reef-like bodies are considered, Fig.4 for example, to lie in this position) for the upper member. Overlying the Rochester Shale is the Lockport Group that lacks any "key" genera but is itself overlain by the Salina Group containing Benthic Assemblage 1 fauna. Summing up these data one brackets the *Striispirifer* Community in the Benthic Assemblage 3 position despite the fact that it lacks the "key" genus *Pentamerus*. Absence of the widespread Benthic Assemblage 2 genus *Salopina* of the *Salopina* Community from the Rochester and western member of the Herkimer is another argument for placing the *Striispirifer* Community in Benthic Assemblage 3. The absence of *Dicaelosia*, common to taxonomically diverse, deeper water communities, is a good argument for excluding the *Striispirifer* Community from Benthic Assemblage 4. In this manner the other communities lacking the "key" genera of Ziegler's (1965) treatment may be assigned to a Benthic Assemblage. Note how use of the Benthic Assemblage concept lends flexibility to the process.

The Use of "Rare", "Common", and Poorly Studied Taxa for Characterizing Communities and Biogeographic Units

Efforts to characterize the contents of ecologic units (communities) and biogeographic units (realms, regions, provinces, subprovinces) should, in principle, employ as much of the known biota as possible. However, there are a number of problems that make the use of the entire known biota either impractical in many instances or downright misleading. Poorly studied groups are present in almost every fossil biota. One can almost support the position that all groups are poorly known, but that some are far more poorly known than others. A "well known" group is merely well known relative to a poorly known group; the paleontologic record is so scanty and even misleading due to the vagaries of preservation and collecting that it is difficult to maintain that we have a really complete understanding of the path followed by organic evolution in detail for any group. However, we do have samples and they may be used to arrive at reasonable conclusions if we are willing to try and understand the limitations imposed by the nature of the samples. Therefore, it is only logical that we place greatest reliance for purposes of ecologic and biogeographic unit definition on those taxa that are best known. The relatively poorly known widespread groups will always tend to smear out differences between units and give a false sense of unity to units that in reality, when studied in more detail taxonomically, are not similar. In other words, there is ample justification for omitting from consideration groups deemed poorly known taxonomically. For example, the taxonomy of the widespread linguloid and orbiculoid brachiopods is very poorly known. Therefore, any ecologic or biogeographic unit depending on the differences or similarities between linguloid and orbiculoid brachiopods will suggest a far greater degree of similarity than may in fact exist, although until these fossils have been studied one cannot be certain of the results. A relatively well studied group like the rensselaerinid brachiopods, on the other hand, may be used with a far higher level of confidence for these purposes. There is no automatic way of determining which groups are well known or poorly known; personal acquaintance with the taxonomy of each group, developed over a period of time, by a competent specialist is essential if the conclusions developed are to have any value.

It is clear as well that "common" taxa, preferably on a worldwide or regional basis, will provide more reliable information to be used in the definition of these units than will "rare" taxa. For example, the Givet-age brachiopod *Enantiosphen* is known from only a few localities in Europe, whereas the coeval, related genus *Stringocephalus* and closely related taxa have an almost worldwide distribution. Both taxa occur in what are most easily interpreted as rough-water Benthic Assemblage 3 positions. With this information in hand one has the possibility of deducing either a relatively widespread, almost worldwide, occurrence of the *Stringocephalus* Community or alternatively a biogeographic unit outlined by the few known occurrences of *Enantiosphen*. Johnson's (1974) recent discovery of a New World occurrence of a form (*Enantiosphenella*) closely related to *Enantiosphen* immediately points up the danger of placing undue reliance on the known distribution pattern of a rare taxon as compared with a common taxon. Decisions to use common taxa are easy once it is realized that their distribution pattern is wide and that they have indeed been recovered at many localities within a number of regions. Decisions as to whether to use rare taxa are far more difficult. It is necessary to find out if the rare taxa are rare both on a worldwide and also on a local basis. A taxon represented by a single specimen known from only a single locality is clearly of little significance in the defining of these units, but a taxon known from only one region where it is present at many localities is another thing. Ultimately the decision whether or not to employ such rare taxa must be made by the student on the basis of experience, with the firm understanding that some misleading results are to be expected from the use of rare taxa in those cases where the decision to use them is made.

Reef and reef-related taxa illustrate this problem very well. They occur py relatively small areas. Reefs tend to be the developmental sites of rapidly evolving, highly endemic taxa that contrast with the far more widespread, coeval level-bottom taxa in this regard. Therefore, one faces examples of high provincialism in reef organisms that are not reflected in the nearby level-bottom organisms. Consideration of this situation must be made when defining biogeographic units.

All of these comments point up the fact that ecologic and biogeographic units are statistical concepts arrived at by a somewhat subjective consideration of the available raw data. Statistical tests may be applied to the data in an effort to make conclusions based on the data more reliable, but ultimately the decision to use or reject a particular category must rest on the informed judgment of the student. This "judgment" may appear to the uninitiated as a very subjective quantity, but experience has shown that although it is difficult to quantify "experience" in a manner satisfactory to the statistician it is also absurd to reject all conclusions not arrived at with the aid of formal statistical procedures. The taxonomic experience of the last two hundred years in defining species as contrasted with that obtained in the last twenty-five years with the application of statistical "tests" has shown again and again that the "subjective" judgment of the "old-fashioned" taxonomist stands the test of "modern" approaches in most cases when it comes to recognizing a species. The appeal to what is "modern" and "mathematical" should not cloud our vision as to what has been shown over the years to be workable in practice despite the value and desirability of quantifying our data wherever possible. The ultimate test and validation of any ecologic or biogeographic unit is provided by the experience of subsequent students attempting to employ the unit.

Level-Bottom Community Classification

As detailed earlier (Boucot, 1970) a community is defined here as a specific set of organisms adapted to a specific set of environmental conditions, irrespective of their separation in time or space. Community is defined in the same way by many paleontologists and biologists, but a fossil community is less clear conceptually because of incomplete preservation. For practical purposes a paleontologist recognizes a community by means of a *recurrent association of taxa*, i.e., *constancy of associated taxa* occurring in fossil collections (Speden, 1966, p.411).

The significance of animal and plant communities, both living and fossil, entails a number of puzzling philosophic problems. The defining and recognition of animal and plant communities is intimately related to the same philosophic problems.

(1) Any taxon will be able to exist within an environmental range.

Insofar as the environmental range is a function of more than one variable it is possible that the range for any one variable may be partly dependent on other variables if the variables do not uniformly vary together.

(2) Following from this first premise is a second, namely that any taxon will be able to exist within an area covered by the environmental range conducive to the existence of the taxon.

(3) Different taxa will have either the same environmental ranges, completely differing environmental ranges, or overlapping environmental ranges.

(4) There should be a complete range among taxa from those that exist dependent only on the physical environment to those which are highly dependent as well on coexisting taxa.

After outlining these four points we are able to consider the significance of communities. Communities may be defined as recurring associations of taxa. For those taxa that are independent of the presence or absence of other taxa a recurring association merely indicates the presence of forms with common environmental requirements that are met with in the area of their co-occurrence. In other words, the recurring association is indicative of a specific set of physical factors in the environment that are expressed by the organisms. The more unrelated taxa incorporated into such a community definition, the more narrowly are the physical conditions specified due to the differing physical requirements of the individual taxa. In such a community situation the occurrence of one taxon is not dependent on the presence of another taxon. Very different is the community situation for which one organism is dependent on the presence of another, or in which there is feedback making possible an interdependence in order to make possible the presence of more than one taxon. There should be gradations in the above matter of biologic dependence or independence.

The elegant and logical treatment of community diversity presented by Whittaker (1972) assumes an interdependence between organisms. If the organisms occurring in the same area do not have completely interdependent relations one must appeal to far more complex and less predictable models for diversity. The paleontologist can only hope to gain some glimmering of the dependence or independence of taxa by accumulating a vast amount of information regarding associations and abundances for various taxa from as many areas and regions as possible. The final conclusions, if based on adequate data, may prove decisive.

The preceding paragraphs were presented in order to provide a background against which to consider the practical question of setting up fossil communities. Specifically, is it valid to consider only selected taxa or must the entire fauna and flora be considered for the community concept to have any utility? In principle, taxa dependent on the physical environment only may be considered individually. The omission of co-occurring taxa in this case will be a loss only in the sense of giving us less qualitative appreciation of the amount of environmental range occupied by a species in terms of other species. The omission of members of dependent or interdependent communities from consideration is a more serious one. Such an omission will prevent our arriving at an understanding of potential dependent and interdependent relations.

In practice, however, the paleontologist proceeds by considering that communities are recurring associations of taxa. One is not initially able to determine whether the members of these recurring associations have any dependence relations on each other. It is only after arriving at an understanding of the actual functioning and requirements of the individual taxa that one can decide whether or not they should be classed as dependent. For example, an association of Silurian graptolites and brachiopods containing *Climacograptus* and *Eocoelia* is known from a number of localities, together with *Monograptus priodon*. This association could be considered as a community. The recognition that graptoloid graptolites are planktic requires us to remove the graptolites from this association and consider them as a separate planktic community not related to the presence or absence of *Eocoelia*, a benthic form. This is an obvious case for which we may provide a quick answer. The association of Pterygotidae-Carcinosomidae eurypterids with a great variety of benthos (benthos that do not co-occur with each other) might indicate a far greater environmental tolerance for the eurypterids than for the benthos. However, the knowledge that the eurypterids are probably vagrant benthos or nektic carnivores not dependent on the same factors that control the distribution of the benthos restricted to small areas of the bottom answers the question. This again is an obvious case. The ordinary situation for most benthos is far more difficult as we have less information about the habits of the animals in question. For example: should we consider associations of brachiopods, gastropods and bryozoans as members of the same community subject to the same environmental controls or should we consider them as separate communities entirely independent of each other? If we consider them jointly we help to specify the number of physical subdivisions of the environment they may represent if they are not dependent on each other. The invariable co-occurrence of taxa with each other indicates either an invariable dependence or interdependence relation or else the invariable co-occurrence of the same environmental range factors. It is clearly desirable to consider the occurrence of as many taxa as possible in either case in order to better understand the environment at any one time, as well as the changing environmental tolerances of organisms. But, it should be made very clear that the co-occurrence of taxa does not necessarily indicate that they are dependent or interdependent. Proof of dependence relations requires an understanding of the functioning of organisms, not just their distribution patterns. Therefore, if we set up a community framework based, as we must, on the recurrent association of taxa we must understand that we define statistical entities, not necessarily functioning biologic entities.[6]

Benthic taxa with non-overlapping areas of occurrence cannot ordinarily be thought to have dependent or interdependent relations. Benthic taxa having largely overlapping areas of occurrence may or may not have dependent or interdependent relations.[7] Filter-feeding or suspension-feeding co-occurring benthos presumably co-exist by partitioning the available food supply in various manners, as it is unreasonable to think that their food intakes would be identical for long if they compete for precisely the same elements in the available supply in the same relative amounts at the same time. It is not clear whether or not the population densities of co-occurring filter and suspension feeders are even partially controlled by each others occurrence *if* the spat of the co-occurring taxa set at the same time. If an abundance of suspended food material is available it may turn out that suspension and filter feeders are not limited by food supply, but by other factors unrelated to food supply. Even in those cases where food supply is a limiting factor it may not be assumed that one among two or more competing species will ultimately completely take over the area (see Ayala, 1972). However, it is conceivable that suspension feeders with the same food requirements may either occupy or not occupy an area on the basis of the first spat to settle in an area, after a wipeout, existing in the area until the next wipeout permits another chance spatfall of the first group to make spat available. In the fossil record one might observe such a chance alternation of taxa as a co-occurrence of the taxa. Carnivore-prey relations are harder to unravel because of the necessarily smaller carnivore populations present in the first place. Minor alternations in key environmental factors at a single locality over short intervals of time may also produce an apparent association of taxa that is an artifact of either collecting or preservation (later mixing misleading the paleontologist as to the original separateness of the taxa during their lives).

In view of the above, what then is the recurrent association of taxa that we view as a community? The best that can be stated at the outset is to treat it as containing at least one community while recognizing that it may contain more than one community, but hopefully no more than the maximum number of taxa present.

Next, let us consider the significance of the relative abundance of each taxon in a recurrent association of taxa. If abundances vary widely and erratically the possibilities for dependence or interdependence are far lower than if they occur consistently in the same proportions and also with the same population densities. However, the occurrence of taxa together with the same abundances and population densities is still no guarantee of either dependence or interdependence, although it is strongly suggestive of temporal co-occurrence under the same environmental regime.

Recognition of an association of taxa in actual living position completely undisturbed by subsequent bioturbation or physical events is the only acceptable proof of co-occurrence during life. Such life assemblages in living position are unusually rare. However, the co-occurrence of articulated bivalved shells with an appropriate population structure may be considered as an approach to living position that is acceptable.

Experience with both living and fossil communities shows that in many instances the taxa within adjacent entities have a gradational abundance relation in large part from one into another (Boucot and Heath, 1969, table 3; Boucot, 1971; Winter, 1971; Fuchs, 1971). The presence of gradational abundance relations between some of the taxa common to adjacent communities, as well as the complete presence or absence of still other taxa in adjacent communities does not detract from the community concept. In a sense communities are statistical concepts. In some instances a complete taxic gradient, both in absolute and relative abundance as well as presence or absence may occur between two community "poles" (poles in the sense of being as mutually exclusive as the adjacent communities in question ever become). There are, of course, many adjacent communities that share nothing or virtually nothing taxonomically due to very sharp, precipitous environmental gradients or changes, but the presence in other instances of very low, gradual environmental gradients ensures that the polar taxonomic groupings will be linked together by a very complete taxonomic and abundance gradient that reflects the environmental gradient (Fig.11A).

See Part II, Chapter 6 concerning the emphasis given here to level-bottom, brachiopod-dominated as opposed to coral-stromatoporoid communities (as well as the non-consideration of reef communities).

Fig.3—10 classify many Silurian and Devonian level-bottom communities. On these figures the boundaries between rough- and quiet-water[8] environments belonging to the same Benthic Assemblage have been indicated as being precisely equivalent, although it is probable, as more detailed information accumulates, that it will be found that few of these boundaries are precisely equivalent despite their approximate equivalence.

The biologic significance of the rough- and quiet-water classification employed in Fig.3—10 is unclear. Despite the correlation observed with sediment type (see footnote) it is unlikely, unless the grain size of the sediment be thought of as the critical, controlling factor, that rough water or quiet water in themselves are the controlling factors. It is more likely, in those cases where grain size of the bottom sediment is not critical, that other factors commonly accompanying rough-water and quiet-water conditions are the key controls for occurrence or non-occurrence of those communities here considered to occur in rough-water and quiet-water conditions. It is likely that factors involving food supply ("resource-rich" or "resource-unstable" of Valentine, 1971b), oxygen supply, metabolites, dissolved nutrients of various types, microtidal or macrotidal regimes, and the like that themselves commonly correlate with rough-water or quiet-water conditions are the more basic controls. However, in the absence of any ability to estimate these possibly more basic variables I am reduced at present to using the rough-water/quiet-water classification. There are enough occurrences of rough-water communities in such a condition of associated sediment, degree of articulation, and shell sorting indicative of quiet-water conditions to suggest that the basic controls are chiefly met with only under rough-water conditions, but that may be rarely met with in quiet-water conditions. Examples of the rough-water *Stricklandia*, *Pentamerus* and *Kirkidium* Communities in an articulated condition, in life position, in fine-grained sediment do exist, although not commonly, to make this point clear.

An attempt has been made to distinguish here between infaunal communities of pelecypods and epifaunal communities in Benthic Assemblage 1. This brings up the basic question whether or not the epifauna and the infauna in a particular spot should be considered as parts of a single community or as two distinct communities. C.F. Calef (written communication, 1972) reports finding a succession of infaunal pelecypod communities in the Arisaig, Nova Scotia McAdam Brook Formation of Ludlow age that are associated with the *Salopina* Community brachiopod community (all presumably belonging in the very shallow part of B.A. 2).

Calef's data suggest that it is proper in at least some cases to consider that the environmental controls governing the occurrence of infaunal bivalves are different from those governing the associated epifauna. We do not know, however, whether the geographic limits of the different infaunal bivalve communities would correspond to the geographic limits of the associated epifaunal brachiopod community. It is logical to infer that in some instances the infaunal communities will have different geographic boundaries than the epifaunal communities. Bader's (1954) study of the intimate relation between the nature of the organic content in at least some marine sediments and the bivalves that consequently either occur or do not occur, makes very clear that separate controls may be acting within the infaunal environment that are absent at the sediment—water interface, i.e., it is necessary in some instances to consider the infaunal taxa as subject to influences not distributed areally in the same manner as those present in the immediately overlying epifaunal environment. This being the case it is probably best for the present to try and keep infaunal associations of shells separate conceptually from epifaunal associations.

An additional complication is the presence during life of more than one community over another in a vertical sense. It is obvious, as discussed above, that infaunal communities may overlap with partly co-occurring epifaunal communities. It is also obvious that various planktic and nektic communities may overlap in a complex manner with epifaunal and infaunal communities (the nektic Pterygotidae-Carcinosomidae Community that occurs associated with many epifaunal communities is discussed in this book as a possible higher trophic-level set of carnivores that occur associated with a variety of epifaunal benthos). What is not so obvious, however, is the possibility for a variety of epifaunal benthic communities existing at different elevations off the bottom as well as directly on the bottom. Communities epifaunal on seaweeds[9] may exist that are different from those present on the bottom, yet after death all may be consolidated into one association concluded by the paleontologist to represent a single community. Different, depth-controlled species of barnacle or periwinkle may exist on rocks, for example, quite separately from each other yet be consolidated together after death into the bottom sediment. These "elevated" benthic epifaunal communities may have an overlapping distribution range relative to the benthic epifaunal communities resting directly on the sediment thus giving rise to a far more complex set of taxonomic associations and abundances. It must also be emphasized that the epifaunal communities discussed in this book have been defined almost exclusively with the brachiopods that form such a large part of the fauna. Little attention or study has been given to the other non-brachiopod taxa in these communities, nor has much consideration been given to the coral and stromatoporoid level-bottom communities except to discuss some of the major subdivisions. This being the case it is likely that careful study of the non-brachiopod, less abundant taxa, in the brachiopod-dominated communities may indicate that there will be considerable overlap of non-brachiopod taxa from community to community, just as there is with some of the brachiopod taxa. More importantly, however, it may turn out that some of the non-brachiopod taxa are distributed in such a manner as to suggest that they belong to separate communities controlled by different combinations of environmental factors, thus making up separate although superimposed communities.

No attempt has been made to consistently consider the effect of bottom sediment type. Present information shows conclusively that there is *no* correlation between the mineralogy of the enclosing rock and the taxa

of brachiopod present; brachiopods obviously lack the ability to discriminate mineral species. This statement about mineralogy is based on the
study of samples from many thousand localities sorted by genus (it is
observed in genus after genus that quartzose sandstone or siltstone, limestone of varying grain size, graywacke, mudstone, even light and dark-
colored volcanic rock as well as chert and dolomite — the last two may,
however, be of secondary origin — and varying composition shales are
associated with the same genus from place to place) with there being little
likelihood that the varying mineralogies associated with the same taxon
can be repeatedly explained by transportation, although it is very common for a particular taxon to be restricted to a particular mineralogy in
any small region (this is because of the far greater chance for correlation
of a mineralogy in a small region with other physical factors that also
correlate more basically with the occurrence of the taxon but are not
obvious to the paleontologist as contrasted with the mineralogy). However, it is probable that adequate petrographic and sedimentologic studies
would show a high correlation between grain size, sorting, sedimentary
structures and taxa present. In other words, those petrographic characters
of the sediment that relate to environment may become very helpful in
trying to understand some of the reasons for presence or absence of
particular communities at a specific locality.[10] Infaunal communities have
not been distinguished separately except in Benthic Assemblage 1 (it is
obvious in Benthic Assemblage 1 that most of the bivalves and linguloids
constitute infaunal communities, indicated in Fig. 3—5 above which live
separate epifaunal communities of articulate brachiopods, chiefly rhynchonellids, gastropods, and some additional bivalve taxa).

It should be emphasized that many of the communities in Fig. 3—10
plus those in Part II, Chapter 6 may contain a variety of smaller divisions
each characterized by a unique fauna. For the present it is reasonable to
consider these smaller faunal units as community subdivisions, although
future studies may prove that they more nearly resemble the communities
of the ecologist. If such comes to pass it will be necessary to devise an
intermediate category, between Benthic Assemblage and Community (a
"Megacommunity"), for the units here termed communities. What is,
however, important is that the rank of the communities employed in any
discussion be more or less equivalent; I have tried to attain this goal but
realize that in some cases (the Bohemian Complex of Communities, for
example, is larger than most of the other communities discussed here) the
units are of greatly differing size. Some of these smaller units may be
intimately related to sediment grade (sand-size or silt-size and finer sediments).

The working out of Benthic Assemblage (i.e., shoreline proximity, depth, or temperature correlated with depth, etc., ranges in which various taxa occur; many being present in more than one such Assemblage) ranges of benthic invertebrates is done by tabulating data from as many collections as possible. Several problems inevitably arise. Collections from steep-slope environments (geosynclinal and reef-related in particular) present the problem of distinguishing between indigenous shells and those that have traveled post mortem for some distance downslope from one Benthic Assemblage into another or occur in small patches. It is easy to ascertain the upper depth ranges of various taxa, as except for the very shallow, rough-water environments there is not very much chance of shells being transported en masse upslope. The most reliable results for lower depth limits are obtained from low-slope platforms characterized by broad reaches having uniform environments, well removed from reef proximity. Geosynclinal collections certainly are employed in such studies, but the results for lower depth ranges must be checked against platform equivalents. Fauna derived from turbidite sequences are obviously displaced, but many other geosynclinal as well as steep-slope trough or basinal faunas may have been displaced for ecologically significant distances either in part or in toto.

Both the Silurian and the Devonian provide examples of quiet-water well-preserved starfish present in positions inferred to be either farther from shore or in deeper water than Benthic Assemblage 5 shelly faunas. E.D. Gill (written communication, 1972) provides typical information bearing on the relatively deep water position of the Starfish Community in the Melbourne region: "At Kinglake, for example, there are thick beds with little stratification and evidence of turbidite activity that are succeeded by the starfish community and then the brachiopod community." It is difficult not to make the comparison between the Starfish Community and the occurrences of abundant starfish found in the deep sea at the present time, although we have every reason to conclude that these mid-Paleozoic starfish lived in deep shelf environments and not in the true deep sea.

Significance of the Low-Diversity Communities

The lower-diversity communities plotted on Fig. 3—10 fall naturally into two groups. The first is comprised almost entirely of single-taxon, rough-water communities occurring chiefly in Benthic Assemblage 3, a few in Benthic Assemblages 2 and 4. The second is comprised almost entirely of communities with from one to four or five taxa that occur chiefly in quiet-water communities in Benthic Assemblage 2, some in Benthic Assemblage 3, and a few in more offshore positions. Quantitative description of diversity in some of these communities is given by Watkins and Boucot (1973). In addition, of course, all of the Benthic Assemblage 1 communities are of very low diversity. The rough-water low-diversity communities are understandable in terms of the very specialized environmental conditions that they reflect. However, the high number of Benthic Assemblage 2 quiet-water low-diversity communities requires more consideration, particularly as there is no suggestion in many cases of a gradation from these low-diversity Benthic Assemblage 2 communities to the high-diversity Benthic Assemblage 3 quiet-water communities.

The restrictive conditions of the shallow intertidal environment reflected in Benthic Assemblage 1 with wide fluctuations in temperature, salinity, and long exposure to the atmosphere account adequately for the low taxonomic diversity encountered for benthic epifauna living in that position (the more stable infaunal environment present in that position permits a somewhat higher-diversity fauna to be present).

E.J. Anderson (oral communication, 1972) has suggested that a rational explanation for the large number of low-diversity, quiet-water communities occurring in Benthic Assemblage 2 is the presence of an environment subject to wide fluctuations in salinity, exposure and temperature such as might occur in nearshore, or intertidal lagoons and back-bay deposits. Jackson (1972) has discussed similar relations from recent examples in Jamaica. E.J. Anderson (oral communication, 1972) suggests that in many instances the rough-water, Benthic Assemblage 3 low-diversity communities may have been associated with offshore bar environments that absorbed much of the turbulent energy generated in the nearshore region, and that they may actually in many instances indicate the presence of shallower environments than present in the more onshore Benthic Assemblage 2 position. Anderson's views are entirely consistent with the available paleontologic and stratigraphic data, although we lack adequate information regarding sedimentary structures (Klein, 1972a) that would bear very significantly on this question. Following Anderson's suggestion, the vertical ordinate in Fig. 3—10 is labeled "Distance from shoreline" rather than "Depth" as although there is certainly an overall trend from shallow to deep as one moves offshore, the real possibility that the rough-water environments present in Benthic Assemblage 3 and 4 may be shallower than some of those present in Benthic Assemblage 2 prevents a 1:1 correlation of distance from shoreline with depth. The quiet-water communities present in Benthic Assemblage 3 may, at least in large part, represent deeper water positions than do the rough-water communities present in that Benthic Assemblage. The quiet-water Benthic Assemblage 3 communities may, in some cases, represent deep water channels through shallower rough-water regions occurring about the same distance offshore. The work of R. P. Sheldon (in: Berry and Boucot, 1970) supports the concept that the *Pentamerus* Community of the Birmingham (Alabama) region lived in shallow water just seaward of an offshore bar region, behind which were deposited more quiet-water lagoonal sediments. There is, however, no reason to conclude that the rough-water communities of Benthic Assemblage 3 invariably lived either in or adjacent to an offshore bar environment. It is possible that the low-diversity quiet-water communities of Benthic Assemblage 2 merely represent, in most cases, a somewhat deeper intertidal position that was subject to far more fluctuating conditions, with corresponding reduction of taxonomic diversity, than the adjacent subtidal region that may have begun with Benthic Assemblage 3 (both rough- and quiet-water communities).[11]

The significance of the quiet-water, low-diversity communities occurring further offshore than Benthic Assemblage 2 is uncertain. Some recent mega-invertebrates show the same pattern (Rowe, 1971). All one may do at present is to call on unspecified, restrictive conditions. The occurrence, for example, in Benthic Assemblage 5 of the single-taxon quiet-water, Notanoplid Community is very puzzling except as caused by some type of unknown restrictive set of conditions.

The final expression of this offshore low-diversity, quiet-water condition is the complete failure of the offshore shelly benthos beyond Benthic Assemblage 6. We currently have no rational explanation for this failure unless one postulates the complete absence of an adequate offshore food supply, a concept that does not appear logical unless one appeals to the absence of offshore plankton, an additional concept that is not very appealing in view of what we know about recent organisms. However, the absence of offshore shelly benthos seaward of Benthic Assemblage 6 appears to be an established fact whether or not it appears logical.[12]

Recognition of Community Mixtures

The recognition and definition of communities within the fossil record depends on the ability to take a series of fossil collections containing one or more taxa characterized by various abundances of the taxa and to distinguish them from collections containing mixtures in varying proportions of these same communities. One deals with a series of statistical arguments aided by a series of geological arguments, all aimed at differentiating mixtures of communities from unmixed communities.

Experience has shown that faunas containing only a single taxon, or a very low number of taxa are more likely to belong to single, low-diversity communities than are faunas containing a very high number of taxa. However, it is important to recognize that high-diversity communities may be recognized and do occur commonly. The critical point here is that a single collection or a small number of collections containing only a single taxon is far more likely to represent a single community than is a single collection containing a large number of taxa. Far more care in sampling and drawing conclusions about mixing or absence of mixing must be taken when dealing with a single high-diversity or a small number of high-diversity collections.

The procedure followed when dealing with high-diversity collections is as follows[13]:

(1) A minimum of thirty to fifty taxonomically similar collections are plotted on a graph (see Boucot and Heath, 1969, table 3) with taxa as one ordinate and localities as the other.

(2) The chart is carefully examined to see what percentage of taxa are common to all collections.

(3) The chart is carefully examined to see what percentage of taxa are not common to all collections.

(4) The original data are examined to see which of the taxa not common to all collections are present as a moderate to high percentage of the collections in which they occur and which are not.

(5) The chart is now replotted with the taxa represented by moderate to high percentage not common to all collections segregated.

(6) The collections characterized by mixtures of the mutually exclusive, moderate to high percentage taxa not common to all collections are now segregated from those which are mutually exclusive.

(7) If a statistically significant number of mutually exclusive collections are segregated from the mixtures it is now time to test the data geologically.

(8) The mixtures are plotted stratigraphically and geographically in order to determine whether they occur in stratigraphic positions intermediate between the mutually exclusive collections and in geographic positions in the same horizon intermediate between mutually exclusive collections.

This procedure will aid in distinguishing and characterizing communities having common boundaries where mixing of taxa has taken place if the communities existed in a relatively low slope region (Fig. 14). The essential point again is to recognize discrete distinctive recurrent associations of taxa. These should occur as mappable distributions in time and space.

If the fossil collections are derived from regions of high slope a different procedure must be followed.

(1) The taxa obtained from the high-slope collections must be referred to communities containing the same taxa defined in low-slope regions.

(2) Such referral will indicate which taxa in a high-slope collection may be referred to communities occurring elsewhere in low-slope environments.

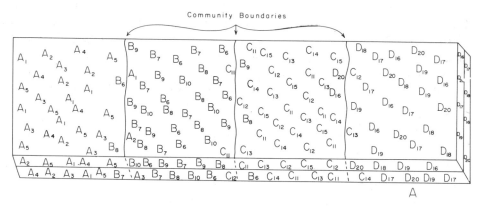

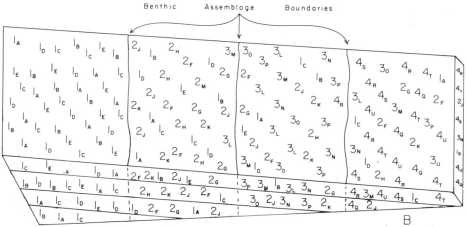

Fig. 14.A. Diagrammatic representation of four communities adjoining each other in a low-slope, platform-type environment. The lateral extent of the figure is in hundreds of kilometers. Note that mixing of the taxa occurs only adjacent to community boundaries. All of these communities belong to a single Benthic Assemblage (the individual communities are lettered with numerical subscripts denoting individual taxa).

B. Diagrammatic representation of four communities adjoining each other in a very high-slope, geosynclinal-type environment. The lateral extent of the figure is in tens of kilometers only (*note* that the lateral extent of Fig. 14A is in hundreds!). Note that mixing of the taxa occurs cumulatively downslope; not just adjacent to community boundaries. Each of these communities belongs to a separate Benthic Assemblage because of the steep slope (the individual communities are given numbers with the letter subscripts denoting individual taxa).

(3) After this procedure has been followed it will be apparent whether the high-slope collections happen, uncommonly I might add, to consist chiefly of single communities, communities derived from the same benthic assemblage or communities derived from more than one benthic assemblage. It follows that high-slope collections from shallow-water will be less mixed than those from deep-water due to downslope, cumulative mixing.

Examples of these problems that come to mind are as follows (Fig. 14):

(1) Low-diversity, single-taxon collections from either high- or low-slope regions: *Pentamerus* and *Kirkidium* Communities (relatively rare in the high-slope environment but common in the low-slope environments).

(2) High-diversity, multi-taxon collections from either high- or low-slope regions: *Striispirifer* and *Dicaelosia-Skenidioides* Communities (relatively rare in an unmixed condition in the high-slope environments but common in the low-slope environments).

(3) Moderate-diversity, multi-taxon collections from either high- or low-slope collections in shallow-water, nearshore positions: *Salopina* Community (relatively common in both high- and low-slope environments).

(4) High-diversity, multi-taxon collections from high- or low-slope in deeper-water, offshore positions: mixtures of *Dicaelosia-Skenidioides*, *Conchidium* and *Gracianella* Communities such as occur in the Roberts Mountains Formation (common in the high-slope environment but uncommon in the low-slope environment).

It must be recognized that mixtures of more than two communities will be rare in low-slope environments, due to the far lower probability of common boundaries between more than two communities occurring, whereas mixtures of more than two communities will have a far higher probability in high-slope regions because of the possibilities for downslope transport and consequent mixing.

Finally, there is the fact that high-slope collections from both the geosynclinal and reef environments are more prone to "mixing" because of the original patchiness and scatter of many communities within a small area. This original patchiness is merely a reflection of the fact that the steep-slope environments offer more opportunity for environmental variability over short distances (the reef environments in addition offer the possibilities of biologically conditioned environments) as contrasted to the far more uniform level-bottom platform environments. Analysis of steep-slope, geosynclinal faunas does not suggest that they reflect environments different from those present on the platforms, but that they do represent far smaller, more randomly scattered patches of similar environments that also offer more possibilities for mixing adjacent to the far greater length of community boundary available plus the possibilities for downslope transport absent on the platform.

Shell Transportation and Mixing

The study of marine invertebrate megafossils for obtaining an understanding of ancient animal communities is highly dependent on the ability
to recognize the amount of transportation, or lack of it, from original
growth sites plus mixing of taxa derived from differing communities. Few
Early Paleozoic marine invertebrates occur in such a manner (original life
position, for example) that one may tell by simple inspection that a
sample contains specimens belonging to a single in situ community. The
rule is to observe collections of shells that have suffered obvious disarticulation of bivalved specimens, breaking of shell margins during movement,
and the like. However, as pointed out by Boucot et al. (1958a) the absence
of shells in life position does not necessarily signify significant net lateral
movement from the growth site.

Studies of shell transport in the modern environment (Martin-Kaye,
1951; Lever, 1958; Craig, 1967; Behrens and Watson, 1969; Warme, 1969)
show that shells are seldom moved for more than a few hundred yards
from their growth site to their depositional site (at least in any abundance) within the beach-near beach region.

Ekdale (1973) has reported on an interesting experiment comparing the
constituents of death assemblages and associated living shells off the eastern coast of Yucatan that indicates movement after death not to be a very
significant factor in shallow-water, subtidal environments. The studies of
Shaw discussed in this section are also confirmatory of this conclusion.
Long-distance transport of shells downslope from the original growth site
does, however, take place through the agency of turbidity currents. There
are numerous examples (for example, the occurrences in the British Basin
Facies Ludlow of *Dayia* Community and Gypidulid Community shells in a
highly sorted, disarticulated condition as well graded constituents) of such
transport in the Phanerozoic record. These downslope transported materials
are normally easy to recognize because of the ecologic (many communities
represented in the same collection) mixtures they include, and their occurrence with sedimentary structures characteristic of turbidite-type sedimentation (graded beds, sole marks, graywackes, etc.).

The "mixtures" occurring in non-turbidite, low-slope type sedimentary
sequences raise serious questions. It is probable, as pointed out here by
Shaw, that most of these mixtures represent changing conditions in time
at the same locality (if amount of non-biogenic sediment is very low) or
areas situated near the boundaries of communities (where movement of a
few meters or less is more than adequate to account for such mixing).
Study of such "mixtures" location in stratigraphic sections, in time, and
for single horizons, in plan, normally indicates very conclusively the

boundary nature of the "mixtures" (Boucot and Heath's, 1969, table 3 presents a typical example). As discussed by Boucot et al. (1958a) there are simple statistical techniques for studying the size-frequency distribution, ratio of opposing disarticulated valves, percentage of articulated valves and the like that aid greatly in determining the likelihood of extensive transport away from a growth site as well as the possibilities for mixing of constituents from more than one growth site at a given depositional site.

The question of shell transportation and resultant mixing in areas of shallow slope (platform environments) has been little investigated. The presence of many distinctive coeval faunas and the good preservation of many such faunas suggest that extensive transportation for many tens of miles is unlikely. A.B. Shaw (written communication, 1972) reports that in the low-slope Florida region ". . . we have found that even the great hurricanes like Donna, which was strong enough to cut one key in half, do not cause any significant mixing of the bottom shells. Transport of shells more than a few yards seems to be unknown. This can be checked readily in the Bay because there is a very strong depth effect, and shells moved from water nine feet deep would be clearly separable from shells in water four feet or one foot deep. The ridges of shells heaped up on the windward sides of keys by the hurricanes all come from the flats a few yards offshore; there is no sign of shells from the bottom of the 'lakes' half a mile away. This observation implies to me that transport of shells on the floor of shallow epeiric seas can be discounted as an explanation of 'mixed' assemblages of fossil shells. Far more likely, in my estimation, as a cause of 'mixing' are the changes in micro-environment that occur over a period of years in any given area. In the areas between the reefs off Florida, where great masses of staghorn coral lie dead on the bottom, off-reef microgastropods are now accumulating among the coral fronds. The 'mixing' of reef and non-reef remains is occurring, but there is no significant transport." Shaw's information plus the circumstantial information provided by platform deposits combine to suggest that lateral transport and mixing are not too important for distances over a few hundred yards, even in areas of very steep slope adjacent to reef topography. Shaw's suggestion that the platform "mixed" communities are probably the result of changing environment rather than transportation fits the data very well.

Macrotidal and Microtidal Consequences

Gill (1973) has summarized some of the effects of both macrotidal and microtidal regimes on the intertidal and nearshore regions. He emphasizes that a microtidal regime concentrates physical activity within a smaller area than do macrotidal regimes; and that microtidal regimes have the property of compressing both biologic and physical zones, in addition to being associated with negligible tidal currents and depressed salt marsh development. Gill emphasizes that the area available for intertidal organisms is much smaller under microtidal conditions, and consequently that the total populations (although not the population densities) are far smaller than on macrotidal coasts.

The properties of macrotidal and microtidal regimes must be considered in connection with the distribution and composition of intertidal animal communities on both steep-slope and low-slope shorelines. In addition, one must consider the possible effect of macrotidal and microtidal regimes on shallow-water, subtidal animal communities.

The first conclusion to be made from Gill's work is that intertidal communities (probably Benthic Assemblage 1 and 2 communities) preserved from a microtidal regime will show far more evidence of mixing due to the higher amount of energy per unit area expended within the microtidal region and the smaller area occupied by the intertidal fauna. The higher the slope present in such a microtidal region the more pronounced these effects will be.

For example, the far greater degree of faunal mixing found in the Late Llandovery of the Welsh Borderland as compared with that known from the Appalachians (north, central and south) has been ascribed to the effects of slope because of the narrower band occupied by shelly faunas in Britain than in the Appalachians. The effects of a microtidal regime, if present in Britain, would also tend to reinforce this effect of steep slope within the intertidal region.

It is, of course, important that the physical effects of the more concentrated, more prolonged intertidal high-energy environment of the microtidal regime also be looked for. This has not been done within either the Welsh Borderland or the Appalachians, although one has the impression that such features as boulder beds, and extensive wave-cut platforms, as well as the absence of salt marsh developments might fit the Welsh region far better than the Appalachian region.

The effects of microtidal and macrotidal regimes on the Lower Paleozoic platforms have not been considered in terms of animal communities. The absence of widespread animal community mixing, as well as of evidence for extensive gullying and the like suggests that strong intertidal currents (as well as subtidal currents) did not effect these platforms. In other words, a microtidal regime over the platforms should favor relative stability of animal community boundaries and lack of faunal mixing in the subtidal environment (the most widespread in the Lower Paleozoic), although encouraging faunal mixing in the theoretically narrower intertidal regions.

In principle, the smaller area occupied by the microtidal intertidal fauna, if very widespread in the world, should encourage more rapid evolution because of the implied smaller population size. Unfortunately, we are not in a position to assess the distribution of microtidal as opposed to macrotidal environments on a worldwide basis in the Lower Paleozoic, although their potential importance should not be ignored when trying to explain relations.

Anderson's Model

Anderson (1971) has skillfully reviewed the available published and unpublished information about Late Ordovician through Late Devonian marine animal communities. In such a rapidly developing field varying opinions are to be expected about interpretation of the same data, and new interpretations will arise based on additional data. Anderson pointed out that Bretsky's (1970) Late Ordovician three-"community"-division (Anderson, 1971, fig. 2) is associated with an offlap situation whereas Ziegler's (1965) Late Llandovery five-"community" division (Anderson, 1971, fig. 1) is associated with an onlap situation. Anderson (1971, fig. 4) went on to extrapolate the five-"community" onlap subdivision as typical of most of the Silurian and Helderberg, followed by a three-"community" subdivision in the Oriskany and its equivalents (he interpreted the three-"community" subdivision of the Oriskany equivalent's as an offlap sequence in a large part of the Central Appalachians Shriver Chert through Ridgely Sandstone sequence).

Anderson proposed two environmental models to explain the presently known community and Benthic Assemblage data for the Middle Paleozoic. The two models are related (1) to a low-slope stable or transgressing regime in which low-, high-, and then low-energy environments are encountered as one approaches the shoreline; and (2) to a higher depositional slope associated with a prograding shoreline regime in which low- followed by high-energy environments alone are encountered as one approaches the shoreline.

There is no question that Bretsky's Late Ordovician "communities" are 3-fold and associated with an offlap situation and Ziegler's Late Llandovery "communities" are 5-fold and associated with an onlap situation. But, in other regions, including Arctic America, the northern Ural-Polar Ural regions, as well as the Baltic region, one finds Ludlow-Pridoli offlap regimes that follow a four-community pattern (*Lingula* Community)—(*Atrypella* or *Didymothyris* Community)—(*Conchidium, Kirkidium, Dayia, Dubaria* or *Gracianella* Community)—(*Gracianella* Community associated with *Dicaelosia-Skenidioides* Community). In the Late Silurian of the Welsh Borderland, an onlap sequence is present with a brackish-water *Hughmilleriidae-Stylonuridae* Community underlain consecutively by (*Lingula* Community)— (*Salopina* Community)—(*Striispirifer, Dayia* or *Kirkidium* Community)—(*Dicaelosia-Skenidioides* Community) sequence of five Benthic Assemblages. In the Malvinokaffric Realm Devonian, an onlap—offlap sequence is present with a symmetrical four-community sequence (both in the onlap and offlap phases). Thus there does not appear to be a 1:1 correlation in the Silurian-Devonian between a unique number of communities for onlap as opposed to offlap. The Llandovery worldwide does represent onlap conditions in its shelly facies (see Berry and Boucot, 1973a, for the glacial-eustatic control of this phenomenon), except for the offlap present in North Africa where only Benthic Assemblages 1 and 2 are present in the Llandovery. However, the North African offlap is interpreted (Berry and Boucot, 1973a) as offlap associated with postglacial isostatic rebound combined with a cold-water Malvinokaffric regime different than that of the North Silurian Realm (see Fig. 25). Anderson's suggestion that the *Stromatolite-Howellella* Community (essentially our *Tentaculites* Community) and a part of the *Mutationella* Community are Benthic Assemblage 1 equivalents (his *Lingula* Community correlatives) is questionable, in my opinion, but does not affect Anderson's conclusions regarding the three- versus the five-community model. However, Anderson (oral communication, 1972) now concludes that the specimens of *Howellella* in question were transported from a more offshore position to their depositional site, which would permit him to place the *Mutationella I* Community in Benthic Assemblage 2 while retaining the *Stromatolite-Howellella* Community (now conceived of as containing transported *Howellella*) in Benthic Assemblage 1. Anderson's model is also questioned because of almost total taxonomic non-continuity between Bretsky's (1970) communities, based on taxa of the highly provincial Late Ordovician North American Realm, and those of the North Silurian Realm. Boucot (1968a) pointed out that the Llandovery and younger Silurian brachiopod fauna was derived from the Old World Realm Ashgill, not from that of the North American Realm. Therefore, Bretsky's (1970) Late Ordovician Communities may overlap the taxonomically unrelated Silurian entities. For example, we lack assurance that Bretsky's farthest offshore communities are equivalent to Benthic Assemblage 5.

Anderson's two models have much merit in that they emphasize that the number of communities encountered across a seaward profile may differ. Anderson's models also have merit in that they link benthic animal community distributions to observed physical characters present in the beds, that is to environmental patterns. One may disagree in detail with Anderson's conclusions but the overall effort to relate animal community distributions to the distribution of physical characters is essential if we are to more fully understand the significance of the animal communities in environmental terms. However, alternate explanations also exist (a fact which Anderson himself partly anticipated) for the varying numbers.

To illustrate, I first attempt to analyze Ziegler's (1965) original concept. There is general agreement that Benthic Assemblage 1 (Ziegler's *Lingula* Community) may be characterized by *Lingula*, abundant bivalves, and certain poorly studied rhynchonellids, although it does not follow that all of these taxa occur together. Benthic Assemblage 1 faunas may be subdivided into several environmentally controlled communities (Fig. 4 is a preliminary attempt for the Late Llandovery—Ludlow). The basic essentials of Benthic Assemblage 2 (Ziegler's *Eocoelia* Community) were recognized by Ziegler on the basis of abundant *Eocoelia*. Other brachiopod taxa do occur within Benthic Assemblage 2, but in the pioneering stage of the work they received little attention. Certain of these, like *Howellella*, are known elsewhere to extend seaward into deeper-water Benthic Assemblages (4 certainly and possibly 5) in situations where their presence cannot be ascribed to transportation. However, Benthic Assemblage 2 contains other taxa (e.g., *Salopina*, *Protochonetes*) that do not extend into deeper-water regions. At least two communities, or subdivisons, of Benthic Assemblage 2 can be recognized: a 'pure' *Eocoelia* Community (the 'pearly layers' of the Sodus Shale of western New York, for example), and the *Salopina* Community, containing abundant spirifers, atrypaceans, chonetids, rhynchonellids, and stropheodontids (it is similar to the *Striispirifer* Community although with a reduced number of taxa). Present information would suggest that the more taxonomically varied community inhabited relatively quiet water, but lack of good data does not permit any conclusion relative to the *Eocoelia* Community. This conclusion accords with the general increase in taxic diversity as one moves offshore.

Benthic Assemblage 3 (Ziegler's *Pentamerus* Community) may also be easily subdivided. As Anderson recognized, the *Pentamerus* Community, the *Kirkidium* Community, and the Gypidulinid Communities of this Benthic Assemblage appear to represent a rather rough-water environment. These rough-water communities contrast strongly with the taxonomically more varied *Striispirifer* Community (the more restrictive nature of the rough-water environment correlates strongly with the lowered taxonomic diversity), as well as the low-diversity *Atrypella*, *Dayia*, and *Didymothyris* Communities of relatively quiet-water environments (as indicated by the common occurrence of articulated shells in an argillaceous matrix). The Trimerelloid Community is another rough-water representative of Benthic Assemblage 3. Similar reasoning is followed for Benthic Assemblage 4, within which the *Stricklandia* Community may represent rough-water as opposed to the taxonomically more varied quiet-water communities (Fig. 4). In areas of steep bottom slope, as in the Welsh Borderland and the Northern Appalachians, these varying communities commonly are mixed in differing proportions, but this occurs in very few places on the low-slope platform (see Fig. 14A), except in boundary regions. There is no good evidence to date for rough-water communities within Benthic Assemblage 5 or 6. These last assemblages probably occur at depths too great for rough water.

Thus, from inspection of Fig. 4 and 6 it may be seen that onlap during the Llandovery might be accompanied by the presence of from five to eleven or more brachiopod communities, and during the Ludlow might be accompanied by the presence of from five to twelve or more brachiopod communities in a traverse seaward from the shoreline. During Oriskany time a similar traverse might encounter from three to six or more brachiopod communities. The precise numbers actually encountered will differ from time interval to time interval and from biogeographic unit to biogeographic unit with strong dependence on the distribution locally of important factors in the environment. Bretsky's (1970) data indicate that the communities within his major divisions have an irregular distribution that may have depended on several variables in addition to depth and shoreline proximity.

Inspection of Fig. 3—10 shows conclusively that there is nothing unique about the numbers three and five when considering the distribution of communities present from the shoreline out to deeper, offshore regions. Ultimately, after consideration of rocky-bottom, biostromal environments, reef environments, etc., the Benthic Assemblage situation must account for a series of communities and subcommunities controlled by various physical and biological factors, with internal feedback. All of these communities are statistical concepts for data that may be divided in a number of ways. The only requirement is that the divisions be consistent with the data.

Finally, there is a basic biologic objection to Anderson's conclusions. Anderson's synthesis suggests that there will be a specific and stable number of communities in both an onlap or offlap regime from the Late Ordovician through the Late Devonian. This conclusion would require that organic evolution progressing during this time interval not affect the niche breadth tolerances of any of the taxa existing during the time interval enough to alter the number of communities. Such a conclusion is very unlikely from a probability viewpoint. It also disagrees with much of the data for the Silurian and Devonian.

Complicating any effort at generalization about systematically changing environments is the knowledge that in some regions, and in some stratigraphic sections, there is good evidence for very rapid environmental changes back and forth that defy any attempt at generalization unless *very* detailed, cm by cm study is carried out. Attempts to generalize overall environmental shifts in such regions from information provided in the older literature are doomed to failure except as they may afford a very gross overall picture that may give misleading ideas concerning the actual relations.

Modern Continental-Shelf Environments versus Ancient Platform Environments

The attempt by a paleontologist specializing in the Early Paleozoic to take advantage, for comparative purposes, of the information gathered by marine ecologists concerned with benthic organisms inhabiting the modern continental shelves is frustrating in many ways. Uniformitarianism dictates that modern examples should throw considerable light on similar situations in the past, but it by no means guarantees that all past situations are indeed similar. The ancient continental platforms with their covering marine deposits laid down in vast epicontinental seas certainly have no modern counterparts (only much smaller, more restricted platforms). Therefore, one must be cautious in trying to compare rigorously the ecologic results and the conclusions obtained from the modern continental shelves to the ancient platform deposits. For example, the vast expanses of Early Paleozoic platform covered by a single community during some time intervals have no modern counterpart, and may require a very different ecologic interpretation than any provided by the modern record. The present day presents no examples of vast areas of hypersaline water occurring in the midst of relatively normal marine shallow water and affecting organismal distributions. One cannot be sure what type of water circulation (probably very localized; not long-distance, short-term) would occur on a vast platform, or what type of cycle would have been followed by various nutrients on these platforms, or of their relation to plankton abundance. One can reasonably conclude that recycling of calcium carbonate, more or less in place, was the rule for the interior parts of the platforms (how else to explain the relatively small carbonate thickness preserved as fossil accumulations? These generally do not exceed 100—200 m for an entire Period). If one assumes any reasonable growth rate for the carbonate secreting invertebrates of those times it implies that the preserved limestone and dolomite or organogenic origin is only a small fraction of the total production. In view of its solubility it is remarkable that so much organogenic carbonate has been preserved on the ancient platforms. An additional and perplexing problem is the lack of solution effects shown by some fossils in these platform carbonate rocks. However, many of these well preserved shells are preserved in relatively uncommon shaly matrix that may have protected them from solution. If one concludes that most of the shells produced on these Early Paleozoic platforms were dissolved, with only a fraction of a percent being preserved, it is comforting to find that some specimens show marked evidence of solution (E.J. Anderson, written communication, 1972) except for those preserved in shales (low permeability). Replacement by dolomite has not altered the situation as even here the shells may preserve all their surface morphology in great detail.

Possibly because of the slower rates of sedimentation, and the far smaller influx of terrigenous material, the carbonate platform deposits are characterized as well by a wealth of stromatoporoidal "reef" communities, and a rich association of level-bottom communities, communities with tetracorals, and also others with abundant calcareous algae. These reef and coral, stromatoporoid, algal-dominated level-bottom community associations account for much of the "difference" between geosynclinal and platform community aspects within the Silurian and Devonian. The temperature distributions present on the platforms (as opposed to modern seas) are unknown. However, it is reasonable to infer that the thin sheet of water present on such platforms would have been far more thermally mixed (although subject to diurnal effects) and uniform than a column of far greater depth in an ocean. The presence of a single widespread community, *Virgiana* for example, during some time intervals, certainly suggests a constant thermal environment over most of a platform. The absence of cooler water at depth may signify that platform water temperatures were uniformly higher than those to be expected in surface waters in the modern oceanic environment.

Shaw (1964) suggested that tidal currents would probably be very ineffectual on the platforms of the past, and that large currents of the type present and effective in modern oceans would have been prevented by friction in the very thin sheets of water on the platforms.

G. de V. Klein (1972b) finds evidence for the existence of tides on the Paleozoic platforms and does not feel that the tidal ranges were different than those present today. All one can do is to inspect the fossils, note their mode of occurrence and distribution in time and space, note something about their depositional environment (rarely about their living environment from life position), and then attempt to draw comparisons with the modern environment. From the above it is clear that one should expect serious problems to arise during the making of such comparisons. Therefore, ecologists using the present as their frame of reference, and paleontologists employing platform data derived from the Early Paleozoic, may find themselves in serious disagreement over basic questions without realizing just why they are disagreeing. At the moment earth scientists certainly know too little about the environments present on the ancient platforms to be able to understand just what all of the significant differences may be, but I certainly feel that all of us should entertain the possibility that significant differences do exist, and that they may alter seriously the ground rules of the game.

Absolute Depth

It is important to try and assess as many of the physical factors in the environment as possible in order to arrive at meaningful ecologic conclusions. Absolute depth is one such variable that should be considered. The strand-line position is determined with reference to the position of the boundary between the brackish-water communities and Benthic Assemblage 1 on the one hand and with reference to the facies change from non-marine to marine-type sediments (largely determined by means of fossils in both cases) on the other hand (Fig.15). Going into deeper environs it is reasonable to conclude that those areas yielding an abundance of algae, calcareous algae in particular, were within the zone of photosynthesis and probably did not extend below about 200 ft. in

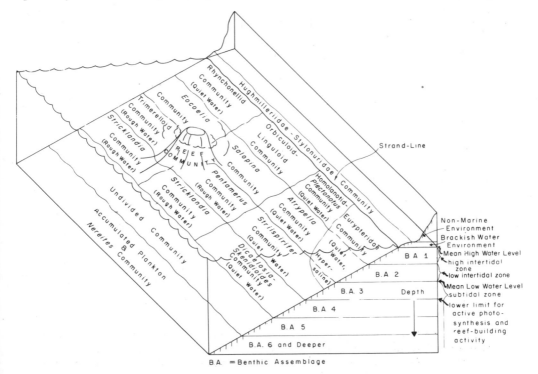

Fig. 15. Block diagram indicating inferred relations between Silurian epifaunal Benthic Assemblages and communities.

maximum depth. This zone of abundant calcareous algae corresponds essentially to Benthic Assemblages 1 through 3 (Adey and Macintyre, 1973, have reviewed the depth significance of modern calcareous algae). Active reef growth does not normally extend much below 200 ft. As the reef complex of communities is largely restricted to Benthic Assemblage 3 it is probable that the outer depth limits of Benthic Assemblage 3 do not exceed 200 ft. in most instances. The relatively narrow ribbon occupied by Benthic Assemblages 4 and 5 in many regions, with little physical evidence for the presence of very steep slopes, suggests that the outer limit of Benthic Assemblage 5 does not exceed more than a few hundred feet additional depth over the outer limit of Benthic Assemblage 3.[14] All in all an outer depth limit for Silurian-Devonian shelly benthos of 500—600 ft.

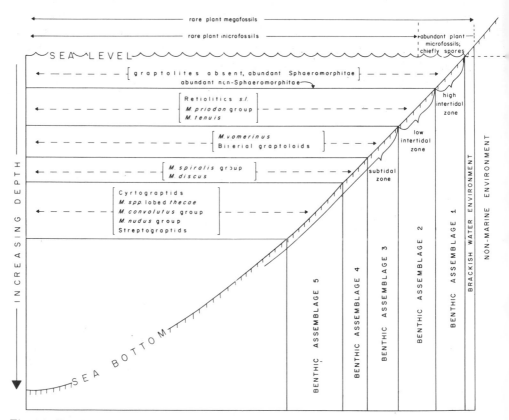

Fig. 16. Depth distribution of Silurian acritarchs (after Gray and Boucot, 1972), Silurian graptolites (after Berry and Boucot, 1972b), and plant spores (after Gray and Boucot, 1972), plotted against Benthic Assemblages (including Kjellesvig-Waering's, 1961, brackish-water Hughmilleriidae-Stylonuridae Community). "Sea Level" in this Figure is equivalent to Mean High-Water Level.

appears most consistent with the present evidence. The abundance of land-plant type spores in Benthic Assemblage 1 pre-Devonian beds is entirely consistent with this conclusion (Fig. 16). Winterer and Murphy (1960) provide an excellent example of the rapid change over a short distance (well under a mile) from the outer limits of Benthic Assemblage 3 (with Pentamerinid Communities known in places) to the outer limits of Benthic Assemblage 5. Winterer and Murphy's evidence does not suggest the presence of excessively steep slopes between the outer limits of Benthic Assemblages 3 and 5.

Depth significance of the Benthic Assemblages

It is critical to try and determine the depth significance of the Benthic Assemblages. Benthic Assemblage 1 with its included *Lingula* Community is concluded to represent an intertidal position because of its paleogeographic position adjoining non-marine facies rocks as well as the common occurrence today of abundant linguloid brachiopods in the intertidal environment. Benthic Assemblage 2 is characterized by single-taxon communities and also by higher-diversity communities. However, these higher-diversity communities have consistently lower taxonomic diversity than those occurring in Benthic Assemblage 3 and deeper positions (not, however, than the specialized rough-water communities occurring in Benthic Assemblage 3 and deeper positions). In addition the Benthic Assemblage 2 communities tend to be characterized by larger specimens (sic. species) of the same or related genera than are present in more offshore Benthic Assemblages. The *Hipparionyx* Community, a Benthic Assemblage 2 community, was originally (Boucot and Johnson, 1967a) termed the 'Big Shell Community' to emphasize this fact. Calef (1972) has pointed out the same thing for Ludlow communities in the Welsh Borderland. It is obvious that most Benthic Assemblage 4 and 5 communities in the Silurian and Devonian of the world contain far smaller shells[15] than do those in Benthic Assemblage 2 (the fauna of Benthic Assemblages 4 and 5 in the Roberts Mountains Limestone in Nevada is a conspicuous example). The only major exceptions to this generalization are the rough-water Benthic Assemblage 3 and 4 communities of the Pentamerinid Communities type. Quiet-water taxa tend, in general, to be smaller than rough-water taxa regardless of Benthic Assemblage. The systematically lower taxonomic diversity of Benthic Assemblage 2 communities suggests, in line with the findings (Jackson, 1972; Stanton and Evans, 1972) for the Recent, that Benthic Assemblage 2 represents a lower intertidal position. This conclusion is consistent with the systematic differences in size. The systematic differences in size may be related to a richer food supply

(Calef, 1972) associated with the more turbulent environment present in both the intertidal and rough-water environments. These systematic differences may also be related to a potentially higher rate of metabolism made possible in warmer, shallower waters than in deeper, colder waters acting in concert with a depth-stratified and/or shoreline-related food supply. The supply of phytoplankton would be predicted to have been richer in shallower depths, and also to have been richer in nearshore regions. The far higher abundance of land-plant type spores in Benthic Assemblages 1 and 2 during the Silurian is consistent with the presence of either a shallow-water, marine flora or of the transportation and deposition of this type of plant material to this position in abundance from the land. Both of these cases are consistent with an enhancement of the food supply in Benthic Assemblage 1 and 2 positions as contrasted with further offshore positions. [16] If Benthic Assemblage 1 is interpreted as "high" intertidal and Benthic Assemblage 2 as "low" intertidal then we have the possibility of estimating something about absolute depth for both. Assuming that most of our material from both of these Benthic Assemblages lived in a macrotidal regime (the lack of microtidal features is consistent with this assumption) then a lower depth limit below mean high water for Benthic Assemblage 2 of between 20 and 30 ft. is reasonable (Klein, 1972b, p. 402, finds no physical evidence in the Lower Paleozoic for tidal ranges different than those present today). [17]

A lower depth limit for Benthic Assemblage 3, as discussed in the earlier part of this section, is taken as about 200 ft. because of the restriction of reef and reef-like structures to this position; the assumption being made that as active reef growth today, as well as algal growth, does not occur below about 200 ft. it is reasonable to assume the same situation during the past. However, it is important to point out that the upper limits of the reef environment probably extended into the intertidal region, as is the case today, although a position offshore is occupied by such reefs. Therefore, it is reasonable to conclude that level-bottom communities belonging to Benthic Assemblage 3 will range from about 20 or 30 ft. as an upper limit to 200 ft. below mean high-water level as a lower limit.

As detailed earlier it is concluded that the lower limits for Benthic Assemblages 4 and 5 do not exceed 500—600 ft., although we lack any good depth indicators as such. [84]

In a general way we have assumed that shoreline proximity is a function of depth, as the physical features associated with more offshore Benthic Assemblages in general suggest deeper water than do those occurring nearshore. However, the rough-water communities in Benthic Assemblages 3 and 4 pose a problem. These rough-water communities may indicate the presence of shallow regions, within the zone of active turbulence, occurring offshore and related or unrelated to bars and bar-like features. The absence of these rough-water communities intertongueing with the non-marine facies certainly shows that they occur offshore, as is also the case for the reef environments, but this does not demonstrate necessarily that they represent deeper water than occurs in all of Benthic Assemblage 2, or even as deep as Benthic Assemblage 2. The absence of Benthic Assemblage 2 communities on the seaward side of rough-water communities belonging to Benthic Assemblages 3 and 4 does suggest that their seaward boundary represents a deeper environment than does their landward boundary.

The conclusions presented here about the relations existing between the Benthic Assemblages, shoreline proximity and absolute depth may, as mentioned earlier, be partly controlled by food supply. Food supply is, of course, for paleontologic purposes a most difficult quantity to evaluate. If the conclusions about the outer limits of Benthic Assemblage 3 corresponding with the lower depth limit of the photic zone are correct we may also conclude that primary food supply would have been far more abundant in a shoreward direction from this point than in a seaward direction. This photic zone effect would, of course, also be influenced in part by the presence of land areas with their potentially increased supply of nutrients made available for nearby primary production and by any local upwelling of nutrient-rich waters. It is, unfortunately, not now possible to produce any reliable data about the benthic biomass characteristic of the various Benthic Assemblages. The overall decrease in benthic biomass encountered from the shallow-water environment to the deep-water environment (Rowe, 1971) might be detectable in transects across the Paleozoic shelf deposits. I have the general "impression" that benthic shelly abundances do go down as one progresses from Benthic Assemblage 2 through 5 and 6. But, this "impression" is very subjective, subject as it is to the inspection of collections made from environments having highly different rates of sedimentation, not to mention the varying preservation problems that would tend to strongly bias the results as well.

Province and Community Extent

In many instances it is clear that all of the communities belonging to a particular biogeographic unit are confined to that unit. However, there are a few examples of communities occurring in more than one biogeographic unit. For example, the *Kirkidium* Community occurs in both the North Atlantic and Uralian-Cordilleran Regions of the Late Silurian. In the Givet, the *Stringocephalus* Community occurs in the Old World and Eastern Americas Realms and also within all of the subdivisions of the Old World Realm.

Caution should be exercized, however, in deciding that a particular community occurs in more than one biogeographic unit *if* the taxa within the community have not been carefully studied. The Orbiculoid-Linguloid Community, common to most of the Silurian-Devonian biogeographic subdivisions in Benthic Assemblage 1, may actually be a motley assortment of communities (a Community Group at best) that are merely lumped together due to lack of study.

The Lower and Middle Devonian is a time of gradual increase in number of reef and reef-related structures together with the number of communities associated with them. The reasons for this increase from a Pridoli-Gedinne low to an Eifel-Givet high, together with the abrupt reef termination at the end of the Frasne are not known. However, the effect on number of communities is pronounced. Study of reef communities will undoubtedly make the present anomaly in increasing number of communities during the time interval even more pronounced.

Number of Level-bottom Communities

The theme of this treatment, that area occupied by a species is a direct function of population size, requires that a reasonable estimate be made of the number of communities containing the species. Fig. 22H indicates that differing numbers of level-bottom articulate brachiopod communities have existed from time interval to time interval through the Silurian-Devonian. Therefore, it is necessary to consider the controls over total number of level-bottom communities that may exist at any one time. Fig. 11B is intended to provide a diagrammatic view of the number of biogeographic units, as well as their relative area, that may exist on a hypothetical continent under conditions of equable climate, steep climatic gradient, and presence of a single, continent-spanning biogeographic barrier. Note the differing number of biogeographic units possible under these differing conditions. Also keep in mind that differing biogeographic units existing under the same climatic conditions may give rise to similar numbers of functionally similar, analogous or homologous (depending on the earlier biogeographic history of the involved taxa) communities. But, there are many controls additional to pure biogeography that result in varying numbers of level-bottom communities. Slope of the shelf or platform will have a serious affect on those communities closely correlating with depth (for example, those that are controlled by temperature, a variable that commonly varies directly with depth, at least locally). It is conceivable to have temperature-zoned situations, with either exceptionally narrow or exceptionally broad platform reaches, that exclude certain communities (or at least make the area of their occurrence so small as to be probably not noticed in the geologic record). Unusually steep slope conditions near the margins of the continental shelf (under conditions of high regression) may also act to exclude certain communities from the fossil record if not from actual presence. During conditions of transgression or regression, high provincialism or cosmopolitanism it is probable that the distribution of bottom sediment types and various physical factors that add up to varying environments will have a very profound control over number of communities. It is reasonable to conclude that the number of such physically controlled environments, as well as their areal importance, in time will vary within very broad limits so as to effectively give rise to apparently varying numbers of communities in the geologic record. The disposition of shallow-water marine currents (Gulf Stream types) over the platforms and continental shelves in time should also have played a very important role, in those cases where not responsible for actual provincialism, in the distribution of nutrients and in adjusting the marine temperature regime locally in a manner that might greatly affect the number of communities present in a region.

Examination of the Silurian-Devonian platform record suggests that during certain time intervals individual communities may cover either a very large portion of a platform or a very small portion so as to give rise collectively to intervals of time during which either many or few communities were present. The analytical problem is satisfied by merely tabulating the number of communities per interval of time, as well as their areal importance, but the genetic problem is something else. For example, during the Early-Middle Llandovery interval a small number of communities have been recognized on the platforms, and of these the *Virgiana* Community appears to have covered well over ninety percent of the available carbonate platform area. Whereas, during the Wenlock an equivalent area of the carbonate-covered platforms affords a far greater number of communities with far less areal dominance of any one community. Clearly some factor, or factors, that controls areal dominance has been at work unevenly in time. Transgression-regression does not appear to hold the answer as intervals of high transgression are known with both a small and a large number of communities (even after normalizing for differences in level of provincialism). We probably will lack any profound insight into this question until such time as the correlations between physical nature of the sediment associated with the communities has been established in enough detail to provide the necessary insight.

This discussion is, of course, concerned with level-bottom communities. During time intervals of widespread reef environments the numbers of reef and reef-related communities may be very high. It is important in any discussion of total number of animal communities existing among the benthos to keep the reef and reef-related units separate from the level-bottom entities.

CHAPTER 3

Rates of Evolution

Point Sources, Marginal Sources, Broad Sources and Rates of Evolution

One should be able to test whether the concept that rates of phyletic, cladogenetic and taxonomic evolution are inversely correlated with size of the area occupied by interbreeding populations (area being viewed as a direct function of actual population size, i.e., similar population densities) has any validity by mapping the areas occupied by populations in successive time intervals. Actual mapping (Berry and Boucot, 1970) of the areas occupied by populations through time for pentamerinids, eocoelids and stricklandids does indicate such a relation as does the large number of chonetid taxa, each restricted to a particular provincial unit during the Early and Middle Devonian as contrasted with the far smaller number during the more cosmopolitan Silurian (Fig. 17, modified from Boucot and Harper, 1968).

A logical next step is to investigate these examples, as well as others, in terms of point sources, marginal sources and broad sources. A point source is here used in the sense of a small population restricted to a single small area of the world. A marginal source is here used in the sense of a small population occurring at the edge or margin of the area occupied by a very large population of the same taxon with the marginal population giving rise to a new, closely related taxon. A broad source is here used in the sense of a large population occupying a very large area.

Mapping of the areas occupied by closely related taxa in time may be accomplished by plotting the localities from which the various closely related taxa are known on a time sequence series of maps. When such mapping is coordinated with the mapping of biogeographic boundaries for the same time intervals, and also of the area of the benthic assemblages and animal communities to which the taxa belong, a fair estimate may be made of the total area in time occupied by the taxa in question.

Point sources are probable for taxa that appear suddenly in the geologic record in a widespread condition, i.e., with no known antecedents that are reasonably close morphologically. Many of the families of Late Llandovery brachiopods (see Berry and Boucot, 1970) appear suddenly over the en-

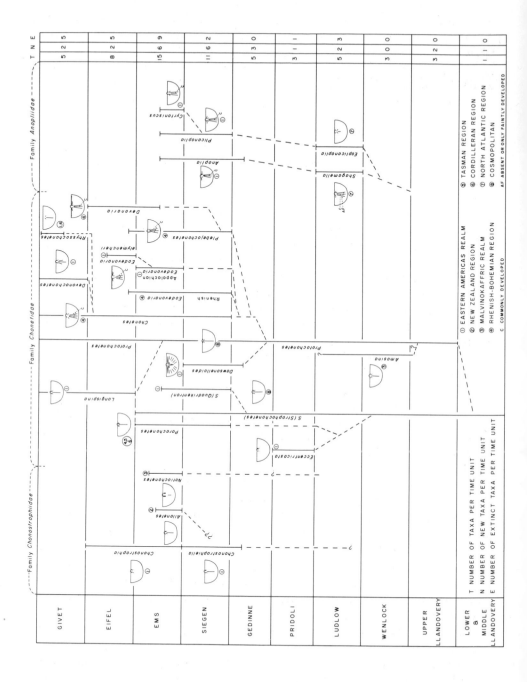

tire,vast area of the North Silurian Realm. It is now known, thanks to the efforts of several Soviet paleontologists (Nikiforova, Ushatinskaya, etc., largely unpublished) that many of these families have representatives in pre-Late Llandovery strata occurring in southeast Kazakhstan. In other words, southeast Kazakhstan forms an adequate point source from which these groups radiated near the beginning of the Late Llandovery throughout the North Silurian Realm.

Neuman (1972) has pointed out the anomalous occurrence within the Northern Appalachians of a number of Early Ordovician brachiopod taxa unknown elsewhere in the world until Middle Ordovician times. These Northern Appalachian taxa occur associated with rocks best interpreted as the remains of volcanic islands. The inference is that relatively small populations of these Middle Ordovician-type brachiopods developed in isolation (point sources) within a restricted part of the Northern Appalachians before becoming widespread with the advent of Middle Ordovician time. The interpretation of a small population size and relatively rapid evolution is easy to make in this instance.

Marginal sources are probable for taxa that appear suddenly in the geologic record, with a very restricted distribution, occurring side by side with closely related taxa possessing a very widespread distribution in the same time interval and during the preceding time interval. The virgianid taxon *Platymerella* is an example of this type; preceded and co-occurring with vast populations of the cosmopolitan taxon *Virgiana*. The many and varied, reef-related genera of the Wenlock and Ludlow may be best interpreted in the same, marginal-source, manner, as may many Eifel and Givet reef-related taxa. Eldredge and Gould (1972) emphasize the possibility for extremely rapid evolution of species (commonly in peripheral locations) as opposed to the more commonly observed slow rate (their "phyletic gradualism") consistent with the data obtained from larger populations (or at least inferred from them). However, the Eldredge and Gould term "punctuated equilibria" appears to be largely, if not entirely, nothing more than rapid evolution of small populations in which the chances for preservation of intermediate forms is low (Mayr, 1970, has already commented on how unlikely it is for rapidly evolving small populations to leave any fossil record), rather than any sudden and radical change in rate or change from an equilibrium condition of essentially no change to one of precipitous change.

Broad sources are of two types: (*1*) very large areas occupied by a taxon; and (*2*) a large number of small areas that communicated reproduc-

Fig. 17. Inferred phylogeny and known distribution of Silurian through Givet Chonetaceans with numeration of the rates of production and extinction of genera.

tively with each other to effectually form a large interbreeding population. The broad sources of these two types will, however, contain a complete range of population size from extremely large (see Fig.18, the pentamerinids) ones characterized by a relatively low (taxonomic in this case) rate of evolution, to far smaller population size (see Fig.18, the stricklandids and eocoelids) ones characterized by a far greater rate (phyletic in this case) of evolution.

In principle one would predict that rate of evolution will be slowest in the broad source type 1 examples, more rapid in the broad source type 2 examples, and most rapid in the point source and marginal source examples, in increasing magnitude of isolated small populations: the key to rapid evolution. Enough data have accumulated for Siluro-Devonian examples to substantiate this conclusion for the broad-source type-1 (pentamerinid, Fig. 18) and type-2 (eocoelid and stricklandid, Fig. 18) situations. By their very nature, however, the substantiation of this point for both marginal- and point-source examples is far more difficult, i.e., the problem of making certain that the known sample is a really adequate sample. There is always the danger that a poor sample of the situation belonging to the broad source type 2 will lend itself to the point-source or marginal-source explanation. There is no way to obviate this potential source of error except continuously to compile incoming data. However, even if this error in interpretation of the data is made, the concept of the inverse relation between population size and rate of evolution (taxonomic, clado-genetic or phyletic) may be adequately tested again and again as the magnitude of the differences between point sources, marginal sources, and broad sources of type 2 will be far smaller than between any of these and broad sources of type 1.

Broad sources may, of course, give rise in time to rapidly evolving populations *if* numerous faunal barriers appear in the area formerly occupied by the broad source so as effectively to cut the broad source up into a number of small to moderate-sized isolated sources (see Fig. 17, chonetid evolution). The recognition and identification of point sources, marginal sources, type 1 broad sources and type 2 broad sources with some degree of reliability is of importance if conclusions drawn about their presence are to be significant. The common and widespread occurrence of a taxon for the first time suggests derivation from a point-source or from a marginal-source taxon. These point sources and marginal sources may never be identified within the older rocks underlying those containing the overlying common, widespread occurrences of a taxon owing to the unlikelihood of hitting on such a small area by chance. The few instances for which the ancestral point source or marginal source has been recognized serve to underline the difficulty. The sudden appearance

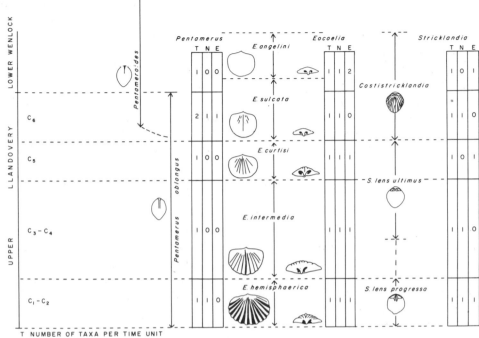

T NUMBER OF TAXA PER TIME UNIT
N NUMBER OF NEW TAXA PER TIME UNIT
E NUMBER OF EXTINCT TAXA PER TIME UNIT

Fig. 18. Ranges of three selected groups of Late Llandovery-Early Wenlock brachiopod lineages. The *Pentamerus* lineage occupies a large geographic area; both the *Eocoelia* and stricklandid lineages small areas. Note the similar numbers of taxa present (number of taxa per time unit) for all three groups irrespective of area occupied as opposed to the significantly differing rates of evolution (number of new taxa per time unit). The species of *Eocoelia* and *Stricklandia* show a rapid rate of phyletic evolution; the pentamerinid genera a slow rate of taxonomic evolution (*Pentamerus* itself presumably has a slow rate of phyletic evolution during this time interval as shown by the lack of change in time). Note too that the morphologic change between *Pentamerus-Pentameroides* is far smaller than the changes present in either of the other two lineages.

throughout the North Silurian Realm of the Eospiriferidae in the Late Llandovery combined with their recognition to date only in the pre-Late Llandovery of southeast Kazakhstan is a reasonable example. It is to be expected that the older, potential point-source or marginal-source location for most of these suddenly widespread, common taxa will not be found.

Widely or narrowly distributed younger taxa having an older broad source, of either type 1 or type 2, will be recognized with ease as having such a broad source because of the far greater likelihood of finding many remnants of the older broad source. Fig. 17 illustrating the evolution and distribution in time and space of the chonetids is intended to make this point. *Eocoelia angelini* is known from only a few areas of northwestern Europe, but the far more widespread occurrence of the antecedent species *E. sulcata* (Fig. 18) leaves no doubt about the broad-source, type 2 origin of *E.angelini.*

The widespread, common occurrence in the Early Llandovery of the North Silurian Realm of *Stricklandia*, with no known reasonably close Late Ordovician antecedent strongly suggests either a point source or a marginal source itself isolated from a broad source for a considerable interval of time (i.e., effectively a point source).

Absolute Time and Relative Time

It is fair to ask how certainly the absolute and relative time duration of the various intervals employed in this treatment have been made. There is, for example, no a priori reason for believing that the Lower Silurian and the Lower Devonian are of equivalent absolute or relative length, nor that the Upper Silurian or Upper Devonian are of equivalent absolute or relative length. If it could be shown that the Lower Devonian represented a far greater absolute or relative time span than the Lower Silurian it would be reasonable to conclude that rate of evolution was actually lower during the Lower Devonian than during the Lower Silurian, necessitating a reversal of the conclusions reached here. There are inadequate absolute dates for accurately calibrating the relative time scale made up of fossil zones. However, the small number of absolute dates does indicate that the Silurian has an approximate absolute duration of about 30 million years and that the Devonian represents about 55 million years (see data given below). Employing this information one may roughly equate the entire Silurian to the Gedinne through Eifel interval and arrive at the conclusion that the Early Silurian and the Early Devonian are probably of about the same absolute time duration and that the Late Silurian and Middle Devonian are probably of about the same absolute time duration.

Consideration of radiometric age determinations for the Silurian and Devonian has not yet provided agreement on information for the limits of the two periods or for their subdivisions. Friend and House (1964) have summarized and interpreted data available prior to 1964 for the Devonian. Their approach to the problem of devising a reasonable integration of the relative and absolute time scale for the Devonian was to plot the absolute age data, together with an indication of the error, against a fossil-based relative time scale based on maximum stratigraphic thickness for the standard subdivisions of the Devonian. The maximum stratigraphic thickness approach was adopted from Holmes' earlier work. The cumulative maximum stratigraphic thickness approach involves the assumption that maximum thickness per standard subdivision of the Devonian, for example, will be proportional to absolute time. There is no good substantiation for this assumption. The assumption based on maximum thickness employs differences of as much as four times (Gedinne 5,500 ft. and Siegen 20,000 plus ft.), differences for which I feel there is considerably less justification in view of the known tremendous variation in rates of sedimentation from place to place and time interval to time interval.

Despite the admitted paucity of adequate information with which to construct a time scale for this interval it is remarkable how much consistency may be obtained, admittedly with a bit of data selection. The following section will outline the limits employed in constructing Fig. 19.

A point for the Frasne-Famenne vicinity is taken as 362 m.y. plus or minus 6 m.y. This Frasne-Famenne data is based on the study of McDougall et al. (1966). They placed their samples near the Famenne-Lower Carboniferous (Tournais) boundary, but J.A. Talent (written communication, 1973) has provided the following information that impels me to move McDougall et al.'s (1966) data down farther in the column:

"The problem here as is so often the case is an assumption that a thing is of a certain age. . .

The sequence, all units conformable, is:

youngest: Cerberean Volcanics — being part of a cauldron subsidence it can be construed as essentially instantaneous geologically

Taggerty Group — volcanics and sediments including *Bothriolepis gippslandiensis*

oldest: Snobs Creek Volcanics

Australian Late Devonian non-marine correlations are highly tenuous, and it is within this capricious context that we attempt appraisal of the age of the Cerberean Volcanics. It all hinges on the age indicated by *B. gippslandiensis*, a form whose distribution does not enable direct correlation with marine sequences. A suggestion of age can be obtained by comparison of the faunal and floral successions in the Mitchell River-Freestone Creek area of eastern Victoria with the Eden-Genoa River area of southern New South Wales. In the latter, the Merimbula Formation has yielded a marine fauna including *Cyphotopterorhynchus* suggesting a Frasnian, perhaps Late Frasnian age (P.J. Sartenaer, written communication). Landward equivalents of these beds (the Genoa River Beds) yield a characteristic flora of *Archaeopteris howitti*, *Cordaites australis*, *Sphenopteris carnei*, and *Pecopteris? obscura*. Substantially the same flora with *A. howitti* and *C. australis* occurs in the Avon River Group 500 m or so above a volcanic-sedimentary succession containing *Bothriolepis gippslandiensis*. Even allowing for the manifest uncertainties attendant on floral correlations and

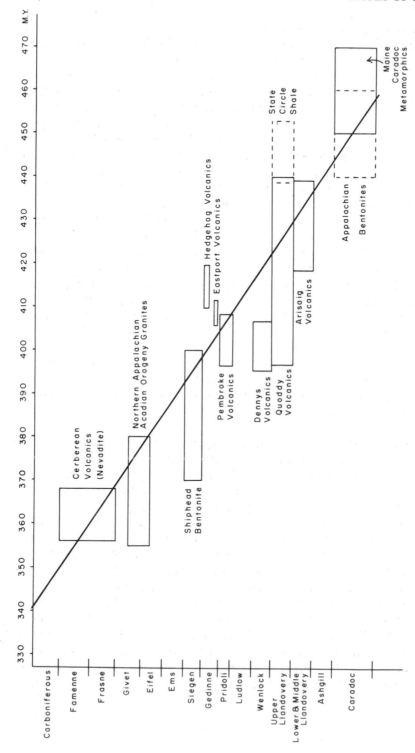

Fig. 19. Plot of absolute age determination data, with estimated errors, for the Caradoc through Famenne against fossil-based, relative time units scaled off through an estimate of the amount of organic evolution estimated for each relative time unit.

uncertainties as to the persistence of *B. gippslandiensis* higher up the column, we have here an indication that the Taggerty Group could well be of Frasnian age and that isotopic dating of the conformably overlying Cerberean Volcanics was more likely to yield an approximation of the Frasnian-Famennian rather than the Famennian-Tournaisian boundary."

A point for the Eifel-Givet boundary is obtained by considering the information presented in Wetherill (1965) for granitic rocks that intrude fossiliferous beds as young as the Eifel and are overlain unconformably by fossiliferous beds as old as the Givet in the Northern Appalachians with an age span of about 355 — 380 m.y. I used Northern Appalachian rocks occurring in northern Maine and adjacent New Brunswick on the one hand and in Nova Scotia on the other hand, but not the granitic rocks of southern Maine and southern New Brunswick which may be somewhat older (about 400 m.y., see Wetherill, for a compilation of data up to 1965). A point for the Siegen is taken from bentonite occurring in Gaspe (Smith et al., 1961) of 385 m.y. plus or minus 15 m.y. A point for the Pridoli is provided by volcanic rocks of the Pembroke Formation of southern Maine cited by Fullagar and Bottino (1970) as 402 m.y. plus or minus 6 m.y. Volcanics from the Dennys Formation of Wenlock age are given an age of 401 m.y. plus or minus 5 m.y. by Fullagar and Bottino. Volcanics from the Quoddy Formation of Late Llandovery age are assigned by these authors an age of 418 m.y. plus or minus 22 m.y. Another Late Llandovery point is indicated by the 445 m.y. plus or minus 7 m.y. age for the State Circle Shale study of Bofinger et al. (1969), that overlaps the Quoddy data in part. A point for the Lower Llandovery or latest Ordovician Arisaig Volcanics of 428 m.y. plus or minus 10 m.y. is taken from Fullagar and Bottino. D. Brookins (in Boucot et al., 1972) assigns an age of 460 m.y. plus or minus 10 m.y. to a point within the Caradoc. The Hedgehog and Eastport Volcanics of Gedinne age from Maine, studied by Fullagar and Bottino, and assigned ages of 414 m.y. plus or minus 5 m.y. and 408 m.y. plus or minus 3 m.y., respectively (Fullagar and Bottino, 1970) are indicated although they do not lie on the plot of the more consistent data (they would, however, be more consistent if of very Early Gedinne age themselves, and if the Pembroke sample were latest Pridoli). The Caradoc age bentonite ages in the span 440 m.y. to 460 m.y. included by Harland (1964, p. 358) are in essential agreement with the information provided by Brookins.

Using these data and these assumptions an upper limit of 350 m.y. and a lower limit of 405 m.y. is obtained for the Devonian, and a lower limit of about 437 m.y. for the Silurian after fitting a line to the data. An absolute duration of 55 m.y. for the Devonian and 32 m.y. for the Silurian is more in accord with my impressions of the fossils and the relative time needed for their evolution than is the commoner practice of assigning the Silurian a time duration of 20 m.y. or less relative to a Devonian of 60 m.y. or slightly more. It is obvious that the assumption about the relative time duration of the Silurian and Devonian subdivisions is ultimately based on the rates of evolution for the fossils. Despite what may appear like completely circular reasoning on my part (this treatment is devoted to the thesis that different groups of fossils provide evidence during the same interval of time for widely differing rates of evolution) one does, after a time, acquire an overall impression about the relative amounts of evolution occurring during different time intervals based on a familiarity with their fossils (both rapidly and slowly evolving forms). Needless to say, these impressions are difficult to assess quantitatively, if they are in serious error all of the subsequent conclusions presented here are not worthy of serious consideration.

Fossil-based relative time length of the Caradoc through Famenne units

Harland (1964) employs the Holmes approach of plotting maximum known thickness of sediment/time unit as the ordinate against which the limited number of absolute ages for the Early Paleozoic are plotted. As detailed in this section earlier I feel that maximum known thickness of sediment per unit is a far less reliable ordinate than rate of evolution. The problem of employing rate of evolution, however, is that each time unit has examples of both rapid and slow rate of evolution. How may an average be arrived at? Initially I made my own estimate, based entirely on experience with a variety of brachiopods (very limited for both the Caradoc-Ashgill and the Givet-Famenne as compared to the Llandovery-Eifel). Table II (column "Reduced numbers") lists my estimate. After preparing this estimate I had second thoughts because of my own inexperience in parts of the interval Caradoc-Famenne and because I feared that other animal groups might show highly conflicting results. Therefore I enlisted the aid of the following colleagues: W.B.N. Berry, graptolites; S. Bergstrom, Ordovician conodonts; J.T. Dutro, Devonian brachiopods; M.R. House, ammonoids; G. Klapper, conodonts; W.A. Oliver, tetracorals; A. Ormiston, trilobites; H.B. Whittington, Ordovician trilobites; A. Williams, Ordovician brachiopods.[18] All of these colleagues were reluctant to commit themselves to making such estimates because of the assumptions involved (each received a blank form on which to list his estimate of the relative time duration of the units arrived at on the basis of his own "experience" and "impressions"). The results are tabulated in Table II. I have taken the relative estimate provided by each paleontologist, averaged them where a range was indicated, and then reduced them using one of the relative numbers provided as unity (an arbitrary procedure, which if altered, could change the overall figure somewhat but not significantly). Next I averaged the reduced numbers to obtain an estimate of relative unit duration. Note on Table II the higher agreement shown by the paleontologists among themselves than is obtained by comparing either the "relative paleontologic unit duration" (the average of all the paleontologic opinions) or the individual opinions with the sediment based estimates provided by Harland (1964).

Finally, it must be pointed out that the scaling off of fossil-based relative time in this manner tacity assumes that long-term variations in radiation flux, food supply, or other significant variables, adequate to seriously change rates of evolution have been at a far lower level than the effects of population size.

Platform versus Continental-Shelf Population Size Estimation

I have earlier discussed some of the potential sources of misunderstandings between ecologists studying the modern continental-shelf biota and paleontologists studying the platform deposits left behind by ancient epicontinental seas. One obvious source of such misunderstanding is the greatly differing slope and total area present in the two environments. Most modern continental shelves are relatively narrow and slope rapidly seaward (at least several hundred feet within several hundred miles). Many ancient platforms had slopes of no more than, and possibly far less than, a few hundred feet in several thousand miles. These differences insure that populations occurring on and related to roughly equal depth intervals on present-day type continental shelves will be far smaller in size (as judged by area) than will those on the far larger platforms of the past. This in turn will make it far more difficult to estimate the relative size of populations occurring in similar depth intervals on modern continental shelves than those in the same depth intervals on ancient platforms. Therefore, it should in principle be more difficult to understand the effects of population size for the modern continental-shelf environment than for the ancient-platform environment. Consequently, factors other than population size may have effects ascribed to them within the modern environment through serious mismeasurement of the size of effective population districts. Similar difficulty is encountered when trying to estimate population district areas for geosynclinal environments of the past.

Some Thoughts About Diversity

Although much has been written recently about taxonomic diversity as

TABLE II

Relative length of time units employed in this book as indicated by paleontologic "experience"

Unit	Bergstrom reduced numbers	Bergstrom raw numbers	Berry reduced numbers	Berry raw numbers	Boucot reduced numbers	Boucot raw numbers	Dutro reduced numbers	Dutro raw numbers	House reduced numbers	House raw numbers
Famenne					1.0		1.4	(10)	1.5	(3)
Frasne					1.0		1.0	(7)	1.0	(2)
Givet					1.0		1.0	(7)	1.0	(2)
Eifel					1.0		0.7	(5)	1.0	(2)
Ems			0.5	(5)	1.0		1.4	(10)	0.5	(1)
Siegen			0.7	(5–8)	1.0		0.7	(5)	1.0	(2)
Gedinne			0.4	(3–5)	1.0		0.7	(5)	0.5	(1)
Pridoli			0.3	(3)	1.0					
Ludlow			0.9	(8–10)	1.0					
Wenlock			0.5	(5)	1.0					
Llandovery			2.0	(20)	2.0					
Ashgill	(1)	1	1.0	(10)	1.0					
Caradoc	(2+)	2+	2.0	(20)	2.0					

TABLE II (continued)

Paleontologist Unit	Klapper		Oliver		Ormiston		Whittington		Relative thickness unit duration (from Harland, 1964)	Relative paleontologic unit duration
	reduced numbers	raw numbers	reduced numbers	raw numbers	reduced numbers	raw numbers	reduced numbers	raw numbers		
Famenne	1.8	(5 or 6)	1.1	(9)	1.4	(14)			1.0	1.4
Frasne	1.5	(4 or 5)	1.0	(8)	1.6	(16)			0.6	1.2
Givet	1.0	(3)	1.3	(10)	1.8	(18)			1.3	1.2
Eifel	1.0	(3)	1.3	(10)	1.2	(12)				1.0
Ems	1.0	(3)	0.6	(5)	2.0	(20)			0.7	1.0
Siegen	0.7	(2)	1.3	(10)	0.4	(4)			2.0	0.8
Gedinne	1.0	(3)	1.0	(8)	1.2	(12)			0.6	0.8
Pridoli	0.3	(1)	0.5	(4)	0.5	(5)				0.5
Ludlow	1.0	(3)	1.0	(8)	1.2	(12)			2.0 or 7.0	1.0
Wenlock	0.3	(1)	1.5	(12)	1.2	(12)				0.9
Llandovery	0.7 0.3	(2) lower	2.0	(16)	2.5	(25)				1.9
Ashgill					1.0	(10)	1.0	(1)		1.0
Caradoc							2.5	(2 or 3)		2.0

Note added in proof: Williams (1957) gives for Ashgill: 1.0 (reduced), 1.0 (raw) and for Caradoc: 1.5 (reduced) and 1.5 (raw).

a criterion of great value for paleoecologic and evolutionary studies (Bretsky and Lorenz, 1970), additional comment is in order. It is indeed obvious, once various post mortem effects are accounted for (chiefly transport of shells and bioturbation, that tend to change the naturally occurring life assemblage diversity) that populations of fossils show large differences in taxonomic diversity. Very little of this mass of information has been synthesized in such a manner as to provide the basis for sound conclusions of the type needed to support or reject the various hypotheses discussed by Bretsky and Lorenz.

Sanders (1969) has pointed out the obvious fact that taxonomic diversity in all environments decreases as any single physical factor departs from the "normal". In other words, as conditions of excessively high or low salinity, low oxygen content, low nutrient content, etc., occur in the environment one may expect progressively less taxonomic diversity and ultimately abiotic conditions. Sanders has also pointed out that great fluctuations in any single physical factor will also decrease the taxonomic diversity regardless of whether the fluctuations are regular in occurrence or irregular. Jackson (1972) presents data for shallow-water bivalves that accord well with the conclusion that high environmental fluctuations of the type found in the intertidal region correlate well with a small number of species. Sanders presents data that support these conclusions. In this book I have used the term "restrictiveness" to say the same thing. For example, rough-water conditions or hypersaline conditions are paralleled by low taxonomic diversity *(Kirkidium* Community and *Prosserella* Community).

Sanders and Hessler (1969) have gone ahead from the above self-evident situations to present data about polychaete-bivalve taxonomic diversity in deep and shallow seas. They collected by means of an "anchor dredge" dragged across and through the upper sediment layer in the deep-sea. They found a far higher taxonomic diversity of polychaetes and bivalves in their deep-sea samples than is known from boreal estuary, boreal shallow-marine, tropical estuary, and tropical shallow-marine environments. Thus they conclude that there is far higher taxonomic diversity among deep-sea polychaetes and bivalves than is present among shallow-water polychaetes and bivalves living on similar substrates ("soft, fine-grained sediments"). They also conclude that there is and has been a "constancy of physical conditions. . . long past history of physical stability in the deep-sea" that "permitted extensive biological interactions and accommodations among the benthic animals to yield the diverse fauna of this region". They then categorize another set of environments as "physically unstable" where "physical conditions fluctuate widely and unpredictably", and "the organisms are exposed to severe physiological stresses". In the physically stable category they include the deep-sea benthic environment, tropical shallow-water environments and tropical rain forests. In the physically unstable category they include hypersaline bays, temporary ponds and "certain shallow boreal marine and estuarine environments" that "approach such conditions". The actual evidence for the "physical stability" of the deep-sea, tropical shallow-water and tropical rain-forest environments is not provided; neither is that for the physical instability of the shallow boreal marine and estuarine, hypersaline and temporary pond environments. They implicitly assume that their deep-sea polychaete samples character-ized by very high taxonomic diversity belong to a single community or set of very similar, closely related communities. Should their "anchor dredge" have been collecting from a great variety of "soft, fine-grained sediment" communities it is possible that their high taxonomic diversity represents the additive effects obtained from a far higher number of communities than present within their shallow-water samples, and that these different communities might represent a great variety of environmental parameters not previously studied in the deep sea but ecologically important there. However, it is critical to point out that none of the Early Paleozoic shelly faunas studied to date shows any evidence of being derived from a true deep-sea environment in the sense of Sanders and Hessler (1969); all of our Lower Paleozoic benthic faunas are derived from what Sanders and Hessler would conclude to be shallow-water shelf environments.

Bretsky and Lorenz (1970) have drawn a number of additional conclu-sions from the Sanders and Hessler (1969) conclusions about high taxo-nomic diversity being directly related to physical stability of the environ-ment and low taxonomic diversity being directly related to physically unstable environments (great extreme of any one physical factor and/or great unpredictability in any one physical factor). Bretsky and Lorenz (p. 532) argue in favor of stable environments (in the Sanders and Hessler, 1969, sense) being characterized by taxa having low genetic variability, therefore characterized by homoselection and low polymorphism, nar-row-niched, provincial, and with a tendency to speciate. The key genetic assumption about low genetic variability in a stable environment, without which the conclusions about homoselection, low polymorphism, lower homeostasis, and tendency to speciate are seriously weakened, is sup-ported by few data. The inverse conclusions for the species present in unstable environments is supported by equally few data. Levinton (1972, 1973) has added more data that support the Bretsky and Lorenz concept of genetic variability correlating well with both phenotypic and environ-mental variability. The Bretsky and Lorenz argument from the experimen-tal data that find a correlation between genetic variability, phenotypic variability and environmental variability does not consider the effects on rate of evolution caused by differing-sized interbreeding populations. Bretsky and Lorenz imply that in populations of the same size, rate of evolution will be governed by the amount of genetic variability. But, in nature we find that populations are of widely differing size. The observational data of paleontology presented in this paper show conclusively that the effects on rate of evolution of population size are far larger than the presumed effects of genetic variability (as judged by interpretations of both phenotypic variability among fossils and of the inferred environmental variability). In principle the Bretsky and Lorenz conclusions could be tested by means of fossils *if* populations of the same size were studied through time; this has not yet been done. There is, however, no reason to suspect from our data that their conclusions would not be valid for interbreeding populations, in time, of the same size. Valentine et al. (1973) and Ayala et al. (1973) have even more convincing evidence that environmental stability does not necessarily correlate with a low genetic variability condition (they find a high degree of heterozygosity in a tropical species of *Tridacna*!).

Gooch and Schopf (1973) have examined the genetic variability of certain deep-sea invertebrates and find examples of high genetic variability. Their conclusions are strongly at variance with those of Bretsky and Lorenz (1970) which suggested that the deep-sea should be a region of low genetic variability because of its presumed high "environmental stability". The key assumption about "high environmental stability" of the deep-sea is, as Gooch and Schopf (1973) are quick to point out, based on assumptions in large part. We know very little about the environmental factors of importance to the benthic invertebrates dwelling on the deep-sea floor. That the deep-sea benthos are many and varied, as recently emphasized by Hessler and Sanders (1967), is beyond dispute, but that their environment is "stable" or unvaried is open to serious question. Therefore, it is still premature to do more than question the assumption that environmental stability correlates well with genetic stability. Gooch and Schopf do not make clear whether their material is from high- or low-diversity communities. If from high-diversity communities their data would suggest that a high degree of heterozygosity does not necessarily correlate with low diversity. In any event, the Gooch and Schopf data call into serious question the conclusions arrived at by Bretsky and Lorenz that the deep-sea fauna (assumed by Bretsky and Lorenz following Hessler and Sanders, 1967, to be high-diversity) will be strongly homozygous, with the deduction of Bretsky and Lorenz that high taxic diversity correlates well, therefore, with a strongly homozygous condition.

The conclusions arrived at in this book (see section entitled "Biogeographic statistics") regarding the far wider longevity of cosmopolitan taxa, i.e., their far lower rates of evolution, suggest that cosmopolitan taxa of the present would form better material for genetic experiments of the sort conducted by Levinton (1973), Gooch and Schopf (1973) and Ayala et al. (1973) if one wishes to test the Bretsky and Lorenz (1970) deduction that high genetic variability will be found to correlate with low rate of evolution. The Gooch and Schopf conclusions question the correlation of low genetic variability being found at depth, although Levinton's (1973) data may not agree with that of Gooch and Schopf. Valentine et al.'s data question the correlation of high genetic variability occurring in unstable environments (their *Tridacna* is thought to live in a very stable environment).[19] It is clear that we lack enough information about genetic variability and various environmental as well as rate of evolution factors to arrive at any hard and fast conclusion at this time.[20]

If the data in this book point out nothing else they do indicate that rough-water communities, characterized by low taxonomic diversity, provide examples of both rapid and slow taxonomic, cladogenetic and phyletic evolution (*Kirkidium* and *Pentamerus* Communities on the one hand as opposed to the *Stricklandia* Community on the other hand). The sudden speeding up in production rate of chonetid taxa (Fig. 17) in the Lower Devonian and the consequent Middle Devonian slowing down do not correlate with any change from an unstable community situation to a stable community situation and then back again. We have no reason to suspect that the low-diversity *Eocoelia* Community environment, characterized by rapid phyletic evolution was any more stable than the low-diversity, co-occurring *Pentamerus* Community characterized by slow taxonomic and phyletic evolution. The low taxonomic diversity of both communities does not accord with the Bretsky and Lorenz thesis that low taxonomic diversity indicates little tendency to speciate (in *Eocoelia* we find to the contrary; in *Pentamerus* we agree!). The Bretsky and Lorenz conclusion that high-diversity (stable) populations are "provincial" whereas low-diversity (unstable) populations are "cosmopolitan" is unsupported by observations. They present no data to support this conclusion; it appears to have been deduced from the genetic variability surmise, and the possible addition of data from Benthic Assemblage 1 (concluded to be an "unstable" environment; Bretsky, 1969) for which some of the taxa are concluded not to have evolved (linguloids and bellerophontids in particular; unfortunately the lack of attention paid to the morphology and taxonomy of Paleozoic linguloids and nearshore bellerophontids makes any conclusion about their taxonomic and evolutionary stability mere inference). An excellent example that directly contradicts Bretsky and Lorenz' (1970) conclusion about low diversity indicating cosmopolitanism and slow speciation is provided by the Devonian mutationellinid terebratuloid

brachiopods of Benthic Assemblage 2. Here we find single-taxon brachiopod communities *(Pleurothyrella, Scaphiocoelia, Cloudella, Mendathyris)*
occurring in a nearshore, shallow-water position, and all with evidence for
rapid taxonomic and cladogenetic evolution (Fig. 20) and high provincialism. The information about the Centronellidae, Rhipidothyridae and
Rhenorensselaeridae summarized in Fig. 21 also contradicts these same
conclusions. Fig. 36 (see p. 242) summarizes a mass of data taken from Part
II, Ch. 6 that shows conclusively that cosmopolitan long-lived taxa occur
as abundantly in an offshore position as in an inshore condition. Many of
the cosmopolitan taxa considered in making up Fig. 36 occur in high-
diversity communities, although many also occur in low-diversity communities. For example, the deep-water Benthic Assemblage 5 *Notanoplia*
Community is both very low-diversity and highly provincial!

If one takes seriously the concept that the more offshore, deeper-water

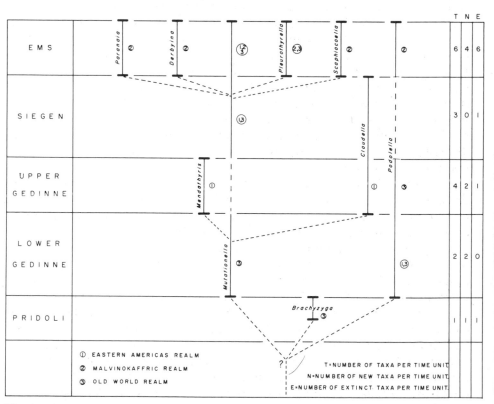

Fig. 20. Evolution and distribution of the Early Devonian mutationellinid brachiopods.

"stable"-environment taxa evolve more rapidly, how could the apparently slow evolution of the offshore Cambrian trilobites used as a source for Palmer's biomeres be accounted for?

E.G. Kauffman (written communication, 1973) states ". . .deep-ocean taxa are *more* cosmopolitan than shallow-water ones", which certainly suggests that the Bretsky and Lorenz conclusions about so-called stability and rate of evolution as indicated by cosmopolitanism do not hold up in the deep oceans any more than they do on the shallow shelf.[21]

The occurrences of Early Paleozoic shallow-water marine invertebrates associated closely with the bottom in one way or another indicate that the key factor in achieving low taxonomic diversity is the restrictiveness (the "unstable" of Bretsky and Lorenz, 1970) of the environment (see Fig. 3—10; Part II, Chapter 6). Restrictiveness of the environment may be achieved in many ways. It should be clear that the degree of restrictiveness may be achieved either by extremes of any one variable (or variability of the variable) such as salinity, turbulence, oxygen content, intertidal exposure, etc., or by a combination of conditions that together spell out a very specialized environment. For example, on a particular type of bottom sediment at a particular depth (depending upon the combination of other factors) either a very specialized, restricted environment with low taxonomic diversity or an unspecialized, unrestricted environment with relatively high taxonomic diversity may be developed.

The Silurian-Devonian data summarized in Fig. 3—10 and in Part II, Chapter 6 suggest to me that faunas living in "normal" marine environments may or may not have a high taxonomic diversity. For example, the *Dicaelosia-Skenidioides* Community of Benthic Assemblage 4 is high-diversity whereas the Benthic Assemblage 5 *Notanoplia* Community is low-diversity. However, we find that when physical conditions departing from the "normal" are specified there is an immediate decrease in diversity whether or not the additional specifications are applied in a predictable or unpredictable way, with or without wide fluctuations. Again, rough-water communities (*Pentamerus*, *Kirkidium*, and others) are low-diversity; hypersaline communities are low-diversity (*Prosserella*); brackish-water communities are low-diversity (Hughmilleriidae-Stylonuridae); Benthic Assemblage 1 communities that may be high-intertidal (Linguloid-Orbiculoid, Rhynchonellid) are low-diversity. We find an overall general increase (not in order of magnitude, however) in taxonomic diversity moving away from the shoreline Benthic Assemblage 1 region *if* the "normal" environment is followed, but not if various restrictive environments are encountered. Whether this general increase in diversity is related to the shallow-sea to deep-sea observations of Sanders and Hessler (1969) made over a far greater depth span is unknown. It is possible that our increase in diversity away from shoreline in the "normal" environment is the result of shoreline- and shallower-water restrictive factors in the environment gradually dropping out and leading to a more "stable" environment, i.e., less restrictive. The information summarized in Fig. 36B might be taken to indicate that there is a general *decrease* in taxonomic diversity from Benthic Assemblage 3 to Benthic Assemblages 4-5, particularly for provincial taxa, but this is not the case because many more of the single-taxon communities occur in Benthic Assemblage 3 rather than in 4 or 5. Turning to planktic depth-stratified associations ("communities"?) of acritarchs (Gray and Boucot, 1971, 1972), graptolites (Berry and Boucot, 1972b), and conodonts (Seddon and Sweet, 1971) we find that there is good evidence for an increase in faunal diversity with increasing depth (all of it relatively "shallow" in the Sanders and Hessler sense) but that there is no order of magnitude increase. Aldridge (1972) has recognized depth zonation among some of the conodonts of the Llandovery in the same, offshore diversity increase, manner. Hallam's (1972) data on the distribution of Jurassic ammonites can be easily interpreted as a depth-stratification phenomenon with a richer assortment of taxa in deeper water than is present in shallower water (as is also inferred by Scott, 1940, for Texas Cretaceous ammonites).

In connection with this apparent increase in taxonomic diversity observed in "normal" environments the recent studies (Berry and Boucot, 1971, 1972b; Gray and Boucot, 1971, 1972; Seddon and Sweet, 1971) indicating that depth-correlated stratification was as marked (see Fig. 16) among the Early Paleozoic planktic organisms as among modern plankton, opens up additional possibilities for understanding faunal diversity in terms of food supply. It is reasonable to predict a possible correlation between increasing diversity of bottom living suspension feeders that utilize material derived from the plankton and increasing diversity going from shallow to deep water of depth-stratified plankton sinking through the water column to the bottom. That is, a greater variety (although not necessarily abundance) of depth-stratified plankton will become available as one goes away from shoreline into deep water, but this potentially greater variety may not necessarily correlate with greater diversity of suspension-feeding consumers unless they become specialized in their tastes.

The role of predators in controlling taxonomic diversity may be very substantial (Paine, 1966, 1971; Murdoch, 1969; Porter, 1972). In the intertidal zone one may expect fewer predators, due to the more rigorous conditions of the environment, than in the subtidal environment. As most predators will have very specific "tastes" (Porter, 1972; Arnold, 1972) it stands to reason that a smaller number of predatory taxa will permit the non-predators to compete with each other more vigorously with the result that the most efficient will tend to take over and lower the number of taxa able to exist. A larger number of predator taxa, on the other hand, will tend to keep the numbers of their prey taxa down and permit other non-predator taxa to exist that would ordinarily be kept out by the competition for resources.

Valentine (1971a) has pointed out that between high or low temperature no consistent diversity trend holds out very well. He emphasized instead that the spatial heterogeneity (essentially the number of niches, i.e., number of communities) within a region is a major factor in regulating the regional diversity. He suggested that the most significant control of diversity is the stability of the trophic resource supply. In short, Valentine (1971b) concluded that low diversity correlates well with resource-rich or resource-unstable environments and high diversity correlates well with resource-poor and resource-stable environments. Valentine's conclusions accord well with the Sanders and Hessler (1969) diversity information for the deep-sea (high-diversity, resource-poor and resource-stable) and shallow-sea (low-diversity, resource-rich and resource-unstable) situations. Without trying to contradict the overall applicability of Valentine's conclusions for large regions, I suggest that in (small-scale) cases of community diversity, adequate consideration should be given to all the factors that may produce restrictiveness within that community's environments.

Gibson and Buzas (1973) have studied the taxonomic diversity problem as it affects Recent and Miocene foraminifers from the North Atlantic region with samples included from northern, southern, deep- and shallow-water positions. They conclude that there is no simple explanation for the observed high- and low-diversity patterns in terms of the time-stability hypothesis or of a climax development. Their data provide additional evidence with which to seriously question the time-stability hypothesis of Sanders (1969) as well as the climax hypothesis of Margalef (1963). They provide data that throw considerable doubt into the Hessler and Sanders concept of increasing diversity indicating stability, if relative stability is characteristic of deep cold and warm shallow water. This information, in turn, throws additional doubt on the validity of the conclusions reached by Bretsky and Lorenz (1970) which depend so heavily on the time-stability concept.

If the environmental significance of taxonomic diversity, both high and low, is to have any utility it will be necessary to tabulate the environmental requirements (and ranges) of as many factors as possible for as many taxa as possible derived from as many faunas as possible. After such tabulations have been made we will then be in a better position to evaluate the environmental significance of the number of taxa present in a community. There is no reason to think that an offhand characterization of the deep sea or the rain forest as "stable" (chiefly I suspect because of their presumed low temperature ranges) as compared to other environments should be accepted.

The most serious defect in the Bretsky and Lorenz (1970) argument that high taxonomic diversity correlates with rapid evolution from the Silurian-Devonian viewpoint is as follows. The most abundant brachiopod taxa in actual numbers of specimens during the Silurian-Devonian are certain cosmopolitan genera like *Atrypa*, *Leptaena*, *Skenidioides*, and *Nucleospira* (plus many others). In addition to being most abundant as regards numbers of specimens these taxa, and others of similar type, are most widespread geographically during large portions of the Silurian-Devonian. Further, these same taxa have the greatest time duration during the Silurian-Devonian. These taxa also happen to be present in *high*-diversity, Benthic Assemblages 3—5, subtidal communities together with taxa that may be less widespread both geographically and temporally. The conclusion is clear that high diversity in the case of these abundant, very cosmopolitan taxa that persisted for very extensive periods of time correlates with very slow evolution (the correlation with large population size and slow rate of evolution is clear). Therefore, the simple 1:1 conclusion suggested by Bretsky and Lorenz is unacceptable as applied to Silurian-Devonian brachiopod studies (Bretsky's studies of Late Ordovician brachiopods have been too restricted both geographically and in time to have shown the above relations for the Ordovician, but a more comprehensive acquaintance with the world Ordovician would show that the Ordovician picture is little different from the Silurian-Devonian picture; the brachiopod leopards have not changed their spots at the Ordovician-Silurian boundary!).

The data presented in Fig. 36A, B (see also section entitled "Biogeographic statistics") indicate that the percentage of cosmopolitan taxa stays about constant in both the onshore and the offshore environments. There is no obvious trend from more cosmopolitan onshore taxa to less cosmopolitan offshore taxa. *If* cosmopolitanism indicates anything about "environmental stability" (as earlier concluded by Bretsky and Lorenz) this should not be the situation. Therefore one is forced to conclude that either there is no change in "environmental stability" from the onshore to the offshore region or that cosmopolitanism is not directly related to "environmental stability".[22]

Whittaker's (1972) comments about plant diversity in tropical rain forests and in the Sonoran Desert make possible some analogies with marine invertebrate situations. The high-diversity tropical rain forest is extraordinarily rich in biologically created niches, just as is the tropical coral reef. In both the tropical rain forest and in the coral reef the concept of floral and faunal interdependence is easier to apply. However, the rich Sonoran Desert flora may have more in common with the level-bottom benthic filter-feeding situation in which biotic interdependence is not nearly so plausible on a large scale as in the tropical rain forest or on the tropical coral reef. We may be looking at different ends of the depend-ence—independence spectrum!

Whittaker's model for organic evolution under conditions of communi-ty diversification, using the rain forest as an example, could be taken to suggest that during times of cosmopolitan stability the level-bottom ben-thic shelly faunas of the world should have become increasingly diverse as resource partitioning through organic evolution progressed. The time avail-able for such a process to have occurred is considerable. Yet we find little evidence that such a progressive diversification occurred, or if it did the upper taxic limit must have been relatively low. The maximum number of articulate brachiopod taxa found in a Silurian-Devonian level-bottom com-munity is about twenty. This suggests, in the absence of any time trends for a marked increase in total number of brachiopod taxa characterizing level-bottom communities with time, that diversification through resource partitioning that accompanied organic evolution was not a major factor then, and possibly may not have been during other time intervals. The level-bottom benthic taxa and the communities they occur in may possi-bly not be too similar in their functioning to the reef-environment com-munities.[23]

Finally, in considering the significance of taxonomic diversity within specific communities of fossils, interpretations commonly are based on observations derived from single beds of fossils, single exposures, or single map areas. This approach — studying a relatively small sample in some detail — parallels the sampling methods of the modern ecologist (although possibly not the "anchor dredge" of Sanders and Hessler, 1969), at least for shallow-marine benthos. However, the paleontologist is in danger of missing and misinterpreting major relations if he does not avail himself of the continent-wide sample, although this is far more poorly defined than that obtained from a single bed or exposure. It is simply the old problem of trying to determine whether or not correlations observed in a small sample represent causal relationships or mere coincidence; this question can best be determined by enlarging the sample as much as possible to exclude coincidence. Surely one must apply both approaches!

What is "environmental stability"?

There has been considerable attention given to the concept of environ-mental stability as a very significant factor in controlling both taxonomic diversity and rate of evolution (see Bretsky and Lorenz, 1970, for a discussion of such conclusions and references to some recent papers in both the ecologic and paleoecologic fields). Some would view long-ranging, cosmopolitan taxa as adapted to an environmentally unstable regime. Fig. 36 makes clear that a high percentage (which is relatively constant) of cosmopolitan taxa occur everywhere from Benthic Assemblage 2 through 5. This being the case one is led to conclude that these regions must include environmentally unstable areas (which are widely distributed in time and space). Yet it is clear that a variety of environmental factors must have a pronounced gradient across this transect! What does this mean? Bretsky and Lorenz (1970), as well as like-minded ecologists, appear to assume that most organisms will react to a specific set of environmental factors and recognize it as either stable or unstable. If cosmopolitanism is a measure of environmental instability then it is clear from Benthic Assemblage 2—5 that a stability gradient did not exist.

In view of this confusing situation it is worth asking just how might organisms react to differing environmental conditions. Some organisms may evolve in the direction of requiring very specialized requirements (the reef biotas afford excellent examples). Other organisms may evolve in the direction of having very unspecialized requirements that exceed or are equal to the environmental variations found over a wide region. Thus, one may conceive of organisms having unspecialized environmental requirements adapted to either a small area (in which the total environmental spectrum is great) and also of organisms having unspecialized environmental requirements adapted to a very large area (in which the total environmental spectrum is far greater). The data summarized in Fig. 36 may indicate that it is a great mistake to assume that environmental generalists will be restricted to areas where a varying, unstable environment is thought to occur. Such environmental generalists may be able to occur over wide regions that incorporate individual areas having a relatively uniform, stable environment that varies little, although the total environmental spectrum may be very broad. Such would appear to be the case if cosmopolitan taxa are interpreted as environmental generalists. There is, in this view, no contradiction in having organisms with narrow environmental tolerances living side by side with others having very broad environmental tolerances. In this view one may not speak of a specific environment as being either stable or unstable rather one must state that it is stable (and favorable) or unstable (and unfavorable) for this organism or for that organism. If we do not adopt such a view we must either conclude that widespread, cosmopolitan organisms are indicators of an unstable environment that somehow co-occurs with others in many places that are indicators of a very stable environment (many of the provincial, rapidly evolving taxa) or that co-occurring cosmopolitan and provincial taxa indicate nothing about environmental stability. Dobzhansky (1970, p. 25) has phrased the question very well in noting that some organisms become specialized for existence in a variety of environments whereas others become specialized for rather specific environments; the cosmopolitans obviously form the first group, the highly endemics obviously the second group, and most taxa fall somewhere in between.

Many provincial organisms also have wide environmental tolerances (although they are unable to cross over the environmental barrier or barriers responsible for the existence of the provincialism) as indicated by Fig. 36A, B. In general, provincial organisms (see section "Biogeographic statistics") evolve more rapidly than cosmopolitan organisms, but this appears to be best correlated with the population size effect than with any environmental effect.

The low taxic diversity of the intertidal region (Benthic Assemblages 1—2) correlates well with a different environment (different than the subtidal). But, should this environment be considered any less "stable" than is the fresh water environment where taxic diversity is still lower. I prefer to think of this progressive lowering of taxic diversity as being related to a progressively more restricted, rather than unstable situation. That is, I conclude that the subtidal region is environmentally less restrictive for aquatic organisms, than is the intertidal environment, and that the estuarine environment is more restricted than the marine, and in turn the fresh water more restricted than the estuarine.[24] Whittaker (1972, p. 240) puts one aspect of the stability problem very neatly in commenting that an environmental parameter that is "predictably unpredictable" can be considered to be stable! It is time that the terms "stable environment" and "unstable environment" were analysed in terms of the specific physical factors present in each one, plus the range in these factors thought to be present on different time scales (daily, yearly, etc.). Some of these physical factors may be deduced from the geologic record. When we finally have such information in hand we may be able to have a better understanding of the correlations thought to exist between many organisms and their environment. Continued usage of the terms stable and unstable appears to generate more heat and fury than light.

Diversity, Rates of Evolution, and Climate Correlation

Stehli et al. (1960) and Stehli and Wells (1971) have synthesized data for both modern and fossil groups (Permian brachiopods, Cretaceous planktic foraminifers, Mesozoic to Recent hermatypic corals) regarding taxonomic diversity. They find a strong correlation between high taxonomic diversity, high temperature, high rate of production of new taxa (rate of taxonomic evolution), and low latitude. From this they conclude that temperature may have an important effect in raising rate of evolution although they are able to show for the hermatypic corals that there may also be an "area effect" of importance (more hermatypic coral taxa present in the larger Indo-Pacific region than in the similar-temperature, smaller area Atlantic region; although their "larger" and "smaller" areas refer to the total sizes of the regions, not necessarily to the actual areas occupied by the various taxa).[25] Newell (1971) goes several steps further than Stehli et al. (1969) in presenting evidence (pp. 25—28) that shows rather conclusively that the area of habitat (not just the size of the two regions) for Indo-Pacific as opposed to Caribbean reef corals, shelled molluscs, echinoids, fishes and cypraeid gastropods is much larger, and that the number of corresponding taxa in the Indo-Pacific as opposed to the Caribbean region is much larger. Newell's data might be interpreted as suggesting a greater rate of speciation in the larger Indo-Pacific region than in the smaller Caribbean region. However, it would be very premature to arrive at such a conclusion until the actual size of population per unit of time had been ascertained for the two regions; this data is not avaiable. What we may be observing here is merely the fact that tropical reef and reef-related faunas are very rich in taxa, and that the presence recently of a Panamanian land bridge as well as a restricted Mediterranean region in the latter part of the Tertiary have chopped this tropical region into discrete parts from what was once a continuum. The smaller size and lower number of taxa in the Caribbean region might be found to correlate reasonably with similar sized portions of the Indo-Pacific region. However, Kauffman (1973) concludes that a Panamanian barrier of some sort began to affect marine organisms as early as the Cretaceous. It will clearly be far more difficult to estimate population size for the tropical reef and reef-related communities compared to the ancient platforms just as it is far more difficult to estimate population size for the ribbon-like continental shelves (themselves far larger and easier to deal with than the far smaller reef environments). The critical question then is the area occupied by each species (area being concluded to be a direct function of population size) *not* the species per area. Stehli et al. (1969) are cautious in not concluding that the correlation with increased temperature in the tropical regions as opposed to the nontropical regions acts to speed up rate of evolution, although it may be a factor. It would be of great interest, in view of the conclusions reached in this paper, to try to estimate the size of the areas occupied by the different populations of organisms involved in this work to see if there are smaller-size populations in the low-latitude, tropical regions than in the non-tropical regions. One would predict that the hermatypic corals of the tropics might be characterized by far small-er-size populations of each taxon than is the case with the non-tropical forms. Stehli et al. also point out the far higher percentage of cosmopoli-tan taxa in the "cold assemblage" (1969, fig. 3) as compared to the "warm assemblage" and suggest slower evolution for the cosmopolitan taxa that also, of course, occupy a far larger area than the endemic taxa. All of this information is consistent with the concept that rate of evolu-tion (taxonomic, cladogenetic or phyletic) is highly dependent on size of interbreeding population (as measured by area), and that other factors play a lower-rank role in determining that rate. In the case of the herma-typic corals it would be critical to try and estimate the actual size of the populations for the various taxa, not just the area of the general region in which they are known to occur, i.e., mapping of the community areas in which each taxon occurs and adding up of the total area.[26]

Valentine (1969) has ably summarized the factors underlying the marked increase in diversity (familial, generic and species level) that oc-curred in the Late Cretaceous to Recent interval. He points out the signifi-cance of an increase in provincialism and increase in climatic differentia-tion, and implies that reef diversity as well as level-bottom diversity in-creased in a parallel manner. Although the actual geography of the Late Cretaceous—Recent shallow-marine regions was radically different than that deduced for the Silurian and Devonian it is comforting to realize that the basic controls of taxonomic, cladogenetic and phyletic evolution rate appear to have been the same.

Level-bottom Diversity, Reef Diversity, Rocky-bottom Diversity and Rates of Evolution

Employing the concept that rate of evolution should be inversely pro-portional to size of interbreeding population it is possible to predict the expected relations to be found in the level-bottom, rocky-bottom and reef environments. Level bottoms should have fewer communities present than either rocky bottoms or reef environments because of the smaller poten-tial number of biologically induced ecologic niches. Rocky bottoms should have fewer communities present than reef environments for the same reason. In a way rocky bottoms are merely a special case of the reef environment in which most of the biologically created reef niches are absent. If these deductions are accepted then it follows that taxonomic diversity will be lowest in the level-bottom environment, higher in the rocky-bottom environment and highest in the reef environment. In general the level-bottom environment covers far greater areas in any region than either the rocky-bottom or reef environment; this is certainly true on a worldwide basis. Therefore one may infer a far lower rate of evolution among level-bottom community organisms, except for those communities that cover a small area, than among either rocky-bottom or reef community organisms. Following the same logic one may predict, because of the larger areas present, that rocky-bottom community taxa will evolve more slowly than reef-community organisms. In a general way this is the experience gained in all parts of the geologic column. Reef and reef-related faunas from different parts of the geologic record are almost invariably more taxonomically diverse than level-bottom communities (rocky-bottom communities, although rare in the geologic record, are known; see Surlyk, 1973). Within the Silurian-Devonian we have very little data about the evolution of reef and reef-related invertebrates. What little we possess is highly consistent (Fig. 22), however, with these inferences (the reef-related pentameroids of the Late Wenlock and Ludlow, for example).[27]

Emphasis is placed here not on correlating high taxic diversity with either high or low rate of evolution, as we have examples of both situations, but on the size of the total interbreeding population being the critical, first-order control regardless of diversity. Tropical reef environments happen to be high-diversity and small-population, thus rapidly evolving. *Eocoelia* Community happens to be low-diversity and small-population, thus rapidly evolving. Examples of the opposite situation giving rise to slowly evolving taxa may be cited.

Consideration of the data in Part II, Chapter 6 together with unpublished counts (Watkins and Boucot, 1973) of specimens occurring in samples taken from different-diversity brachiopod-dominated communities shows conclusively that the taxic diversity of brachiopod communities decreases markedly as the abundance of one or of a few species increases markedly.

Reef Environments in Time

The disappearance of reef environments periodically has something to tell us about the environmental tolerances of their organisms as opposed to those of the adjacent level-bottom communities. The reef environment was badly disrupted or entirely wiped out at the end of the Ashgill and at the end of the Frasne, as well as towards the end of the Permian. Therefore, it stands to reason that the abundant level-bottom organisms that survived these extinction events must have had wider environmental tolerances than the varied reef organisms. After each wipeout of reef organisms a new group subsequently developed from level-bottom antecedents. The smaller size populations of reef organisms will also tend to make them more susceptible to extinction when faced with environmental changes which they do not cope with successfully by evolutionary changes of a rapid sort.[28]

Worldwide and Regional Taxic-Diversity Gradients

Worldwide and regional differences in taxic diversity have been obvious to travelers since the beginning. The lowered numbers of plants and animals commonly observed as one travels from a warm climate to a cold climate, whether in climbing a high mountain or in going from the Mediterranean region to the Far North aroused the interest of observers from the beginning. The same is true regarding the flora and fauna of various deserts, small islands and islands large or small far removed from the continental land masses. In more modern times biologists and paleontologists have begun to analyse and attempt to gain an understanding of the mechanisms that must underlie these systematic changes in diversity.

Several basic questions are involved. The first is to merely prepare worldwide and regional syntheses of raw differences in diversity. Examples of this type are Fischer's (1960) census of bivalve taxa occurring along the eastern coast of North America with a significant increase in taxa from north to south. Stehli (1971, and references included therein) has prepared a worldwide summary of changing taxic diversity for Permian brachiopods that indicates greater taxic diversity in the non-polar region. Stehli et al. (1967) have prepared a summary of modern worldwide marine bivalve diversity showing greater diversity in the equatorial region than in the polar regions. Other work by Stehli and his collaborators on hexacorals and Foraminifera is referred to elsewhere in this book. MacArthur's (1972) classical work on the differing taxic diversity found on islands of varying size and position relative to the mainland is summarized with references to earlier work. MacArthur (1965) has discussed the additional problem of analysing diversity in terms of changes in number of taxa present in the same or similar niches as one travels from region to region, and the introduction of new niches which, of course, materially increase the overall regional diversity gradients.

The application of MacArthur's principles of island biogeography to marine benthos will probably be in largest part an exercise in planktic larval ecology. For example, data from the Late Llandovery of New York indicates that planktic larvae of depth-correlated brachiopod communities existed in water shallower than did co-occurring depth-stratified graptolites with the result that deep-water brachiopod communities occurring in topographic depressions are accompanied only by shallow-water graptolites. Taylor's (1971) data concerning Indian Ocean reef-associated bivalves indicate the importance of larval ecology in the distribution of these

Fig. 22. Fig. 22AA through EE are plots of the same data presented in 22A—E normalized for the estimated relative duration of the various time units. The relative time estimates are taken from Table II.

A. Total number of genera per time interval (Caradoc through Famenne). Data for taxa not occurring above the Ashgill, or not shown as possible ancestors (Fig. 27) of post-Ashgill taxa were taken from many sources, and are not plotted.

B. Total number of newly appearing genera per time interval (Ashgill through Famenne).

C. Total number of genera becoming extinct per time interval (Caradoc through Frasne).

D. Percentage rate of appearance of new genera per time interval (total number of genera per time interval divided into total number of new genera per time interval).

E. Percentage rate of extinction of genera per time interval (total number of genera per time interval divided into total number of genera becoming extinct per time interval).

F. Diagrammatic representation of continental marine transgression and regression trends during the Caradoc through Famenne interval.

G. Diagrammatic representation of relative changes in degree of faunal cosmopolitanism and provincialism during the Caradoc through Famenne interval.

H. Number of level-bottom articulate brachiopod-dominated communities per time interval during the Early Llandovery-Frasne time interval. The solid line sums the brachiopod communities considered in Part II, Chapter 6; the dashed line provides an estimate of communities not considered (primarily from Devonian biogeographic entities not known in enough detail; for example, Uralian Region and Uralian-Cordilleran Region).

I. Diagrammatic representation of relative orogenic intensity and duration during the Caradoc through Famenne time interval.

J. Diagrammatic representation of the duration and intensity of continental glaciation during the Caradoc through Famenne time interval.

K. Diagrammatic representation of the intensity of marine dolomite production during the Caradoc through Famenne time interval.

L. Diagrammatic representation of the intensity and duration of platform marine evaporite production during the Caradoc through Famenne time interval.

M. Diagrammatic representation of the varying importance in time of reef and reef-like structures—both their abundance and geographic scatter from the Caradoc through Famenne time interval.

animals. Ability of planktic larvae to survive for varying time intervals will probably be one of the chief factors, together with the location of transporting currents, in determining diversity gradients between far removed shallow-water regions either on platforms or in the oceanic environment.

If we are to understand worldwide and regional diversity gradients we must first make every effort to analyse differing taxic diversity in what appear to be similar niches occurring repeatedly from region to region, and keep these separate in our compilation from niches not present in every region. In other words, the observation that polar marine bivalve diversity is far lower than that present in the shallow equatorial waters although true does not bring out the fact that the level-bottom bivalve diversities do not change nearly as much from the poles to the equator as the raw data might indicate, because one of the biggest equatorial increments is the addition of the varied reef bivalves occurring in a variety of communities and corresponding niches totally absent in the polar regions.[29] The biggest effect in worldwide diversity gradients is the presence in marine shallow waters of reefs with their rich faunas in the equatorial, warm-water regions and the corresponding absence of these structures elsewhere. The presence of the equatorial rain forest similarly affects the terrestrial diversity story. The recent discovery of high diversity present in certain groups living in the deep seas has extended the problem of the controls governing diversity. MacArthur's penetrating analyses of the problem of faunal diversity on islands, which may be applicable to island-like isolated regions like mountain tops and Paleozoic reef environments occurring in a scattered manner on platforms has enabled part of the diversity story to be analysed in terms of extinction rate, immigration rate and the like to these isolated habitats. However, there still remains the nagging problem of why the overall latitudinal diversity gradient is present on land and in the shallow seas. Possibly these diversity gradients are related to the physiologies possible under equatorial conditions as opposed to polar conditions, but the story is probably one of a number of interacting factors that add up to the observed result. Valentine's (1971a) thoughts about food supply and stability of that food supply may be deeply involved. Pianka (1966) has summarized the bulk of the possibilities used in explaining the latitudinal gradients.

The Silurian and Devonian afford excellent examples of both worldwide and regional diversity gradients that parallel those of the present. The level-bottom communities of the Malvinokaffric Realm (see Figs.6 and 9 for diagram of some of the important communities, and Chapter 6 for a listing of taxic diversity of brachiopods) are significantly lower in taxic diversity than those of the North Silurian Realm or of the Eastern Americas and Old World Realms of the Devonian. In addition, worldwide

Silurian-Devonian diversity is strongly affected by the presence of the reef environment with its rich faunas during many time intervals, although not all, in the non-Malvinokaffric Realm areas. This information affords some of the best evidence for concluding that the Malvinokaffric Realm was a cold-water region during both the Silurian and Devonian (physical evidence summarized elsewhere in this book strongly supports this conclusion). Regional diversity gradients in the non-Malvinokaffric Realm exist between reef environments with their very high-diversity faunas and level-bottom environments with their far lower diversity faunas. The taxic diversity of Silurian-Devonian reef environments have not been summarized in this book, community by community, but the reader should be assured that the differences between reef and level bottom in the Silurian-Devonian are substantial.

Insofar as marine benthos is concerned there has been little attention paid, to date, to diversity trends between subdivisions of the level-bottom environment (sandy substrates, muddy substrates, substrates covered with algal growths of any type or another, etc.), although it has been noted that the intertidal benthic diversity appears in general to be lower than that of the subtidal. The diversity of rocky bottoms could be compared with profit to that of muddy and sandy bottoms, as well as to that covered with shelly debris of one type or another. Study of the fossil record can afford a wealth of information concerning taxic diversity trends correlated with these bottom-type materials.

Summing up the problem of latitudinal species diversity one may conclude that the more rigorous environments found progressively away from the tropical regions permit fewer physiological and biotic types to exist because of the increasingly restrictive nature of the physical environment. In other words, the environments closer to the poles tend to be more physically conditioned, and the organisms more closely related to the physical parameters than to the biological parameters of the environment. Conversely, in the more tropical regions more diverse physiologies and biotic types may be expected to occur, and the environment is more highly determined by the biota itself. The lowered restrictiveness of the tropical environment also permits a greater variety of predators, parasites, diseases and scavengers which in turn all act to raise diversity and lower the effect of competition for resource and space as a primary factor in controlling diversity. Lowe-McConnell (1969) and MacArthur (1969) have both provided careful discussions of the tropical situation. My own comments are only intended to emphasize the more physically conditioned nature of the lower-diversity, non-tropical regions (and regions of any extreme environmental types such as hot springs, conditions of excessive aridity and the like). Both Lowe-McConnell and MacArthur emphasize that it is the interaction of a number of factors that must be used to account for the observed diversities in the tropics and elsewhere. Viewed in this context Valentine's (1971a) concept that resource-poor stable environments are highest taxic diversity, resource-rich unstable environments are lowest-diversity, resource-poor unstable environments have low diversity and resource-rich stable environments have high diversity, must be modified to read that environments in which numbers of predator, parasitic, scavenging and disease taxa are high may outweigh resource as a taxic-diversity factor by raising taxic diversity above the level deduced from resource supply alone. In this context Valentine's terms "stable" and "unstable" should be read as "normal" or "optimum" as contrasted with highly "restrictive". It may be very difficult to decide whether or not high diversity in any specific example should be ascribed to the character of the resource supply or to the character of the diverse predators, parasites and disease taxa. It is also clear that when resource supply falls below a certain level diversity is sure to follow.

Biogeographic Units, Reefs and Islands

It has been noted that the scattered reef faunas of the Silurian-Devonian appear to show a higher degree of endemism (hence the usage "microprovince") than associated level-bottom faunas. In other words, the level-bottom communities of the Silurian-Devonian platforms show less evidence for endemism from one part of a platform to another part, or on adjacent platforms, than do reef communities occurring in reefs separated from each other by a great distance on even a single platform. In a way one may consider that reefs occupy almost the same biogeographic position in the shallow-water marine scheme of things as island biotas do in the terrestrial sphere. The difference is that the reefs occurring on a single platform, to carry the analogy farther must be considered as submerged islands occurring on rather than away from the continent.

This distinction between reef communities, level-bottom platform communities and level of endemism is apparent when dealing with the Early Paleozoic where the vast platforms do, during many time intervals, bear reefs. However, during parts of the post-Devonian, including the present, this distinction is not always apparent. At the present time the reef communities are not associated with shallow-water level-bottom communities covering vast platforms (it is not reasonable to compare the deep-sea level-bottom from this point of view with the Early Paleozoic shallow-water level-bottom communities). Therefore, many tropical and subtropical biogeographic units defined at present with the aid of shallow-water benthos are essentially reef-community-related, whereas those of the

Silurian-Devonian are largely level-bottom plus reef-community related. It is important in defining biogeographic units, because of this problem, to keep level-bottom and reef communities separate. It is reasonable to infer a higher level of endemism in the reef faunas over the same distance than would be expected in level-bottom faunas covering the same distance. One may try to apply the concepts of island biogeography developed by MacArthur and Wilson (1963) to the reefs occurring on a platform whereas the level-bottom communities that co-occur may not behave in the same manner.[30]

It is reasonable to infer that there were oceanic islands during the Silurian-Devonian just as during the Cretaceous and Cenozoic. Therefore, it is also reasonable that endemic shallow-water marine island and reef communities may have existed during the Silurian-Devonian about which we have little inkling at present. Possibly such islands served as source areas for some of the point source taxa that appear suddenly on the platforms and adjoining geosynclines. As our biogeographic understanding improves we may be able to make inferences about the likelihood of such possibilities.

Whittaker's (1972) rain forest resource partitioning model for the evolutionary process suggests that in an environment like the reef where biologically induced niches are very important one should encounter a progressively increasing number of taxa with time. The geologic record does not appear to support this surmise. The reef environment, as well as the level-bottom environment, appear to be characterized by relatively fixed numbers of taxa during extended intervals of time. Apparently, after the initiation of the reef environment following any earlier and widespread wipeout there is a very rapid, almost geologically instantaneous, evolutionary diversification following which a relatively steady state of taxic diversity is reached rather than a geologically slow diversification during the process of progressive reef partitioning. In terms of the reef environment this makes sense if one concludes that initially small level-bottom marginal populations diverge into the reef environment (essentially point sources) and evolve very rapidly. Following this geologically rapid diversification induced by small population size there is a leveling off in evolutionary rate. I have not yet thought of any logical explanation for this leveling-off phenomenon. An upper limit for number of taxa will certainly be afforded by the minimum number of individuals/species necessary to maintain the species. Thus the geologic record would appear to present us with a very stable number of reef taxa during any given interval of reef occurrence. Otherwise, one should expect to find an almost infinite number of reef and reef-related taxa developing during a relatively modest interval of time. This conclusion about the rapid attainment of maximum taxic diversity in the reef environment, as well as in the level-bottom environment of the Silurian-Devonian, is no different than that arrived at by Buzas (1972) for Cenozoic benthic foraminifers or by R. Watkins (written communication, 1973) for Cenozoic bivalves: "Deep-shelf, low-diversity bivalve communities of the California Oligocene in some instances contain essentially the same number of preservable species as communities of similar genera and families in the Recent outer shelf environment, although some do show an increase in the Recent over the Oligocene. Within the past 40 million years, no increase in number of niches in these communities appears to have occurred". These taxic diversity conclusions appear to disagree with those arrived at by Bretsky and Lorenz (1970) which are based in large part on the time-stability hypothesis.

Competition and Rates of Evolution

Stanley (1973c), in a most provocative paper, has advanced the concept that competition between allied forms is one of the most important controls over rate of evolution (both phyletic and cladogenetic). Stanley advances the concept that intensive competition for food and space, with the latter possibly being ultimately related to food supply, is the most important control over the rate of evolution of mammals, bivalves, hexacorals and possibly other groups such as ammonites and trilobites. If this concept of competition intensifying rate of evolution, as a first-order control over rate of evolution, could be substantiated from the fossil record as well as from living materials, we would have a powerful tool for understanding many relations. However, as will be discussed below, the data discussed by Stanley can be equally well if not more satisfactorily correlated with size of interbreeding populations. The evidence that Stanley has produced to support the concept of increasing competition correlating well with increasing rate of evolution is unsupported by experimental or observational data from the present,

Stanley points out that overall taxonomic rates of evolution (compiled from various sources) for bivalves are far lower than for mammals at the generic level. He makes a good point that suspension feeders, filter feeders, and deposit feeders do not appear to compete for space or for food supply, and certainly do not display aggressive behavior towards each other. He then goes on to conclude that the territoriality shown by many mammals, as well as that shown by members of many other animal groups, may be interpreted in terms of competition for food supply. However, no experimental or observational data is really put forward to prove that the territoriality of the cited organisms is strongly related to

competition for food as opposed to many other possibilities (sexual prero-
gatives, protection of young, protection of a breeding or nesting structure,
protection of a living area, etc.).

Stanley points out the widely differing rates of evolution met with in
hippuritid bivalves (rapid) as opposed to most level-bottom bivalves. Stanley ascribes this differing level to competition for food and place as hippuritids commonly occur in densely packed masses of shells. However, other
bivalves as well as non-bivalve groups (barnacles) also occur in densely
packed masses (one thinks immediately of some mussel species and of
many oysters) and yet do not apparently give evidence for rapid rate of
taxonomic evolution. This hippuritid situation vis-a-vis other bivalves may
be easily interpreted in terms of varying population size. Hippuritids are
essentially tropical forms, occurring during a time of relatively high provincialism (relative to the earlier parts of the Mesozoic) restricted to certain hippuritid reef-type environments. It is not surprising that they
should show rapid evolution as compared to level-bottom taxa because of
the far greater likelihood of far smaller populations, despite their wide
geographic scatter in the tropical zone (Coates,1973). It is difficult to
ascribe rapid evolution for hippuritid bivalves living under crowded conditions to competition for food while at the same time ignoring crowded
conditions when explaining some other slowly evolving bivalve groups
(many mussels and oysters). However, in all fairness it must be pointed
out that cases of competition for food and even of aggressiveness among
some attached marine invertebrates have been made on a very sound basis
by Connell (1963), but unfortunately we have no information for rates of
evolution shown by these groups that may be compared to bivalves.

Stanley has compiled data on taxonomic evolution rates of hermatypic and ahermatypic corals that are very similar to conclusions reached
earlier by others making the same calculations. From the fact that hermatypic corals show higher rates of taxonomic evolution than do ahermatypic corals Stanley deduced that they compete more intensively for food.
However, as discussed separately I feel that the correlation with size of
population is a relation that explains these differing rates of evolution
more satisfactorily. Stanley's surmise that the rapid rates of taxonomic
evolution shown by trilobites and Mesozoic (although not Paleozoic) ammonites is related to competition for food is unsupported except as an
analogy with his conclusions regarding corals, bivalves and mammals. [85]
The conclusion that Cenozoic mammalian evolution has been rapid because of a high level of competition for food, as opposed to the far lower
levels for marine bivalves, is also to be questioned. Cenozoic mammals are
characterized by a high degree of provincialism induced by the geography
of the time interval plus the climatic regimes superimposed on the geo-

graphies. It is probably premature to ascribe to competition for food the premier place in controlling mammalian rate of evolution during the Cenozoic in the absence of more compelling evidence (experimental or observational) that would show size of interbreeding population to be a far less significant factor. It would be interesting to develop tests for Stanley's concept using birds. One wonders if the large numbers of tropical birds really compete more intensively for food than do the relatively smaller numbers found in non-tropical regions? Or can the relative numbers of bird species in these regions be correlated more strongly with population size?

Rates of Evolution and Extinction, Biomass and Species Diversity; their Relation to Nutrient Supply

There is no reason to assume that the controls governing total biomass and maximum species diversity are identical or even operate in the same manner. Total biomass is probably controlled by the interaction of many factors of which available energy, available nutrients, available living space both for adults and young are considerations, as well as the capacity of the system for recycling both metabolites and dead organisms.

Assuming that the bulk of the nutrients available in the marine realm is derived from the continents, then intervals of widespread, continental exposure like the Late Permian—Early Triassic and Late Tertiary should be effective in providing an excess of nutrients over those intervals when the bulk of the continents were covered by shallow seas as was the case in portions of the Lower Paleozoic. Gibbs (1967) has shown how important is the effect of high relief in controlling both suspended and dissolved material in river waters. Gibbs also has pointed out that the effect of physical weathering in high-relief areas is even more important than tropical weathering in low-relief regions for generating a high supply of nutrients (see also Livingstone, 1963). Combined with this effect of continental exposure is the effect of warm climate as contrasted with cold climate (assuming equivalent relief, area and topography). Time intervals of widespread warm climate favor more rapid production of available nutrients through chemical weathering when compared with intervals of widespread cold climate. Accordingly, time spans of widespread warm climate favoring chemical weathering, if combined with time spans of widespread continental uplift resulting in high relief maximize the supply of available nutrients, whereas widespread cold conditions combined with widespread continental submergence minimize the nutrient supply. The effects of a land plant cover, probably present since the beginning of the Devonian, may be significant but probably are not the most significant factors in determining nutrient availability in the sea. An evaluation of the effects of climate in terms of nutrient supply makes it very important to consider the role of topography as emphasized by Gibb's (1967) studies. During intervals of low relief on the continental land area, nutrient supply will be at a lower level than during intervals of high relief. The warm-climate weathering of a low-relief surface should result in rapid removal of surface nutrients followed by an interval during which nutrient supply remains very low. A high-relief surface (which must, of course, be renewed at a constant rate by the effects of uplift) will permit comparatively higher levels of nutrient supply to be continuously available. The topographic factor may result in a minimizing of climatic effects during time spans of very low continental relief that approach peneplanation. The depletion of available nutrients in the sea effected by the removal of various ions due to their incorporation in bottom-sediment minerals is undoubtedly very significant on a long-range basis. Also, this depletion may be partly controlled by the composition and texture of the terrigenous material supplied, which in turn will be partly controlled by climatic regime.

Total species diversity does not appear to be a direct function of the factors cited in the first paragraph although species cannot exist without the available energy, etc., listed in that paragraph. The geologic record suggests that there are slow changes, as well as rapid perturbations, in total species diversity from lows like those encountered in the Lower Triassic to highs like those present in the Upper Cretaceous. It is unrealistic to conclude that the high diversity of the Recent is necessarily unique because of the preservation problem that has biased the paleontologic record against the recovery in significant numbers of many groups of organisms, as well as of organisms from many terrestrial environments where preservation of even the most massive hard parts is unlikely (upland floras and faunas, for example).

Consideration of the paleontologic record indicates that total worldwide species diversity is mainly a function of degree of provincialism versus cosmopolitanism supplemented by community diversity plus conditions of regression-transgression. However, it is instructive to consider total species diversity per small area per individual community. There have been evident swings in total species diversity of organisms possessing hard parts per community per small area in time. For example, Silurian-Devonian level-bottom, megascopic, benthic-marine invertebrates per small area per community seldom exceed about forty or fifty species whereas those of the Permo-Carboniferous commonly reach eighty to one-hundred and those of the Early Triassic less than twenty or thirty. What can be the controls over the upper numerical limit?

If the Time-Stability Hypothesis functions through extensive intervals of geologic time one would expect far larger numbers of fossil species than are actually encountered. Therefore, one must deduce that the Time-Stability evolutionary mechanism, if it functions at all, operates only over a very brief interval of time. This interval is probably so brief as to be virtually undetectable in the geologic record. In other words, maximum species diversity in an area is probably reached in a geologically almost instantaneous time interval that is well under a million years.

If resources are considered to be a seriously limiting variable it may be that the upper limit of species diversity in a small area is controlled by the resource specificity of each species, as well as the efficiency of each species in reproducing and utilizing the resources it is capable of employing. Low species diversity may indicate an interaction between available species having low resource specificity (relatively omnivorous) and a physical environment providing the necessary resources for nutrition. High species diversity may indicate an interaction between available species having very specific tastes (high resource specificity) and an environment in which the basic physical resources have been greatly modified through many trophic layers to provide a wide variety of food types.

It is clear from the work in the deep-sea benthos that total amount of resource does not limit species diversity as Valentine (1971a) has so ably concluded. The upper limit imposed on species diversity by the need to maintain a population size capable of perpetuating itself is clearly not reached by most taxa, which suggests that room exists for the presence of far more taxa than actually occur from the viewpoint of the reproductive consideration alone. Biologically modified environments like the rain forest and the coral reef are very rich in species per small area, but the high number of deep-sea species discussed by Sanders and Hessler (1969) from an environment that has not been thought of as being biologically modified is puzzling. Also puzzling is the problem of why the coral-reef and rain-forest environments are so biologically modified whereas biologically modified analogs in the low species diversity regions of the world including both the cold and hot deserts, tundras, and continental shelves, etc., do not occur. Possibly the presence of a heavily biologically modified environment depends on the presence of an optimum set of physical conditions that enables more varying physiologies to co-exist as contrasted to physical environments characterized by one or more extremes (from the organisms physiological viewpoint).

Further consideration of Valentine's (1971) suggestion that environments poor in resources and possessed of relative stability favor highest diversity is called for. In an environment low in total resources it may be in the direction of increased taxic diversity to specialize for resource utilization as suggested by the high diversity present in the deep-sea benthos, the modern coral reef and the rain forest, all of them being environments relatively poor in available nutrients. In other words, natural selection in an environment poor in resources might favor high diversity in the interests of intensive recycling of available, limited nutrients at all trophic levels. Hessler (1974) has presented a succinct summary of the large increase in taxic diversity of the benthos from the continental shelf into deep-ocean floor, and also commented on the inverse relation along the same transect between species diversity and biomass. Dayton and Hessler (1972) have provided an interesting alternative to the possibility that low resource supply drives evolution in the direction of highly specialized consumers by suggesting that a large number of varied predators (their "croppers" of both living and dead organisms) will so disturb the environment as to permit a low-density, highly-diverse group of "cropped" organisms to develop that utilize the same basic resources. In some ways the Dayton and Hessler suggestion is merely an extension of the observation that there is a good correlation between diversity of predators and herbivores as indicated by Paine (1966) and others combined with an explanation for the increased diversification. The Dayton and Hessler (1972) conclusion about disturbance helping to maintain diversity may be viewed as a paraphrasing of Hutchinson's (1957, p. 420) comment "... that there will exist numerous cases in which the direction of competition is never constant enough to allow elimination of one competitor". Inspection of the fossil record in terms of highest number of taxa per area per similar community both spatially and temporally, if this deduction is correct, should provide a clue as to relative abundance of nutrient supply in time and space. Hessler's (1974) observation that the deep-sea benthic faunas most removed from continental regions, which occur below oceanic gyres and have the lowest food supplies known, happen to have a somewhat lower diversity (although still high as contrasted with those of the continental shelves) as compared to deep-sea faunas lateral of them that are provided with somewhat higher food supplies, opens up the possibilities that either the inverse correlation of food supply with taxic diversity breaks down here or that there may be some lower limit below which further diminution of food supply results in taxic diversity going down as well rather than continuing to rise as had been the case from the continental shelves down into the deep oceans (it is worth noting that the Time-Stability concept fails to account for this deep-sea situation in any straightforward manner!).

Using this assumption one might conclude that the low taxic diversity of Cambrian brachiopod communities as contrasted to Silurian-Devonian brachiopod communities reflected differing levels of nutrient supply.

An attempt to correlate taxic diversity per community per small area in time with climate and continental exposure (regression-transgression and relief) is then in order. The Silurian-Devonian record indicates low taxic diversity per community per small area in the colder climate (Malvinokaffric Realm) as contrasted to the North Silurian Realm and the Old World and Eastern Americas Realms of the Devonian. If this correlation is taken to indicate a greater supply of available nutrients in the Malvinokaffric Realm than elsewhere one must conclude as well that the cold-climate Malvinokaffric Realm had a far greater exposed area of high relief than did the far warmer, low-relief and low-elevation regions present elsewhere. This situation is possible in view of our geologic information (widespread clastic units), but certainly has not been investigated in enough detail to be considered proved. There is no very clear correlation between taxic diversity per community per small area for the North Silurian Realm level-bottom faunas, which existed during an interval of high transgression (especially during the Wenlock-Ludlow) and those of the Eastern Americas or Old World Realms during the Lower Devonian, an interval of relatively high regression and higher relief. However, this non-correlation in terms of available nutrient supply might be a result of the higher diversity for the Lower Devonian due to conditions of far higher provincialism which more than cancelled out the effects of potentially increased available nutrient supply. The generally higher taxic diversity of Permo-Carboniferous level-bottom benthos per community per small area over that of the Silurian-Devonian might be taken as a consequence of lowered available nutrient supply, but it is difficult (outside of the cold-climate Malvinokaffric and Gondwana Realms) to correlate this situation with any inference about widely differing climates, conditions of overall relief and elevation, or conditions of transgression—regression (inferred to have been comparable in both instances). Possibly our data for these two time intervals will have to be analysed in more detail before a meaningful conclusion may be drawn.

The relation between taxic diversity and potential nutrient supply is apparently not simple as viewed by the geologist working in any specific time interval. For example, geosynclinal areas of North Silurian Realm rocks containing benthic faunas may be compared with areas of platform carbonate containing benthic faunas. The geosynclinal area, rich in terrigenous material (of either volcanic or basement complex source) yields brachiopod faunas with taxic diversities similar to those derived from the areas of platform carbonate. One would initially deduce that the geosynclinal regions should have been richer in available nutrients insofar as the bottom sediment is any indication. Possibly bottom sediment is not a good indicator of nutrient supply. Possibly factors of water circulation are more important than local bottom sediment. What, then, may be used as an independent measure of available nutrient supply other than the taxic diversity? Relating taxic diversity to available nutrients provides only a circular argument. Long-term perturbations in taxic diversity are clearly related to many variables which may be normalized only after making a number of assumptions that may, in specific instances, be unwarranted. An additional difficulty is the unlikelihood that faunal barriers will necessarily correspond with sharp nutrient gradients if the possibilities for upwelling and various currents trending in unknown directions are considered.

If these concepts have much utility in understanding the control over taxic diversity per community per small area they may then be combined with the information bearing on the smallest-size population capable of being reproductively viable (thus favoring stable environments as commented on by Valentine,1971) to conclude that we should try to correlate the taxic diversities in time with the geologic factors in a better attempt to understand changes in available food supply, climate, topography, relief and the like. This could take the form of a plot of maximum taxic diversity for marine benthos per community per small area against geologic time interval analysed in terms of climatic regimes, changes in topographic relief and elevation and regression—transgression.

Finally, it is important to understand that marked reductions in available nutrient supply may result in a marked rise in terminal extinction for some groups while at the same time accounting for a marked rise in rate of cladogenetic evolution, due to the evolutionary consequences of decreased supply of available nutrients, for other groups. Low nutrient supply should also favor a heightened rate of phyletic evolution at the same time that a heightened cladogenetic rate is favored, due to the smaller population sizes possible with a reduced supply of nutrients.

Rohr (oral communication,1973) has suggested that one possible independent measure of nutrient supply over time would be non-detrital phosphorus present in sedimentary rocks, particularly those of the carbonate platform well removed from geosynclinal regions where a higher level of detrital phosphorus might be expected. His initial examination of readily available data indicates a certain degree of correlation between high-phosphorus—low-diversity and low-phosphorus—high-diversity although far more study is indicated before this conclusion may be accepted without reservation.

Trophic structure position and rates of evolution

As mentioned earlier it is reasonable to conclude that, in general, organisms like carnivores which exist at a higher trophic level might be expected to be characterized by higher rates of evolution due to inferred smaller population sizes. Examination of this argument in more detail suggests that it may be a valid generalization. However, it is obvious that carnivores having relatively broad preferences in prey may be characterized by larger populations than many if not all of their prey. On the other hand, highly prey-specific carnivores may be characterized by far smaller populations than their prey. Therefore, one must deduce a complete spectrum in prey specificity of carnivores as well as in their consequent rates of evolution relative to their prey. Without detailed information about the prey specificity of a specific carnivore it is probably not possible to conclude whether it has a higher or a lower rate of evolution than an organism occupying a somewhat lower trophic position. These same arguments may be applied to scavengers as well as to carnivores.

Turning to the fossil record of the level-bottom, marine benthos we find that the great bulk of the preserved organisms are relatively low in the trophic structure; including such organisms with filter-feeding, suspension-feeding or deposit-feeding habits as bivalves, tetracorals, brachiopods, pelmatozoan echinoderms, some groups of gastropods, tabulate corals, stromatoporoids, trilobites, bryozoans, benthic foraminifers, etc. On the contrary the groups suspected to have been vagrant, benthic carnivores are far more limited as to number of specimens and number of species, including such items as cephalopods (ammonoids, nautiloids, belemnoids), some groups of gastropods, eurypterids, and some groups of eleutherozoan echinoderms including the various starfish.

Our knowledge of the taxonomy and evolution of most of the carnivore groups, with the exception of the cephalopods, is far poorer than that of the corresponding filter feeders, suspension feeders and deposit feeders due to the combined rarity of specimens and the lack of attention that this rarity brings with it. For example, it is entirely possible that the subtidal gastropods of the Middle Paleozoic, particularly those groups represented by rare specimens (in other words, not the coprophagous platyceratids, or the euomphalopterids, or oriostomatids that might have been herbivores), may have been carnivorous. But, owing to their rarity as specimens one finds that their taxonomy is so poorly known and neglected that little of consequence can be said about rates of evolution one way or the other.

The record of the cephalopods is somewhat ambiguous. During the Paleozoic both nautiloids and ammonoids are rare to moderately abundant, although seldom very abundant. Their taxic diversity appears to be relatively high as contrasted with their abundance. Belemnoids are extremely rare during the Paleozoic. However, during the post-Paleozoic the nautiloids are characterized by a very much lowered taxic diversity and abundance with the present-day situation being a logical continuation and sample. This situation for the post-Paleozoic nautiloids can best be interpreted as a series of limited populations that have been unable to expand their numbers (except for a minor interval in the Early Tertiary; see Miller,1949) in the direction of more individuals or more taxa. The Mesozoic ammonoids, on the other hand, exploded into a tremendous development of both taxa and individuals during several post-Paleozoic intervals that are consistent with the concept of population size being inversely related to rate of evolution. The ammonoid situation during the Mesozoic is paralleled by that of the belemnoids. Therefore, one is led to conclude that, assuming cephalopods of the Phanerozoic to have been higher-trophic-level carnivores, most ammonoids, belemnoids and Paleozoic nautiloid groups were characterized by rapid evolution and smallish populations, but that the post-Paleozoic nautiloids are anomalous in that they display a combination of relative rarity and low number of taxa suggesting a low rate of evolution which may only be explained by appealing to special mechanisms.

Rates of Evolution in "Higher" and "Lower", Marine and Terrestrial, Mobile and Sessile Forms as Viewed from the Population-size Position

From the beginning of thinking on the subject of organic evolution there has been serious concern about the controls governing rate of evolution (see Rensch, 1960, for a summary of this thinking). There has also been concern about whether evolutionary rates have remained constant in geologic time, have changed in one direction, or have varied back and forth in time due to influences of one sort or another. At the generic and specific level many workers have been concerned about whether or not, in general, lower organisms have evolved less rapidly than higher organisms, marine organisms less rapidly than terrestrial forms, and sessile organisms less rapidly than mobile forms. There has been a tendency to conclude that placental mammals have evolved more rapidly than marine organisms (as exemplified by the placental mammal-marine bivalve examples discussed in Stanley, 1973c). Thought about the relative rates of evolution characteristic of higher- and lower-type organisms earlier concluded that higher types evolved more rapidly, undoubtedly influenced by the far more comprehensive taxonomic knowledge concerning the higher-type placental mammals than of other lower groups, but has long since arrived at the conclusion that most animal groups high and low have examples of both slow and rapid evolution. However, as held earlier in this section on the possible importance of competition in governing rates of evolution, I conclude that population size is the most important variable responsible for determining the observed rates in both marine and terrestrial organisms. The observation that placental mammals display an apparently far higher rate of evolution than is the case for marine bivalves correlates well, as discussed previously, with population size. The work on the sessile brachiopods of the Middle Paleozoic discussed in this book should make clear that a sessile mode of life need not necessarily correlate with a low rate of evolution at the generic level.

One of the important factors that has probably influenced the discussion of evolutionary rates in the major groups has been the greatly differing levels of study they have received. It is hard for the non-specialist to realize how uneven has been the attention given to the taxonomy of the various groups. Why would one suspect that the taxonomy of fossil sponges is far less well known than is the case for foraminifers? Why would one suspect that the taxonomy of the foraminifers has been improved many times during the period 1920—1970 as compared with most other groups of animals? Therefore, it is not completely meaningful to prepare tabulations of genera per unit of geologic time for various major groups with the intent to compare them and extract meaningful conclusions about different as well as changing rates of evolution. The uneven nature of the fossil record, the uneven attention given to various fossil groups by specialists, the uneven collecting coverage, the changing position of biogeographic boundaries, the changing complexities of community structure, and a host of other variables have all conspired to make raw compilations dangerous for comparative purposes. Illustrative of this problem is Brodkorb's (1971, p.21) statement: "The reason for the imperfect state of our knowledge is not the alleged dearth of fossil birds — it lies in the scarcity of paleo-ornithologists to study them." The situation referred to by Brodkorb for birds is also true of many other potentially important groups, important if we are to have records of evolutionary rates that may be compared with any degree of confidence that we compare like quantities.

Total Number of Species in the Record

As Mayr (1970; see quotation in preface, this book) has pointed out, the fossil record consists chiefly of taxa characterized by low to moderate rates of evolution. Therefore, when considering the implications of Fig. 29 it must be remembered that the bulk of the taxa characterized by very rapid rates of evolution (comparable to those found in many of the endemic fresh-water fishes discussed by Lowe-McConnell, 1969) will probably not be preserved in the fossil record. In other words, the fossil record provides an exceptionally poor, statistically unreliable count of the number of rapidly evolving forms developed in the past, probably a fairly reliable record of those forms with moderately rapid rates of evolution (those discussed in this book as rapidly evolving forms), and a very reliable record of those forms with slow rates of evolution (all of this being a function of degree of overall cosmopolitanism and local abundance that correlates with it). In view of these relations one clearly may *not* regard the number of species recorded from the fossil record per time interval as a reasonable approximation of the number of species living during any time interval of the past. Rather, one must regard the number of cosmopolitan species per time interval of the past as a reasonable approximation of cosmopolitan species abundance, the number of taxa characterized by moderately rapid rates of evolution as considerably lower than the real number, and the number of taxa characterized by rapid rates of evolution as far, far below the actual number. If one expects that rapidly evolving taxa, during most time intervals of the past, should have exceeded in number the moderately rapidly evolving and slowly evolving forms combined, it is clear that estimates of overall number of species per time interval of the past must be adjusted. Any raw compilation of total number of species recorded from the fossil record per time interval of the past will certainly not be a direct function of actual number of species living during any time interval of the past. By appropriately weighting the numbers of cosmopolitan, endemic and rare taxa per unit of time for the past we may be able to provide a more realistic estimate of total number of taxa per time unit of the past (present estimates may well be far too low). Time intervals of high and low cosmopolitanism must be appropriately weighted if a reliable estimate of actual number of species per time interval is desired. Cosmopolitan time intervals will provide a far more reliable estimate of actual number of species due to the greater percentage of cosmopolitan species present in the first place. Time intervals characterized by a high level of endemics, the highly provincial parts of the geologic record, will presumably yield an unusually low estimate of the actual number of species originally present due to the far higher probability for larger numbers of very rare, endemic taxa. For example, a time interval of high cosmopolitanism like the Early Triassic should provide a far more reliable estimate, albeit low total number, of species present than intervals like the Miocene, Early Devonian or Lower Ordovician with their high levels of provincialism despite the far higher number of species recorded to date from these latter time intervals. In other words, normalizing our data for changing numbers of species requires that we pay very careful attention to the effects of changing levels of provincialism per unit of absolute time. Even casual consideration of the fossil record in terms of changing provincialism suggests that the actually recorded swings in number of species present per time interval have been highly underestimated. Real figures for highly provincial time intervals are probably much higher than recorded. All such normalizing estimates must await a better compilation of the relation between numbers of cosmopolitan, provincial and rare species in living groups. After abundance curves have been developed in terms of living organisms and their degree of provincialism, we will be in a better position to weight the evidence of the fossil record more realistically.

Raup (1972) has carefully considered additional factors capable of having altered the numbers of fossil species and genera. He has concluded that all of these factors have badly biased the available record, and decreased the apparent numbers of genera and species recorded from older rocks. The information discussed above by me tends in the same direction as Raup's conclusions, but tends to expand them manyfold as well as giving another base from which to extrapolate a more meaningful figure than that provided by a raw compilation of numbers of genera and species currently published.

Silurian-Devonian Evolutionary Rates and Patterns for Brachiopods

Areas occupied by potentially interbreeding populations through geologic time may aid in the fuller understanding of the evolutionary implications of population size and of faunal diversity within communities and within biogeographic entities. Present knowledge of the Silurian and Devonian is reviewed from this viewpoint. Sufficient preliminary data are available for a serious attempt at correlating these variables, although not all of these correlations may prove to be causal.

Measuring rate of evolution

It is desirable for several reasons to try to measure rates of taxonomic, cladogenetic and phyletic evolution (Simpson, 1953). It is clear that rate of evolution is a complex quantity that results from the operation of many factors. Various approaches can be made to measure rate of evolution using fossils. In the past some paleontologists have obtained a "rate of evolution" by counting the number of taxa present, for either a single group or for many major groups lumped together, in successive time intervals and reducing this information to a rate (Total number of taxa, Fig. 22A, 23).

Rate of taxonomic evolution is essentially the result obtained by summing up phyletic *and* cladogenetic evolution. However, in many cases the information summed up in the taxonomic summary is very fragmentary. This fragmentary nature may bias the result so that it does not evenly sum up both cladogenetic and phyletic evolutionary rates.

The above assumes, of course, that estimating rate of evolution by adding up totals for taxa per units of time is a "natural" approach unsullied by "monographic" effects or migration from unknown regions. Williams' (1957) work on generic compilations of Paleozoic brachiopods

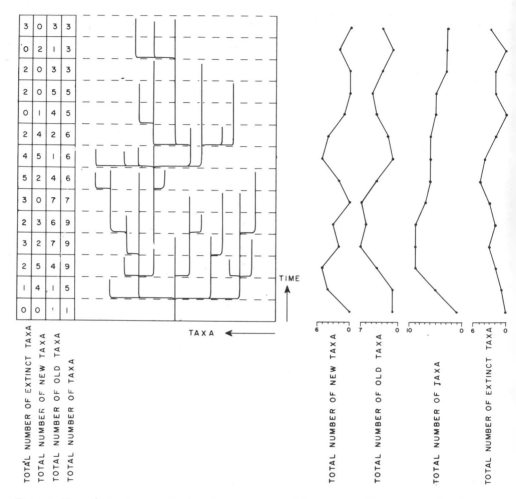

Fig. 23. Hypothetical example showing summary of taxa per unit of time analysed as to new taxa, extinct taxa, through-going taxa, and total taxa. Note that taxonomic rate of evolution as shown by introduction of new taxa is not a direct function of the total number of taxa. All taxa are of equal rank. It is important to note that only cladogenetic evolution has been indicated on this diagram (phyletic evolution in the individual "lineages" shown as unbroken lines could have been indicated, but would have complicated the diagram). If phyletic evolution had been indicated in each "lineage" then the overall taxonomic rate of evolution would have been altered appropriately and would not necessarily have corresponded with that obtained by considering cladogenetic rate only.

shows how misleading a raw summary of genera per unit of time may be, as it changes significantly whenever the shells in any particular time interval receive more attention in publication than has been the case previously relative to other groups. Finally, a mere summation of the number of taxa per unit of time makes no distinction between intervals of high evolutionary or low evolutionary rate combined with high or low extinction rate (Fig. 23. Total number of new taxa and Total number of old taxa).

It is desirable to consider the information bearing on rate of evolution provided by the total number of new taxa per time interval and also that provided by the percent of new taxa per time interval (that is the total number of taxa divided into the number of new taxa for the time interval) in order to have some idea of the production rate of new taxa and the relative production rate of new taxa (the relative production rate permits some insights into rate of evolution divorced from numbers of taxa existent during any one time interval; see Silurian and Devonian examples in Fig. 22).

It is crucial to any consideration of taxonomic evolutionary rates that one understands the following. We actually deal with the *appearance* rate of new taxa in the geologic record. We then may interpret this appearance rate as accurately reflecting either rate of taxonomic evolution, as sudden migration and dispersal of taxa unknown previously (sources not located in older rocks) or as a combination of the two. In other words, one cannot assume that appearance rate of taxa is synonymous with taxonomic evolutionary rate. The discussion of Fig. 22 makes clear that in some instances appearance rate and rate of taxonomic evolution are almost identical, but that in other instances appearance rate definitely is a reflection of migration in largest part, not of taxonomic evolutionary rate. Each case must be decided on its own merits (the introduction of a large number of taxa without known antecedents makes migration from an unknown, previously existing source the explanation of choice).

It is very desirable to analyze this raw rate of evolution, into as many of its components as possible (Simpson, 1953). I find that the real distribution patterns shown by fossils may be employed to evaluate the factor concerned with the differing size of interbreeding populations and their apparently very profound effect on overall rate of evolution. The reader must remain aware that we assume population densities of the same magnitide for the taxa under consideration; in detail this assumption about equivalent population densities will certainly not hold up, but as an order of magnitude estimate it is reasonable.

If we are able to normalize the rate of evolution for varying size of interbreeding populations, it is possible to analyze for other factors formerly obscured. In this manner one may be able to determine whether differing physical environments have an effect. Then one also should be able to determine whether different time intervals are characterized by differing rates that might be ascribed to reversals in the earth's magnetic field and differing fluxes of cosmic radiation or to a host of other biologic and physical factors.

Just as we can measure phyletic, cladogenetic and taxonomic rates of evolution following Simpson's (1953) lead, so it is possible to measure both phyletic and taxonomic rates of extinction (as pointed out by Newell, 1967, p. 75). For the latter case, however, it is obvious that phyletic extinction rate is a pure synonym of phyletic evolution rate. Taxonomic extinction rate will be a pure summary of number of taxonomic disappearances per time unit just as its counterpart is a pure summary of appearances per time unit. A novel quantity, here termed terminal extinction rate, may be measured for the termination of phyletic groups. It may be measured by counting the number of final terminations per time unit for family trees. The measurement of terminal extinction rates will give a more precise measure of the environmental factors tending to affect extinction than will the raw taxonomic extinction rate. The taxonomic extinction rate incorporates, unfortunately, the results of both phyletic extinction and terminal extinction; in other words, taxonomic extinction rate is a mixed quantity just as taxonomic rate of evolution is a mixed quantity incorporating the effects of phyletic evolution, evolutionary diversification (cladogenesis), and even migration (the appearance rate problem of this treatment) from unknown source regions.

Phyletic evolution (= "bioseries" of many paleontologists) rate may be considered as related to the time duration of taxa belonging to an evolving lineage. Phyletic evolution rate provides a measure of the speed with which taxa are evolving. Short-lived taxa belong to rapidly evolving lineages; long-lived taxa belong to slowly evolving lineages. Phyletic evolution rate is independent of number of communities.

Cladogenetic evolutionary rate is a measure of the amount of branching (cladogenesis), either increasing, decreasing or remaining steady. In a single-taxon lineage the cladogenetic evolutionary rate is zero as no net increase in number of taxa per time interval occurs. Cladogenetic evolutionary rate is related to, but independent of phyletic evolution rate. Different branches derived from an ancestral taxon may have differing rates of phyletic evolution. Cladogenetic evolutionary rate is largely if not entirely dependent on number of communities available, the latter closely related in turn to number of biogeographic entities.

Neither phyletic evolution nor cladogenetic evolution rates may be calculated from a single taxon whose antecedents or descendents are unknown. Such isolated single taxa, if morphologically far removed from either potential descendents or ancestors, are of no use for such calculations. If such taxa are relatively close morphologically to potential relatives one may estimate the time position of direct ancestors, descendents and even unknown morphological intermediates to provide rates.

Phyletic evolution rate is dependent on population size, but cladogenetic evolution rate is dependent on number of communities and biogeographic entities at any one time (unless sympatric species belonging to a single community occur!). These two quantities, as pointed out earlier by Mayr (1965), vary independently. Therefore, one may find examples of high or low cladogenetic rate of evolution that are also characterized by either high or low rate of phyletic evolution. Thus, *Eocoelia* species are widely distributed in the North Silurian Late Llandovery. They are characterized by a high rate of phyletic evolution, but a zero rate of cladogenetic evolution. Pentamerinid genera in the Llandovery and Early Wenlock are characterized by a very low rate of phyletic evolution and also a low rate of cladogenetic evolution. During the Late Wenlock-Ludlow, correlated with an abundance of isolated reef environments, the pentamerinids are characterized by a rapid cladogenetic rate of evolution. During the late Lower Devonian the mutationellinid brachiopods are characterized by a moderate rate of phyletic evolution, but by a high rate of cladogenetic evolution.

Both cladogenetic evolution rate and terminal extinction rate are closely related to the presence of faunal barriers and the reef environment, the former only being a special case of provincialism aided by distance between reefs as the barrier ("microprovinces" if you will). Cladogenetic evolution rate will increase rapidly with the introduction of new faunal barriers and then taper off rapidly. In other words, rapid diversification is almost a function of the introduction of faunal barriers. Increase in terminal extinction rate may be evidence of the removal of faunal barriers, tending to increase rapidly with their removal, and following which it tapers off rapidly, plus cataclysmic events. Both, of course, may also be considered in terms of changing numbers of communities that are also related to faunal barriers.

It is clear that all of the evolution rates considered here will be related to the level of study. It is difficult to be certain that the subspecies, species, genera or families employed by one worker for one group of animals are strictly comparable to those employed by another worker for an unrelated group. However, it is probable that the taxonomic level used by any taxonomist will not greatly exceed that used by another, i.e., the genus of one man will probably not be any different in rank than the species of another or the subfamily of a third. In any event the various rates discussed here will clearly be affected by the taxonomic rank employed, i.e., species will show as great or greater rates of evolution as genera and so on. The same question arises for biogeographic subdivisions; the smaller the taxonomic category considered, in general, the finer the biogeographic subdivision possible.[31]

Rate of Extinction

Rate of extinction is defined here as the decrease in Total number of old taxa (Fig. 22C, 23) per unit of time (Phyletic extinctions plus terminal extinctions = taxonomic extinctions). Percent rate of extinction is defined here as the change in percentage obtained by dividing the total number of taxa into the number of taxa becoming extinct per time unit (Fig. 22). They undoubtedly represent a number of complex factors, some dependent on each other, others not. It is notable that for the Silurian-Devonian brachiopods there is a correlation between raw taxonomic extinction rate and number of biogeographic entities. When the number of biogeographic entities goes down, as near the boundary between the Ems and the Eifel, the taxonomic extinction rate goes up, i.e., the large number of taxa (see Fig. 17, 21) present during a time of high provincialism (the Siegen-Ems interval) goes down during a time of significantly decreased provincialism (the Eifel interval). This correlation may be interpreted as indicating that the removal of faunal barriers permits competition between previously isolated taxa for the same positions and resources, with the inference that the competition is only of short duration. How else are we to explain the evident reduction (Fig. 17) in number of chonetid genera near this boundary, a reduction paralleled in many groups of Devonian brachiopods at or near this boundary between the Lower and Middle Devonian (see Fig. 27).

The removal of the barriers to faunal migration that leads to cosmopolitanism may either coincide with an amelioration of climate or not. If there is an amelioration of climate, Valentine (1967) has suggested that the more uniform climate will lead, naturally enough, to a heightened terminal extinction rate that will complement the terminal extinction rate increase brought about by the removal of barriers. In the event that the barriers are climate-controlled or dependent these two effects will be one and the same.

During the Middle Llandovery—Late Llandovery boundary interval there is an increase in rate of taxonomic and terminal extinction of brachiopod taxa that corresponds with the introduction of a number of new taxa, many of the latter known in older beds only from southeast Kazakhstan. Many other explanations and mechanisms must certainly be considered when trying to understand extinctions, but this sudden competition for place and resources after the removal of faunal barriers appears worthy of serious consideration as a major mechanism. Evolutionary change in niche breadth is another important consideration in understanding extinction rate. The elimination of the reef environment from time to time is yet another important factor.

The remarkable increase in rate of terminal extinction of brachiopod genera on a worldwide basis encountered near the Ordovician-Silurian boundary (Boucot, 1968a, p. 34), that may reach more than 60% (Fig. 22) on a worldwide basis, is also correlated with the continental glacial event of the Late Ordovician-Early Silurian and following deglaciation (Berry and Boucot, 1973a). This remarkable physical event involving the withdrawal of sea water from vast reaches of the continental platforms presumably had a pronounced effect on the benthic marine fauna, involving as it did the effects of changing temperature distributions and withdrawal (regression). Hutt et al. (1972) indicate that there was also a marked diminution in numbers of graptolite taxa between the Ordovician and Silurian corresponding in time to this same glacial event.

A summary of the worldwide extinction event occurring in the Late Ordovician indicates that both biotic and physical controls may be involved. The marked regression of Late Ordovician times, correlated with a glacial maximum and lowering of sea level, should have lowered the size of interbreeding populations and tended to speed up the rate of phyletic evolution. However, this tendency favoring increased rate of phyletic evolution may have been accompanied by the presence of a far more rigorous climate (a gradient from glacial climate in the south to a warm-humid, laterite producing climate in the north), plus the lowering of provincial barriers near the Ordovician-Silurian boundary that allowed extensive competition for place and resources between groups with similar requirements that resulted in a far lower rate of taxonomic evolution (not as great, however, as for the climatically equable Permo-Triassic extinction event).

The apparently unipolar climatic picture indicated for the Silurian and Devonian (a cold Southern Hemisphere Malvinokaffric region as contrasted with a warm region elsewhere) contrasts markedly with the bipolar picture that appears to be present for the post-Devonian (possibly for the post-Givet or post-Frasne). This contrast might be more apparent than real *if* one assumes a continental-drift model in which the present continental masses are brought together into a configuration leaving the north polar region free of any land area on which a cold climate record could have been left. Needless to say, such a model is speculative and unproved at present leaving us still to consider the possibilities inherent in the unipolar model.

In connection with these major Silurian and Devonian extinctions in the marine realm it is only fair to state that the correlation made here between removal of faunal barriers, i.e., return of cosmopolitan, less provincial conditions, works for the Ordovician-Silurian boundary region, the Middle Llandovery-Upper Llandovery boundary region, the Lower Devonian-Middle Devonian boundary region, but may not be applied to the very important extinctions occuring between the Frasne and the Famenne in the Upper Devonian (McLaren, 1970). McLaren makes a very strong case for the worldwide extinctions affecting many groups (about 60% of the brachiopods; see Fig. 22) near the Frasne-Famenne boundary being best explained by some type of as yet unknown catastrophic event (the extinctions at the end of the Ordovician are different in that they coincide with both a glacial, climatic event and a lowering of provincial barriers). Urey (1973) outlines the theoretical possibility of calling on cometary collisions to effect rapid biotic changes; we lack enough data to evaluate this possibility for the Frasne-Famenne event.

The high number of terminal extinctions occurring at the end of the Pridoli (Fig. 22) coincides with the disappearance of most reefs.

In considering rate of extinction it is important to note the difference between termination of a taxon (terminal extinction) that does not give rise to descendents (i.e., the bulk of the chonetid taxa of Fig. 17) as contrasted with those that do give rise to new taxa (the eocoelids shown in Fig. 18). The high extinction rate among the eocoelids is clearly just another way of expressing their high rate of phyletic evolution, whereas the high terminal extinction rate among the chonetids is inferred to be related to the removal of faunal barriers near the Lower Devonian-Middle Devonian boundary. These two separate phenomena should not be confused.[32] Many examples of both types (at the generic level) are illustrated in Fig. 27.

Any attempt to evaluate rate of extinction information must try to cope with the problem of terminal extinction rate as separately from co-occurring phyletic extinction rate (i.e., phyletic rate of evolution) as possible, since phyletic extinction rate represents a quantity very different from terminal extinction rate. The fact that taxonomic extinction rate incorporates the sum of phyletic extinction rate and terminal extinction rate demands that the significance of changes in taxonomic extinction rate be considered very carefully. We have no reason to believe that terminal extinction rate and phyletic extinction rate vary together in the same direction much less at the same rate!

During times of high phyletic extinction rate, such as times of high provincialism and high regression the taxonomic extinction rate will be consequently higher than during times of high cosmopolitanism because of the population size effects. These differing taxonomic extinction rates related to differing phyletic extinction rates may co-occur with terminal extinction rates that are unrelated to the phyletic extinction rates. Fig. 22 should be studied very carefully with these questions in mind.

Major terminal extinction events are of great enough magnitude to completely overshadow differences in phyletic extinction rate occurring both before and after, as the terminal extinction rate effect is overwhelmingly dominant. Therefore, it is during times of only "average" extinction rate that the significance of differing phyletic extinction rates must be subtracted carefully from the taxonomic extinction rates in order to get at the underlying, more important terminal extinction rate data. For example, in view of the above comments it is easy to envisage a time of differing taxonomic extinction rates, induced by differing phyletic extinction rates, underlying which is a constant terminal extinction rate. Many other such anomalies may be conceived.

Worldwide Extinction Events

The three massive, worldwide extinction events at the end of the Ordovician, end of the Frasne, and end of the Permian have certain features in common. Because of the small sample these common features may be, of course, mere coincidence. They all share a very large reduction in number of taxa at every level from species upwards.[33] The event at the end of the Ordovician probably accounts for about 60% of the previously existing taxa, that at the end of the Frasne for a good 60% percent[34] and that at the end of the Permian for more than 95%! In all three cases a physical event[35], or events, is commonly postulated as the precipitating cause, although only for the Ordovician event have we as much information from the geologic record to form a plausible case as one might wish (widespread Southern Hemisphere glaciation, a large climatic gradient from north to south, widespread marine regression). All three cases are followed in the earlier Llandovery, Famenne, and earlier Triassic respectively, by the rapid development and spread of a very low to moderately diverse, cosmopolitan fauna made up of developments from a few previously existing units. All are then subject to a rapid proliferation of both taxa and animal communities that rapidly brings things back to a more "normal" condition of taxonomic richness. It is clear that the Permian event was the most intense, followed by the Frasne event, followed by the Ordovician event, but the similarities in all three cases are striking. Each does away with a large number of previously existing taxonomic, animal community and biogeographic entities, followed by an interval of cosmopolitanism, lowered community number and lowered taxonomic diversity. These three massive extinction events are of far higher importance, particularly in a percentage of previously existing taxa wiped out, than the extinction maxima present between them, many of the latter being ascribable to less profound physical and biological factors. None ·of them acts to either speed up or slow down the rate of evolution except insofar as the size of interbreeding populations is concerned. The smaller number of taxa present after the event, and the wide dispersal of these taxa tend to form larger populations that evolve more slowly, except for those widespread communities that occupy a relatively small area with consequently smaller populations. The worldwide monotony and ease of correlation with these faunas is notable, contrasting markedly with the more normal, more provincial faunas. It is far easier to have a worldwide command of Llandovery, Famenne or Early Triassic shells, for example, than to have a worldwide command of Ashgill, Givet-Frasne or Late Permian faunas. So, these events indicate:

(1) Reduced rate of taxonomic and cladogenetic evolution. (2) Increased rate of phyletic evolution. (3) Increased rate of terminal extinction.

It is clear, however, that there is no necessary correlation between periods of extinction worldwide being followed by the presence of an extremely cosmopolitan fauna. The data just discussed does indicate that such cosmopolitan faunas did follow major extinction events occurring at the ends of the Ashgill, Frasne and Permian respectively, in the Early Llandovery, Famenne and Early Triassic respectively. But, this fact coincides with the occurrence of both climatic and geographic conditions conducive to relative cosmopolitanism. Following the extinction event of the Late Cretaceous it is clear that the "new" fauna is highly provincial because of the various barriers to migration and interbreeding that existed.[36] In other words, it is not a "rule" that worldwide extinction events are followed by intervals of cosmopolitanism, or that immediately prior to worldwide extinction events we have periods of high provincialism (the Frasne is immediately prior to an important extinction event, but it certainly is not a provincial time interval).

The association of the Late Ashgill and Permo-Triassic extinction, low-diversity events with worldwide regression requires comment. The Permo-Triassic is a time of worldwide regression comparable only to the Late Tertiary-Recent, the Late Ashgill not as marked in this regard. The extinctions and accompanying diversity lows of the Permo-Triassic are the most marked, and are accompanied by relative climatic uniformity (Kummel, 1973). The Late Ashgill is a time of large, worldwide unipolar climatic gradients from a glacial southern region to a warm northern region. The Late Tertiary-Recent is a time of large north—south bipolar climatic gradients. The Permo-Triassic is a time characterized by a small number of widespread, marine animal communities (in the Early Triassic), the Late

Ashgill by a far higher number of marine animal communities, and the Late Tertiary—Recent by a very high number of far less widespread animal communities. Thus, it is possible to explain the low total diversity (see Kummel, 1973, for the Early Triassic) and high extinction rate of the Permo-Triassic as resulting from widespread regression combined with a very uniform climatic regime that lent itself to the occurrence of very widespread, larger population marine animal communities with a correspondingly low total diversity. The higher degree of climatic diversity in the Late Ashgill permitted the existence of far more marine animal communities of smaller size, that made possible a much higher faunal diversity than was present in the Permo-Triassic. Finally, the possibly higher degree of provincialism present in the Late Tertiary—Recent coupled with many more faunal barriers (including the existence of major land barriers crossed by isotherms) permitted a tremendous number of communities to exist with correspondingly higher faunal diversity. These regression-linked, community-, biogeographically-, and climatically-related extinction events do not help to explain the Late Frasne extinction event (widespread transgression, with moderate number of animal communities, cosmopolitan conditions). The Late Cretaceous extinction event involves both a high terminal extinction rate, widespread regression and high diversity. It differs from the Late Ashgill, Late Frasne and Permo-Triassic extinction events in not involving co-occurring post-event cosmopolitanism. One may infer for this Late Cretaceous event that the widespread regression and increasing climatic gradients were responsible for the faunal crisis that led to a high terminal extinction rate and also for smaller populations that led to continuing rapid phyletic and cladogenetic evolution that produced new taxa. This may indicate that there is a certain threshold effect occurring during regression which, if passed, leads to massive terminal extinctions induced by a number of factors. It is also possible that this same threshold region may permit the existence of small enough populations among the remaining taxa to permit rapid phyletic evolution. In this view a high terminal extinction rate and a high rate of phyletic evolution may coexist under certain conditions with a lowered rate of taxonomic and cladogenetic evolution. There must be additional factors not considered here that are very important to an understanding of both extinction events and faunal diversity.

Viewing the Permo-Triassic extinction high, low diversity and incoming of many new, higher, suprageneric level taxa from the view of population size it is possible to conclude that the shallow-water, uniform environments, containing a relatively small number of communities were sites of very rapid evolution during the "hidden, unfossiliferous", boundary interval. These communities, during the regression maximum, may have been isolated in such a manner as to favor very rapid evolution (almost of the type discussed by Mayr, 1963, for isolated populations of fishes in lakes and ponds). This conclusion helps to account for the "sudden" appearance of so many new, suprageneric level taxa within a short time interval—taxa absent in the highest Permian marine faunas, but present and widespread in the lowest Triassic marine fauna. This interpretation for the controls over the narrow marine shelf fauna rates of evolution and extinction are consistent with a contemporary slow rate of evolution affecting terrestrial organisms.

Extinction Survivors

Bretsky (1973) suggests that the survivors of a major extinction event (such as that following the Permo-Trias event or the Ordovician-Silurian event) will tend to be chosen largely from previously existing cosmopolitan taxa. In addition he suggests that a higher level of provincialism is to be expected prior to a major extinction event than following such an event. Examination of the articulate brachiopod record suggests that Bretsky's conclusions are oversimplifications of the true state of affairs.

Major extinction events may occur in a sequence that is largely cosmopolitan on either side of the event (Frasne-Famenne event), provincial prior to the event and cosmopolitan after the event (Ashgill-Llandovery event) or provincial both before and after the event (Late Cretaceous-Early Tertiary event). In principle we should also have examples of a cosmopolitan condition prior to the event followed by provincialism, but I am not familiar with any examples.

The factors responsible for causing major extinction events are varied and many. The geologic record suggests that the factors responsible for the presence of provincialism and cosmopolitanism do not correlate in a 1:1 manner with the factors responsible for major extinction events.

Of the forty-five articulate brachiopod taxa surviving into the Llandovery from the Ashgill 27% (12 taxa) are rare. The 27% is almost the same percentage as are present in the entire Early Llandovery fauna (including new forms) as rare taxa. Although the Llandovery fauna is far more cosmopolitan than that of the Ashgill there is no indication that the major extinction event has played a first-order role. The steady decrease in level of provincialism encountered in the Early Ordovician continues into the more cosmopolitan condition of the Llandovery. Fig. 26L shows that the Malvinokaffric Silurian (including the Llandovery) is highly provincial which indicates that the barriers to faunal migration were just as intense during the Silurian as prior to the Silurian, although the major extinction event affected this region as well. The extinction of the Malvinokaffric

Silurian fauna is followed by the introduction of another endemic, unrelated fauna!

The twenty-three brachiopod taxa that span the Frasne-Famenne boundary are cosmopolitan both before and after the event, as are most of the taxa that fail to span the boundary.

The situation regarding Late Cretaceous-Early Tertiary shallow-water molluscan faunas certainly suggests that strongly provincial conditions were present both before and after the major extinction event.

The Frasne-Famenne major extinction event is preceded by a major interval (Frasne) of cosmopolitanism that was arrived at following a gradual decrease in provincialism following the highly provincial Lower Devonian. The same, of course, as pointed out in the preceding paragraphs is true of the Ordovician where a steady decrease in Ordovician provincialism preceded the major Ashgill-Llandovery extinction event.

The Permo-Triassic major extinction event has possibly colored Bretsky's thinking unduly, being the major Phanerozoic event of this type, as it certainly is an example of provincialism prior to the event followed by cosmopolitanism. However, the information presented here suggests to me that it is erroneous to conclude that the controls governing major extinction events coincide with those governing the levels of cosmopolitanism and provincialism. The controls may be related but they do not operate in an identical manner. Bretsky (1973) in essence concludes that the controls responsible for a major extinction event result in the absence of any barriers to faunal migration and mixing, i.e., provincialism, following the event. This is not the case.

CHAPTER 4

Conclusions about Evolution and Extinction

Overall Synthesis of Population-size Effects

The major conclusion reached in this book is that worldwide population size is inversely related to rate of evolution, whether phyletic, cladogenetic or taxonomic (morphologic rates of evolution have not been specifically discussed but would follow the same pattern as the others). All of the data from the record affording any measure of population size (number of individuals belonging to the taxa being considered) provide this same conclusion. Therefore, it seems unavoidable that worldwide population size forms the first-order control related to rate of evolution. This is not to say, however, that there are not a multitude of lower-order controls over rate of evolution. The lower-order controls are probably masked in most instances, by this overall effect of population size. Population size for marine, shallow-water invertebrates is strongly correlated with a number of factors that may be estimated from the geologic record including the following: size and number of biogeographic entities; size and number of communities; number, size and scatter of reef environments; strength or weakness of climatic gradients acting as one of the major barriers involved in presence or absence of biogeographic units; area of the continental platforms covered by shallow seas (a major control over the size of shallow-water invertebrate communities) during intervals of transgression and regression. There are undoubtedly other critical controls correlating with population size such as food supply, but I have been unable to deal with them at this time.

Rates of terminal extinction are also closely related to these same variables in those instances where competition for place and resource are thought to have been largely responsible for extinction.

Consideration of these variables makes clear that cladogenetic rate of evolution and phyletic rate of evolution may vary independently in large part. The Permo-Triassic extinction event for which a high rate of phyletic evolution is coupled with a high rate of earlier terminal extinction and a low rate of cladogenetic evolution is an excellent example.

The primary effect of widespread regression is to reduce population size

overall. As long as this population reduction effects all communities evenly there should be a general increase in rate of evolution related to the
smaller population size.[37] Ultimately, however, as regression continues there
will also be a reduction in number of communities due to the removal of
certain environments from the marine realm. This removal of certain communities may make it possible for others to actually expand their population size during a time of overall reduction in population size, while
additional communities are evolving at a higher rate due to smaller population size. When regression reaches an extreme stage at the very edges of
the continental shelves one may predict the presence of a much smaller
number of communities than during transgression and also of small population size with resultant high rate of phyletic evolution. Decrease in
number of communities leads to resulting terminal extinctions; therefore
we should infer an increasing rate of terminal extinction to accompany an
increasing rate of phyletic evolution during widespread regression. However, the effects of cosmopolitan and provincial developments, climatic
differentiation or its absence, and of the presence or absence of the reef
environment must be considered.

The presence of a highly differentiated climate and of a highly provincial[38] situation caused by a number of faunal barriers will increase the
number of potential communities greatly. This increase in number of
communities will greatly decrease overall population size per community
and thus greatly increase rate of phyletic and cladogenetic evolution. The
same result should be expected from the presence of widespread reef
environments.

Fig.1 graphically outlines these trends. It is important to consider that
during conditions of uniform climate, cosmopolitanism and absence of the
reef environment the number of communities will be far lower, size of
populations will be higher, taxonomic diversity will be lower, rate of
taxonomic evolution will be lower, rate of cladogenetic evolution will be
lower, rate of phyletic evolution will be lower and rate of extinction will
be lower than under the reverse set of conditions. The reverse set of
conditions with higher numbers of communities, smaller populations,
higher taxonomic diversity, higher cladogenesis and corresponding rates
during both regressive and transgressive conditions has profound consequences. It means that times of high regression may involve either high or
low taxonomic diversity, either a large or a small number of communities,
a high rate of extinction, a high rate of phyletic evolution, either a high or
low rate of taxonomic evolution, either a high or low rate of cladogenetic
evolution and widely differing size populations. These varying circumstances, related to the variables considered above, help us to place the
differing rates of evolution and extinction shown by the Phanerozoic

record into a population dynamics size context that does not require inferences about unknown extraterrestrial agents, sea-water composition changes or the like.

In other words, the actual form of the curves shown on Fig.23 is determined by the interaction of all the factors influencing population size that are shown diagrammatically on Fig.1. It is important to understand that each one of these factors interacts with all of the others to produce the resulting population size.

By varying these factors appropriately in time we may produce an infinite number of combinations entirely adequate to explain the observed record of changes in evolutionary rate, extinction rate, taxonomic diversity, number of communities and size of populations.

All of the above considerations are viewed as continuous, or almost continuous variables (number of communities is not strictly continuous but the large numbers involved permit its being treated as continuous). These are not "catastrophic" quantities. Catastrophic circumstances must be invoked to explain the Frasne-Famenne event that frustrates any simple explanation (McLaren, 1970). A worldwide catastrophic event will immediately raise terminal extinction rate, lower rate of taxonomic evolution, lower rate of cladogenetic evolution, lower size of populations, will probably lower number of communities considerably *if* the different communities have differing susceptabilities to the specific catastrophe, and increase rate of phyletic evolution by reducing population size. A large number of physical and biological catastrophes may be conceived, but we have difficulty in developing the geological evidence so as to really demonstrate these catastrophes. Urey's (1973) cometary collisions are, however, a possibility worth considering. Great climatic differentiation during glacial intervals has been demonstrated, but this is not really a catastrophe in the sense of being almost instantaneous. Local volcanic eruptions are certainly catastrophic but their total effect on a worldwide basis should be very minor.

The geologically "sudden" appearance of hard parts in many major groups of animals at the beginning of the Cambrian argues for an excessively high rate of phyletic evolution immediately prior to the appearance of these hard parts as we lack any transitional forms as would be expected had the evolution been more gradual. The basal Cambrian, and the Lower Cambrian as a whole for that matter, are times of major regression for most platforms. This condition of major regression should have materially assisted in raising the rate of phyletic evolution during this time interval, as is also concluded for similar reasons in the Late Permian-Early Triassic transition interval, but it is doubtful if regression alone is adequate to explain the observed phenomena. Regression should probably be viewed merely as a contributory factor tending in the same direction as other factors of even higher importance in producing an excessively high rate of phyletic evolution. We have no reason to suspect that the climate, number of animal communities or conditions of provincialism—cosmopolitanism were far different during the critical time immediately beneath the base of the fossiliferous Cambrian than they were later in the Cambrian. Therefore, an appeal must be made to other environmental factors not considered in this treatment of factors governing rates of evolution in a first-order manner.[39]

Silurian

Lower and Middle Llandovery

The areal distribution of Lower and Middle Llandovery communities is not known in detail except for the *Virgiana* Community (this community consists of almost 100% *Virgiana*). The data for the virgianids is both abundant and potentially significant. Known only from carbonate rocks, none are known with certainty from the geosynclinal environment (the few exceptions that might be found would certainly not significantly alter the generalization).

The plicate virgianid genera *Platymerella* and *Virgiana* are known in North America (Berry and Boucot, 1970). *Platymerella* occurs in a narrow band extending from northern Illinois south to central Kentucky, through a relatively small part of Early and Middle Llandovery time. It is thus a highly endemic, provincial taxon present for a relatively short period of time, possibly in a specialized environment or possibly just reflecting marginal isolation, on the eastern side of the widespread sheet of North American Platform carbonate rocks carrying *Virgiana* itself. Its morphology and stratigraphic position argue strongly for its derivation from *Virgiana*, for *Platymerella* is essentially a flattened *Virgiana* that originated in a marginal position from a broad source taxon.

On the other hand, *Virgiana* is widespread in North America (see Berry and Boucot, 1970), ranging from El Paso, Texas, to northern Baffin Island, and from Anticosti Island to central Nevada. Several species can be recognized within the North American Platform, suggesting that morphologic types were differentiated from one region to another, but the fantastic abundance of *Virgiana* must be seen to be believed. These "species" (discriminated chiefly on the basis of relative size of plication on the exterior as well as of cardinalia internally) probably approximate the species of the neontologist. However, clines may well exist from one "form" to another. This raises the possibility that the "species" might better be termed subspecies segregated from each other geographically, or that they might even be merely "races" segregated geographically in a gradational manner. In any event their morphologies are remarkably close, requiring the services of a specialist provided with very large samples before identification can be made. Beds of many centimeters thickness (20—40 cm), extending for mile after mile along strike, commonly are packed with millions of valves. The potentially interbreeding *Virgiana* population on the North American Platform during much of Early and Middle Llandovery time must have been one of the largest, if not the largest, ever to cover so much of this continent. With the sole exception of *Platymerella* in North America plus a few Uralian genera, a multiplicity of *Virgiana*-derived taxa did not develop, thus arguing for a very slow rate of morphologic change, i.e., slow morphologic evolution. Trettin (1971) collected pentameroids from Prince Charles Island in the Foxe Basin, northern Canada, that may contain a few new virgianid taxa, but the material is too limited to serve as a basis for conclusions (if found to be new they would be interpreted as small, marginal type populations on the edge of the large *Virgiana* population). The differentiation of several *Virgiana* species on a worldwide basis argues for the presence of some broad gradients of variability, a reasonable phenomenon for a population of such size and extent.

Virgiana is also widespread on the Siberian Platform (sensu lato, including the Kolyma Massif Silurian, and the Sette Daban Silurian). Connections between the North American and Siberian Platform *Virgiana* are not presently known as far as their matrix rocks are concerned, but the morphologic similarity of the *Virgiana* and the similar stratigraphic position argue in favor of interbreeding connections. *Virgiana* is also reported from the North Uralian—Novaya Zemlya region, but the areas involved are small.

The virgianid of northern Europe is the smooth taxon *Borealis* (the old *Pentamerus borealis*). Widely distributed through the Baltic region (Norway, Sweden, Esthonia), it may occur as far as the eastern and northern limits of the Russian Platform. However, it has a far more limited geographic and stratigraphic range than has *Virgiana* itself. *Borealis* appears to have been the ancestor of several smooth pentamerinid taxa during later Middle Llandovery time. The more restricted area and shorter time range of *Borealis* are consistent with its having more rapidly given rise to other taxa than did *Virgiana* itself. The relative size of the areas involved may differ by several hundred times or more, or possibly as little as 10-20 times depending on the assumptions made in lithofacies and paleogeographic restorations; in any event, the size of potentially interbreeding populations shows a correlation with rate of evolution. Nikiforova and Sapelnikov (1971) described several bizarre virgianid genera

from the west slope of the northern Urals. These taxa can probably best
be regarded, like *Platymerella*, as very localized, provincial taxa occurring
near the geographic limits edge (marginal sources) of the group's range,
and indicating little about the size or rate of evolution of the parent
population (broad source).

During this time interval the three successively occurring stricklandid
taxa in the areally restricted *Stricklandia* Community, further support the
thesis presented here.

Late Llandovery

World data for the Late Llandovery Benthic Assemblages and their distribution is abundant. Many fewer data concern the distribution of the
communities within the Benthic Assemblages. The salient feature of the
Late Llandovery is the widespread occurrence of Benthic Assemblage 3,
and of the *Pentamerus* Community within this Assemblage, on the North
American, Russian and Siberian platforms. Opposed to this is the very
limited occurrence areally of both Benthic Assemblages 2 and 4 as well as
5. The relative areas occupied by Benthic Assemblage 3 as opposed to the
other assemblages is on the order of 100 times. Benthic Assemblage 1 also
covers a very small area during this time interval as compared to 3.

The widespread *Pentamerus* Community is very low in diversity, the
genus *Pentamerus* comprising the bulk of most collections. The other
communities within Benthic Assemblage 3 are similar to the *Striispirifer*
Community, with a far greater taxonomic diversity, but more restricted
distribution than the *Pentamerus* Community. Benthic Assemblages 4 and
5, occurring seaward of Benthic Assemblage 3, are its equal or exceed it in
diversity whereas the more landward Benthic Assemblage 2 has relatively
low diversity. Benthic Assemblage 1 probably has the lowest brachiopod
diversity of all, although it may have numerous bivalve taxa.

Rate of change of form (i.e., evolution) also differs between these
Benthic Assemblages. The well-known *Eocoelia* lineage of the Late
Llandovery is rapidly evolving phyletically, as is the *Stricklandia* lineage
within Benthic Assemblage 4 (Fig.18). The widespread and abundant
genera *Atrypa*, *Leptaena* and *Nucleospira* in the Benthic Assemblages 3-4
position evolved very slowly. Information is scanty about the taxa within
Benthic Assemblage 5, but suggests that the dicaelosid brachiopods may
have evolved rapidly during this time interval whereas *Skenidioides*
evolved very slowly. There is no evidence for rapid evolution within either
Benthic Assemblage 3 or Benthic Assemblage 1. This appears to be a fair
statement of the Benthic Assemblage 3 situation inasmuch as these shells
have been studied about as carefully as those in the other Benthic Assemblages. However, study of the taxa in Benthic Assemblage 1 is not yet

sufficient to assure us that the apparent lack of evolution reflects a biologic situation rather than a lack of attention. Summing up this information, rate of change of form in the Late Llandovery brachiopods correlates well with the size of the potentially interbreeding population, although this conclusion is based almost entirely on platform information. If one considered only the geosynclinal data, which areally speaking compose only a small fraction of that for the platforms, no correlation would be apparent between size of interbreeding populations and rate of evolution, a fact which emphasizes the importance of having a large worldwide sample. This follows because in the geosyncline it is far more difficult to estimate areas occupied by taxa.

Wenlock-Ludlow

Knowledge of Early Wenlock faunas and communities suggests that those present within Benthic Assemblages 1 into 3 merely represent a continuation of their Late Llandovery counterparts. However, from part of Benthic Assemblage 3 through 5 we have the beginnings of increasing endemism in the Early Wenlock. Information for the Middle Wenlock is inadequate, but Late Wenlock-Ludlow data are excellent. During the latter time interval, Benthic Assemblage 3 is the most widespread, and neither the relatively rough-water *Kirkidium* Community nor the more quiet-water *Dicaelosia-Skenidioides* Community show much evidence of rapid evolution (phyletic, cladogenetic or taxonomic). Data for Benthic Assemblages 4 and 5, plus a portion of 3, has not been adequately analysed to determine whether evolution was rapid or slow during this time interval, although their relatively small area as contrasted with the bulk of Benthic Assemblage 3 might suggest more rapid evolution (for example, the phyletic production rate of the highly provincial *Gracianella* species [Fig.24], restricted to a narrow ribbon, appears to have been high during this interval) except for a widespread cosmopolitan form like *Skenidioides*. Benthic Assemblage 1 and 2 of this time interval have not been studied carefully to determine whether evolution proceeded rapidly.

Interestingly, during this time interval, several sets of provincial, taxonomically and cladogenetically rapidly evolving pentameroid taxa developed from more widespread, very slowly evolving antecedents (Fig.27). The centers of development for these localized pentameroid taxa are adjacent to reef bodies in the middle of the North American Platform, and scattered about the area of the Uralian-Cordilleran Region of the Late Silurian. In these two widely separated regions several endemic genera appeared during the Late Wenlock-Ludlow interval, yet all became extinct by the end of the Ludlow. The reasons for this development are unknown, but the geographic restriction in both cases is notable, correlating well

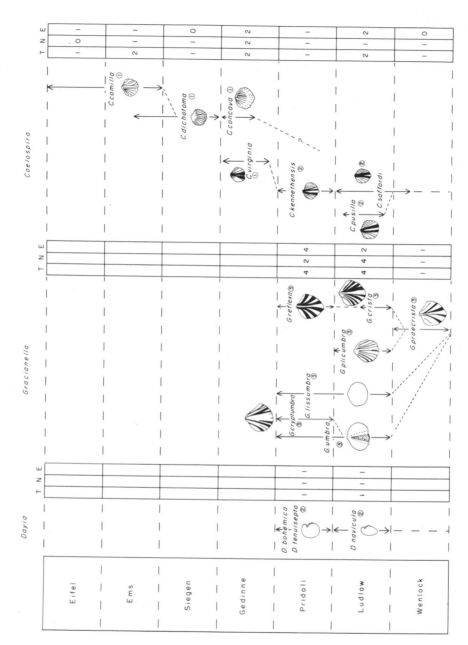

T = NUMBER OF TAXA PER TIME UNIT
N = NUMBER OF NEW TAXA PER TIME UNIT
E = NUMBER OF EXTINCT TAXA PER TIME UNIT

① EASTERN AMERICAS REALM
② NORTH ATLANTIC REGION
③ URALIAN–CORDILLERAN REGION

with the rapid rate of evolution, and proximity to reef-like bodies and coinciding with the existence of large bodies of hypersaline water on the platforms. It is possible to consider these localized developments as marginal sources derived from broad-source Benthic Assemblage 3 taxa. Their extinction at the end of the Ludlow coincides with the spread of gypidulids and may be interpreted as analogous community replacement by the gypidulids and might also have been affected by the lowering of reef importance. The abundant, taxonomically rapidly evolving subrianinid pentameroids of part of Benthic Assemblage 3 and possibly 4 were restricted to narrow ribbons within the Uralian-Cordilleran Region, a position consistent with their rapid evolution.

The higher rate of taxonomic, cladogenetic and phyletic evolution characterizing the Late Wenlock-Ludlow interval correlates well with the abundance of reef and reef-like structures. The creation in some abundance of scattered structures of this type markedly increased the number of available ecological niches which appear to have been promptly filled by rapidly evolving organisms.

Pridoli

Information about Pridoli-age communities, their areas, and respective rates of evolution is currently too poor to permit any generalization. However, inspection of Fig.25 indicates the development of a somewhat higher level of provincialism during the Pridoli. Inspection of Fig.22 and Fig.27 shows that the production rate of taxa during the Pridoli was reasonably intermediate between the Ludlow and Gedinne. The higher terminal extinction rate of Pridoli time (Fig.22) may reflect a marked diminution of the reef environment.

Early and Middle Devonian

The Early Devonian is characterized by the highest provincialism of the Silurian-Devonian, with the Middle Devonian not far behind, in contrast to the Silurian and the very cosmopolitan Late Devonian. The Early Devonian is also a time of marine regression and of marked increase (Fig.22) in number of animal communities. As would be predicted, the great provincialism of the Early and early Middle Devonian regression and the variety

Fig. 24. Family tree for the species of the Atrypacean genera *Dayia*, *Gracianella* and *Coelospira*. Note the relations between number of taxa per unit of time and the times of highest provincialism (Early Devonian highest, Pridoli moderate, Ludlow lower, Wenlock lowest). *D. bohemica* of the Pridoli may be identical to *D. tenuisepta* Shirley (nomen nudum) but a decision in this regard awaits Shirley's publication on the subject.

Fig. 25. Diagrammatic relations between Late Ordovician to Devonian biogeographic entities. Solid lines indicate barriers; dotted lines indicate free flow of taxa either laterally or vertically with time.

Fig. 25 is an attempt to illustrate the relations between the Late Ordovician through Middle Devonian marine biogeographic entities. The "facies" boundaries drawn solidly are either Realm or Region boundaries across which little migration and mixing have occurred; the dashed horizontal boundaries are those across which considerable interbreeding and mixing are concluded to have occurred. Note the irregular pattern, in time, of zoogeographic unit expansion and contraction. The circumstances governing these relations are still poorly known. It also is not known whether the Late Ordovician Malvinokaffric area faunas gave rise to the Silurian Malvinokaffric Realm faunas. It is reasonably certain that the bulk of the Malvinokaffric Late Lower Devonian brachiopods arose from Eastern Americas Realm taxa, although largely modified under conditions of isolation, rather than from Malvinokaffric Silurian taxa. The Late Ordovician North American Realm ceased to exist at the end of the Ordovician. The Old World Realm Ordovician gave rise to both the Mongolo-Okhotsk Region and North Silurian Realm faunas. The North Silurian Realm certainly gave rise to the North Atlantic Region and Uralian-Cordilleran Region Silurian. The North Atlantic Region Late Silurian gave rise to the Appohimchi Subprovince Late Silurian and Early Devonian in turn. The Amazon-Colombian and Nevadan Subprovinces are clearly geographically terminal developments from the Appohimchi Subprovince of the Eastern Americas Realm. The Uralian-Cordilleran Region of the Late Silurian appears to have given rise to all of the Old World Realm Devonian subdivisions, except for the Mongolo-Okhotsk Region. The Rhenish-Bohemian Region during the Eifel is represented only by Bohemian Complex of Communities, no Rhenish Complex being known, but the presence in the Late Eifel-Givet age Eastern Americas Realm Hamilton fauna of Rhenish-type elements argues clearly for the existence somewhere in Eifel time (possibly to the north of the calcareous Eifel present in northern Europe; areas now stripped of Devonian by later erosion) of a Rhenish-type complex of communities. The Eastern Americas Realm during Eifel time (Onondaga Limestone and its correlatives) is clearly derived from Early Devonian Appohimchi Subprovince antecedents.

of communities present and recognized within most of the zoogeographic entities resulted in a far greater production of taxa per unit of time (increased rate of cladogenetic and phyletic evolution) than during either the Silurian or the Late Devonian. The size of interbreeding populations correlates well with the rate of taxonomic evolution for the Early and Middle Devonian.

The relatively well-known Early Devonian terebratuloid brachiopods provide one example of this correlation. The Eastern Americas Realm Centronellidae and Rhipidothyridae (Fig.21) are represented by a number of genera extending from the Early Gedinne through the Ems, except for *Amphigenia* and *Centronella* plus the rhipidothyrids that extend through the Givet as well, and are restricted to that realm (except for *Globithyris* and a few occurrences of *Amphigenia*). The relatively high production rate following the initial sudden appearance from uncertain ancestors for generic-level taxa, many of which have limited distributions within the realm, is consistent with the small area occupied by the realm. The rapid terminal extinction of centronellid genera and drop-off in numbers of new genera near the Ems-Eifel boundary (Fig.21,22) coincides with the opening up of a new Eastern Americas Realm area in northern South America and also with the lateral spreading of Eifel-age Onondaga Limestone equivalents in eastern North America. Thus increase in size of interbreeding population coincides with decline in rate of taxonomic and cladogenetic evolution and in increase of terminal extinction rate. Mutationellinid terebratuloids similarly illustrate this relationship (Fig.20); they are widely distributed within the Early Devonian, occurring from at least Podolia to the Appalachians and throughout the Malvinokaffric Realm. The restriction of the mutationellinid genera occurring in the Malvinokaffric Realm to Benthic Assemblage 2 (Fig.9), a small area, and of the genera occurring in the Eastern Americas Realm to Benthic Assemblage 2 (Fig.7), a small area, is an additional example of the correlation between high production rate of taxa and limited size of potentially interbreeding population. The Givet terebratuloid genus *Stringocephalus*, and several of its less well-known allies illustrate the opposite situation of an almost worldwide distribution correlated with a relatively low rate of taxonomic evolution (the other stringocephalid genera are, without exception, very localized in their occurrence which indicates that they may easily be interpreted as having been derived from the broad-source *Stringocephalus* near the margins of the latter's distribution).

The brachiopod Superfamily Stropeodontacea (Harper et al., 1973) provides another excellent example of rapid taxonomic and cladogenetic evolution for the Silurian and Devonian. Most genera of the Llandovery through Wenlock have a worldwide distribution, the genera of the Ludlow through Pridoli show a slightly more limited geographic range and some-what increased number of genera per unit of time, as contrasted with the endemic, far more numerous genera of the Early and Middle Devonian. Information about chonetacean brachiopod genera (Fig.17) shows the same relation: a few pre-Late Wenlock genera having a worldwide distribu-tion, a moderate number of Late Wenlock-Pridoli genera showing limited distribution, and many more Early and Middle Devonian genera showing a still more limited geographic distribution pattern (following the various zoogeographic entities defined for the Devonian). The fall-off in number of Middle Devonian taxa after the Lower Devonian is marked in both cases, as it was for the Centronellidae. Gypidulid brachiopods follow the same route, although the information has not yet been carefully compiled. By Givet time the limited number of gypidulinid generic taxa has an almost worldwide cosmopolitan distribution. The other major groups of Silurian-Devonian brachiopods appear to follow the same distribution pat-terns.

Total and percentage rates of appearance, evolution and extinction for the Silurian and Devonian

Fig.22 presents the information obtained by summarizing many groups of Silurian and Devonian brachiopods in terms of total number of taxa per time interval[40] (Fig.22A), new taxa per time interval (Fig.22B), number of taxa becoming extinct per time interval (Fig.22C), total number of taxa divided into new taxa per time interval (percent of new taxa, Fig.22D), and total number of taxa divided into number of taxa becoming extinct per time interval (percent of extinctions, Fig.22E).

The information in Fig.22A, total number of taxa per time interval indicates three peaks, one within the Late Ordovician, another in the mid-Silurian (essentially the Wenlock and Ludlow) and the other at the end of the Lower Devonian (Siegen, Ems, Eifel). The Devonian peak is above the Silurian peak, with one low present in the Early-Middle Llandovery another in the latest Silurian (Pridoli), and another at the end of the Frasne (early Upper Devonian). It is very important to realize that Fig.22, total number of taxa per time interval, does not provide a direct measure of changing rate of taxonomic evolution. Rather it summarizes changing rate of taxonomic evolution (influenced by such changing fac-tors as provincialism, number of animal communities, presence or absence of the reef environment, and transgression—regression that influence popu-lation size), appearance rate of taxa from unknown regions (migration and dispersal), and extinction rate. Appearance rate is a term that must be introduced to account for taxa at the generic and family levels that have no good evolutionary ties with older taxa (they appear to come from nowhere; in other words, their previous place of origin is unknown and their earlier rate of evolution cannot be estimated).

The information in Fig.22B, new taxa per time interval, clearly shows that there is a marked increase in appearance rate (although not necessarily of production rate!) of new taxa in the Late Llandovery and Wenlock portions of the Silurian, after a marked low in the Ashgill and Early through Middle Llandovery, followed by a decrease in the Ludlow, and then a pronounced Early Devonian increase culminating in the Ems followed by a marked downward trend through the Frasne followed by a Famenne increase. However, in terms of percentage rate of change in appearance of new taxa (total number of taxa divided into new taxa per time interval of Fig.22D) it is clear that the Late Llandovery through Wenlock shows a significantly higher rate than the Devonian high present in the Gedinne through Eifel, both peaks being separated by a low in the Pridoli interval (exaggerated as the Pridoli is a shorter time interval), with the Silurian peak preceded by a Late Ordovician low and the Devonian high followed by a Frasne low, followed by a high in the Famenne. The unusual height of the Famenne high reflects the severity of the terminal Frasne extinction event that must have decimated the shallow-water, shelf invertebrate populations to such an extent that an exceptionally high rate of evolution involving relatively small populations took place among the survivors. This recovery is all the more remarkable as many major groups had no post-Frasne survivors.

The information in Fig.22C, number of taxa becoming extinct per time interval, clearly shows a steady increase in rate of extinction through the Ludlow, as measured by total number of taxa, a low in the Pridoli, then an increase to a maximum in the Eifel, followed by a decrease in rate to the Frasne after the initial precipitous extinction event at the end of the Ordovician. Fig.22E shows, however, that the percentage rate of extinction gives a far better estimate of the extinction picture than does the change in absolute numbers of taxa going to extinction. Fig.22E makes clear that the Late Ordovician is a time of very high percentage rate of extinction, that the Ludlow is a similar time, and that the Ems through Frasne is a time of increasingly high percentage rate of extinction culminating in the Frasne.

The next question is to assess the biologic and physical factors capable of explaining these profound anomalies! The precipitous decline in total number of taxa at the end of the Ordovician, with a correspondingly low rate of production of new taxa and exceptionally high total number of extinctions as well as high extinction rate, correlates well with the Late Ordovician climatic rigor connected with Southern Hemisphere glaciation occurring at the same time that laterites were forming in Nova Scotia, and significant regression. The Late Ordovician trend away from high provincialism would tend to slow down rate of taxonomic and cladogenetic evolution, the highly differentiated climate would tend to increase rate of taxonomic and cladogenetic evolution, whereas the Late Ordovician regression would tend to speed up rate of phyletic evolution in terms of population size alone.

The sharp increase in number of taxa, after the Late Ordovician decline, culminating in the Wenlock-Ludlow (Fig.22A) is not a simple matter. Inspection of Fig.22B shows that this sharp increase in total numbers of taxa is underlain by a very slow increase in number of new taxa from the Late Ordovician through Early and Middle Llandovery followed by a sharp increase in the Late Llandovery that culminates in the Wenlock to be followed by a decline in the Ludlow. During this same Llandovery-Ludlow interval the total number of extinctions is rising from a Late Llandovery low to a Ludlow high. The rate of production and appearance of new taxa during the Silurian reaches a maximum in the Wenlock followed by a sharp decline to the Pridoli low. Extinction rate during this time interval is low compared to that occurring at the end of the Ordovician. Extinction rate increases from a Llandovery low through a Ludlow high that is followed by a Pridoli low. This information must be understood against a background of low provincialism during the Llandovery followed by an increase in the Wenlock-Pridoli, factors tending to increase rate of phyletic evolution in terms of population size as well as taxonomic and cladogenetic evolution rate, occurring at the same time as a gradual and widespread marine transgression that would tend to decrease rate of evolution in terms of population size. The Llandovery increase in percentage appearance of new taxa is best understood in terms of the almost worldwide spread of a complex of pre-Late Llandovery taxa from pre-existing source regions of a restricted nature (southeast Kazakhstan, for example). In other words, the Llandovery increase does not necessarily indicate an increase in rate of taxonomic or cladogenetic evolution, but a sharp increase of appearance rate (migration in this case). The marked increase in total number of new taxa during the Late Llandovery-Ludlow is probably a combined result of the increase in appearances during the Late Llandovery taken together with an increased rate of taxonomic and cladogenetic evolution for the Wenlock-Ludlow made possible by the presence of a larger number of animal communities (see Fig.22H) that is best interpreted in terms of evolution having produced a group of organisms with narrower environmental tolerances. The abundance of reef and reef-related communities in the Late Wenlock-Ludlow is consistent with this conclusion. The Pridoli decline in total number of taxa, of total number of new taxa, and in rate of production of new taxa might be interpreted as the result of further evolution having produced additional taxa with far wider niche breadth (note the smaller number of communities recognized for the Pridoli as compared with the Wenlock-Ludlow, Fig.22H); this change correlates well with the marked increase in extinctions at the end of the Ludlow, and also with the marked decline in abundance of reefs and reef-like structures at the end of the Ludlow.[41]

The Gedinne through Ems interval sees increasing production of new taxa culminating in the Ems. This increase in rate of evolution may be understood in terms of rapidly increasing provincialism, increased number of communities, increasing abundance of reefs from a Gedinne low to an Eifel-Frasne high, and also by regression that culminates in the Siegen, all factors tending toward an increased rate of evolution because of the possibilities for smaller interbreeding populations existing in isolation. A part of the Gedinne increase is, however, due to the appearance of new forms (terebratuloids and cyrtinoids, for example) from regions unknown.

The Givet through Frasne interval sees a steadily decreasing production rate of new taxa as well as a decreasing percentage production rate of new taxa followed by a burst in the Famenne. This information correlates well with the decreasing provincialism of the time interval, the decreasing number of animal communities, and the renewed marine transgression (beginning in the Ems and culminating in the Famenne), all factors favoring a decrease in rate of evolution through the Frasne, followed by the Famenne burst correlated with rapid phyletic and taxonomic evolution rate increases induced by the massive Frasne extinction event. Underlying the high production of new taxa in the Lower and Middle Devonian is the increase in abundance of reef and reef-like structures that reaches a peak in the Eifel and Givet. This increase in reef activity helps to explain the relatively steady rate of production of new forms in the Gedinne through Givet. Towards the beginnings of this time interval, and extending through the Ems, we have the effects of increased provincialism and regression tending to increase rate of evolution. As provincialism decreases towards the end of the Ems, and transgression progresses again, we have the incoming reef high that helps to keep up the production of new forms by encouraging the formation of small interbreeding populations. In other words, a number of factors are involved here to result in the observed production of taxa.

Fig.22D indicates that the Llandovery faunal replacement was a more marked event as measured by percentage than the Devonian event, despite the greater number of taxa involved in the Devonian event.

The high percentage extinction rate in the Late Ordovician is most easily explained in terms of the changing climate, plus marked regression related to glaciation.

The large number of taxa becoming extinct per time interval during the Ems-Eifel coincides with a marked decrease in level of provincialism which suggests that competition for place and resource is a reasonable explanation. The Ems-Eifel is a time of relatively high provincialism and of moderate marine transgression, neither factor being capable of explaining the high number of extinctions. The very high percentage rate of extinctions (Fig.22E) encountered from the Ems through the Frasne coincides with progressively lower levels of provincialism and continuing marine transgression. However, the marked low at the end of the Frasne probably reflects a major physical event (see McLaren, 1970) of whose nature we are uncertain.

Fig.22F and 22G summarize the overall trends of marine regression versus transgression on the continents, and of faunal provincialism versus cosmopolitanism during the Caradoc through Frasne intervals of the Late Ordovician to Late Devonian. Note the changing tendencies for increasing population size during the latest Ordovician into Llandovery with progressively increasing cosmopolitanism, from Late Ordovician provincial conditions, accompanied by first regression and then widespread marine transgression. Note also the congruence in the Lower Devonian of marine regression and provincialism; both tendencies tending in the direction of smaller, more isolated populations. Note finally the post-Lower Devonian congruence of increasing transgression and cosmopolitanism tending to increase population size.

Fig.22H strongly suggests that the increase in percent extinction rate of the Ludlow should be correlated with a consequent decline in number of benthic communities. In other words, the increased extinction rate may be interpreted as a result of organic evolution progressing in the direction of permitting some taxa to occupy a greater niche width than previously with a consequently smaller number of communities and taxa.

Fig.22I through M suggest a high degree of correlation between the climatic factors and regression involved in Late Ordovician-Early Silurian continental glaciation and both appearance rate plus taxonomic rate of evolution, but no correlation observable between orogenic intensity, evaporite production or dolomite production and rate of taxonomic evolution.

Fig.22M indicates the relative abundance of reef and reef-like structures during the Silurian and Devonian. The Wenlock-Ludlow high in production of new taxa correlates well with this information, as does the Pridoli fall-off. The Devonian high in the Eifel, Givet and Frasne helps to explain why in the face of declining provincialism and increasing transgression the rate of production of new taxa remains relatively high until the massive wipe-out at the end of the Frasne. The low number of brachiopod taxa, as well as other megafossils, known from the Famenne is partly a function of the almost total absence of the reef environment with its possibilities for a multitude of small, isolated, interbreeding populations leading to rapid phyletic and cladogenetic evolution.

All of this information has been compared with the information for certain groups of articulate brachiopods. It is logical to inquire whether or not the correlations here espoused apply only to articulate brachiopods. Despite the lack of compilations in stratigraphic detail for other groups of marine invertebrates my own experience strongly indicates that the articulate brachiopod correlations accord very well with data for other groups. Trilobites, for example, during the Silurian as a whole are present in only moderate taxonomic abundance (far below their Cambro-Ordovician high). During the Early Devonian through Eifel trilobites are present in far larger taxic abundance than during the Silurian. During the Givet trilobite abundances steadily decrease and most groups disappear at the end of the Frasne, to remain as very minor parts of the fauna during the remainder of the Paleozoic. Corals, bivalves and gastropods appear to follow a similar pattern of relatively low Silurian taxonomic abundance (except for the Wenlock and Ludlow), relatively high Lower Devonian abundance, and only moderate Middle Devonian abundance, followed by a terminal Frasne marked diminution in numbers (see McLaren, 1970, for the Frasne data). Acritarch genera follow the same general pattern (A. Loeblich, written communication, 1973).

This summary for the Silurian and Devonian makes clear that summaries of taxa per time interval whether intended for understanding the problems of extinction or evolution have a better chance of affording understanding if done with relatively small time intervals and with as much biogeographic comprehension as possible. Taxonomic summaries based on the *Treatise of Invertebrate Paleontology*, or like sources, not intended for this type of study will provide only a very limited and far overgeneralized understanding of the problems involved. It is critical that the interested student of these questions have the patience to carefully synthesize the primary information rather than trying to employ the *Treatise* or similar sources as a substitute. Failure to do so will result in a far lower level of understanding than is now possible with the information on hand.

Finally, the conclusions presented in this section are dependent on the assumption that the time intervals chosen for comparison are of approximately the same absolute duration. If, for example, the Lower-Middle Llandovery time interval should be only one-tenth as long as the Late Llandovery the conclusions presented here would be completely erroneous. The information we have from absolute ages is very limited. However, our knowledge of the faunal succession based on a number of independent groups (graptolites, conodonts, various co-occurring brachio-pod groups) strongly suggests that the time intervals are at least of ap-proximately the same absolute time duration. For the Pridoli we do have the problem that although it appears of shorter duration in terms of conodont and graptolite zones than the Ludlow and Gedinne, its fauna may also be partly confused with those other two intervals in such a manner as to tend to lower the number of taxa assigned to it as well as the number of communities.

Effects of normalizing total and percentage rates of appearance, evolution and extinction for the stages with estimated relative stage duration

Fig.22A-E and the preceding conclusions drawn from them are based on setting stage lengths, or time divisions used here, at unity. If, following the conclusions shown in Table II, the data presented in Fig.22A—E are normalized for relative stage duration then the results shown in Fig.22AA—EE are obtained. In those instances where the relative stage duration is estimated to be unity no change is obtained, but changes do occur for those stages and time units that depart from unity. The number of Cara-doc taxa appearing per unit time is significantly reduced (Fig.22AA) from the unnormalized figure. The normalized peaks for total abundance reach a maximum during the Pridoli-Ems interval with significant lows occurring in the Llandovery and Famenne. The Pridoli low of Fig.22A is obviously an artifact introduced by the Pridoli representing a smaller time interval than the units on either side.

Fig.22BB shows that a distinct Wenlock high in appearance of new taxa is followed by a Ludlow-Pridoli low followed immediately by a marked Devonian increase that persists from the Gedinne to the Eifel. Following the Eifel there is a marked decrease, a Frasne low, and then a Famenne increase of modest proportions. However, in terms of the normalized percentage of new appearances (Fig.22DD) there is an Ashgill to Wenlock increase, a low in the Ludlow, and then an increase to the Gedinne follow-ed by a decrease to the Frasne, followed by a marked Famenne increase. The normalized data suggest that the Famenne increase in new appear-ances is not unusually marked.

Fig.22CC shows that the Pridoli low of Fig.22C is probably an artifact introduced by the shorter length of the Pridoli. Fig.22EE shows that the normalized percentage extinction figure for the Ashgill exceeds that for the Frasne and that the Llandovery-Siegen extinction rate is fairly stable until the marked increase in the Ems and later intervals.

The normalized data of Fig.22AA—EE suggest that the Pridoli "lows" do not indicate a need for such explanations as evolution in the direction of greater niche breadth and a smaller number of communities. The normalized data indicates the critical importance in any taxic compilations of trying to employ intervals of about the same length. Note carefully the differences in conclusions possible if one does not employ the normalized data!

Needless to say, this process of normalization for relative time duration of the units employed in this treatment entails the introduction of errors based on the paleontologically-based relative time estimates. Any conclusions based on the normalized data must be considered in this light.

Biogeographic statistics

I have outlined the areas occupied in time by various biogeographic units of the Silurian-Devonian and discussed their nature in some detail. In the section "Total and percentage rates of appearance, evolution and extinction for the Silurian and Devonian" (p. 146) I have discussed the significance in terms of rates of evolution and extinction of the total number of articulate brachiopods per unit of time during the Silurian-Devonian. However, all of these approaches still leave unanalyzed a number of statistical relations involving changing biogeography and rates of evolution and extinction.

Fig. 26A—L and Table III indicate the number of endemic taxa (chiefly genera or subgenera), number of taxa shared between two biogeographic units (chiefly adjoining each other geographically), and number of taxa shared by three or more biogeographic units for the time units of the Silurian through Middle Devonian employed in this treatment. This analysis, despite its unsophisticated form, makes readily apparent a number of

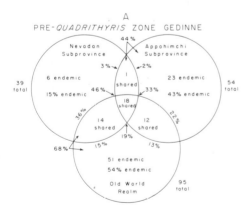

Fig. 26A. For caption, see p. 155.

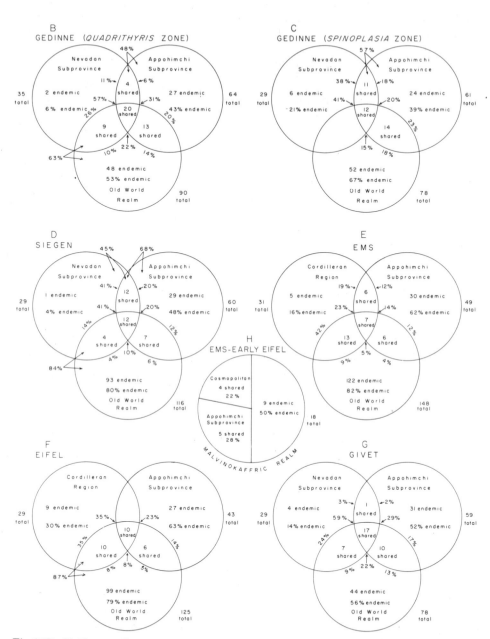

Fig. 26B—H. For caption, see p. 155.

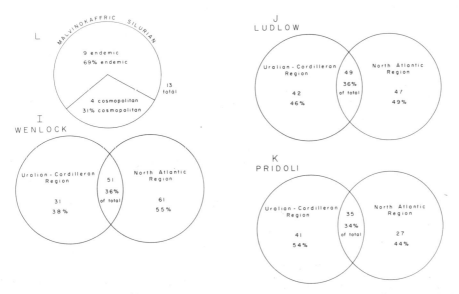

Fig. 26. Numbers of brachiopod taxa (with calculated percentages) known only from Nevada, from the Appohimchi Subprovince of the Eastern Americas Realm, and from the Old World Realm (the Appohimchi data is essentially comparable to that for the entire Eastern Americas Realm).
Composition of the Malvinokaffric Silurian and Devonian faunas:
A. Gedinne (pre-*Quadrithyris* Zone in Nevada); B. Gedinne (*Quadrithyris* Zone in Nevada); C. Gedinne (*Spinoplasia* Zone in Nevada); D. Siegen (*Trematospira* Zone in Nevada); E. Ems (*kobehana* and *pinyonesis* zones in Nevada); F. Eifel (*Leptathyris* and *Warrenella kirki* zones in Nevada); G. Givet; H. Malvinokaffric Devonian fauna (endemic, shared and cosmopolitan taxa); I. Wenlock; J. Ludlow; K. Pridoli; L. Malvinokaffric Silurian fauna (endemic and cosmopolitan taxa).

Note that for Fig.26 I—K the percentage labelled "of total" refers to the taxa endemic to each Region plus the cosmopolitans, whereas the unlabelled percentages for each Region are percentages of the endemics for that Region plus the cosmopolitans (not a percentage of the endemics for both Regions plus the cosmopolitans).

fundamental relations concerning sampling-induced anomalies, changes in degree of provincialism through time, and the difference between highly cosmopolitan and highly provincial taxa during all time intervals as regards rate of evolution and extinction. A few of the "cosmopolitan" taxa undoubtedly represent units that have been so little studied that they actually represent more than one taxon (the basic units of which may turn out to be highly provincial); the item listed as *"Schuchertella"* is an excellent example.

TABLE III

Numbers and percentages of Cosmopolitan and Provincial articulate brachiopod taxa in the Silurian—Devonian

Famenne

Frasne

	Cosmopolitan													
	O.	L.—M.Ll.	U.Ll.	Wen.	Lud.	Pri.	Ged.	Sieg.	Ems.	Eif.	Giv.	Fr.	total	%
	—	—	6%	2%	—	4%	8%	2%	8%	22%	20%	27%	51	55%
			3	1		2	4	1	4	11	10	14		

Givet

	Cosmopolitan												
	O.	L.—M.Ll.	U.Ll.	Wen.	Lud.	Pri.	Ged.	Sieg.	Ems.	Eif.	Giv.	total	%
	4%	—	15%	7%	—	4%	11%	—	15%	15%	30%	27	23%
	1		4	2		1	3		4	4	8		

Eifel

	Cosmopolitan											
	O.	L.—M.Ll.	U.Ll.	Wen.	Lud.	Pri.	Ged.	Sieg.	Ems.	Eif.	total	%
	7%	—	13%	13%	—	—	13%	7%	27%	20%	15	9%
	1		2	2			2	1	4	3		

Ems

	Cosmopolitan										
	O.	L.—M.Ll.	U.Ll.	Wen.	Lud.	Pri.	Ged.	Sieg.	Ems.	total	%
	8%	—	17%	8%	—	17%	17%	25%	8%	12	7%
	1		2	1		2	2	3	1		

Siegen

	Cosmopolitan									
	O.	L.—M.l.	U.Ll.	Wen.	Lud.	Pri.	Ged.	Sieg.	total	%
	5%	—	19%	10%	5%	19%	29%	14%	21	13%
	1		4	2	1	4	6	3		

Gedinne

	Cosmopolitan								
	O.	L.—M.Ll.	U.Ll.	Wen.	Lud.	Pri.	Ged.	total	%
	9%	—	27%	18%	3%	15%	27%	33	24%
	3		9	6	1	5	9		

Pridoli

	Cosmopolitan							
	O.	L.—M.Ll.	U.Ll.	Wen.	Lud.	Pri.	total	%
	8%	8%	42%	31%	6%	6%	35	35%
	3	3	15	11	2	2		

Ludlow

	Cosmopolitan						
	O.	L.—M.Ll.	U.Ll.	Wen.	Lud.	total	%
	12%	14%	43%	22%	8%	49	36%
	6	7	21	11	4		

Wenlock

	Cosmopolitan					
	O.	L.—M.Ll.	U.Ll.	Wen.	total	%
	19%	13%	42%	25%	51	36%
	10	7	22	13		

Upper Llandovery

	Cosmopolitan				
	O.	L.—M.Ll.	U.Ll.	total	%
	31%	20%	49%	74	83%
	23	15	36		

L.—M. Llandovery

	Cosmopolitan			
	O.	L.—M.Ll.	total	%
	67%	33%	52	67%
	35	17		

TABLE III (*continued*)

Famenne

	O.	L.—M.Ll.	U.Ll.	Wen.	Lud.	Pri.	Ged.	Sieg.	Ems	Eif.	Giv.	Fr.	Fa.	Total
	—	—	—	—	—	—	3%	1%	5%	6%	1%	10%	74%	76
							2	1	4	5	1	8	58	

Frasne — Rare

	O.	L.—M.Ll.	U.Ll.	Wen.	Lud.	Pri.	Ged.	Sieg.	Ems	Eif.	Giv.	Fr.	total	%
	—	—	—	2%	2%	—	5%	—	12%	7%	17%	55%	42	45%
				1	1		2		5	3	7	23		

Givet — Eastern Americas

	O.	L.—M.Ll.	U.Ll.	Wen.	Lud.	Pri.	Ged.	Sieg.	Ems	Eif.	Giv.	total	%
	—	—	—	—	—	—	3%	3%	27%	21%	45%	33	28%
							1	1	9	7	15		

total — 60   %endemic -- 55%

Eifel — Eastern Americas

	O.L.—M.Ll.	U.Ll.	Wen.	Lud.	Pri.	Ged.	Sieg.	Ems	Eif.	total	%
	— —	4%	4%	—	4%	7%	4%	48%	30%	27	17%
		1	1		1	2	1	13	8		

total — 42   %endemic — 64%

Ems — Eastern Americas

	O.L.—M.Ll.	U.Ll.	Wen.	Lud.	Pri.	Ged.	Sieg.	Ems	total	%
	— —	5%	3%	—	8%	18%	13%	54%	39	21%
		2	1		3	7	5	21		

total — 51   %endemic — 76%

Siegen — Eastern Americas

	O.L.—M.Ll.	U.Ll.	Wen.	Lud.	Pri.	Ged.	Sieg.	total	%
	— —	8%	3%	—	5%	54%	31%	39	25%
		3	1		2	21	12		

total — 60   %endemic — 65%

Gedinne — Eastern Americas

	O.	L.—M.Ll.	U.Ll.	Wen.	Lud.	Pri.	Ged.	total	%
	—	—	5%	8%	3%	5%	79%	39	29%
			2	3	1	2	31		

total — 72   %endemic — 54%

Pridoli — North Atlantic

	O.	L.—M.Ll.	U.Ll.	Wen.	Lud.	Pri.	total	%
	4%	4%	19%	23%	19%	31%	27	25%
	1	1	5	6	5	8		

total — 62   %endemic — 42%

Ludlow — North Atlantic

	O.	L.—M.Ll.	U.Ll.	Wen.	Lud.	total	%
	9%	6%	13%	34%	38%	47	34%
	4	3	6	16	18		

total — 96   %endemic — 49%

Wenlock — North Atlantic

	O.	L.—M.Ll.	U.Ll.	Wen.	total	%
	15%	5%	20%	60%	61	42%
	9	3	12	36		

total — 112   %endemic — 54%

Upper Llandovery — Rare

	O.	L.—M.Ll.	U.Ll.	total	%
	40%	13%	47%	15	17%
	6	2	7		

L.—M. Llandovery — Rare

	O.	L.—M.Ll.	U.Ll.	total	%
	46%	54%		26	33%
	12	14			

TABLE III (*continued*)   Sum

														Sum
Famenne														76

		Sum
Frasne		93

Givet — Old World

O.	L.—M.Ll	U.Ll.	Wen.	Lud.	Pri.	Ged.	Sieg.	Ems	Eif.	Giv.	total	%	Sum
2%	—	2%	4%	—	—	2%	4%	5%	20%	54%	57	49%	117
1		1	2			1	2	3	16	31			

total — 84   %endemic — 68%

Eifel — Old World

O.	L.—M.Ll.	U.Ll.	Wen.	Lud.	Pri.	Ged.	Sieg.	Ems	Eif.	total	%	Sum
—	1%	6%	4%	3%	3%	5%	10%	17%	51%	119	74%	161
	1	7	5	4	3	6	12	20	61			

total — 134   %endemic — 89%

Ems — Old World

O.	L.—M.Ll.	U.Ll.	Wen.	Lud.	Pri.	Ged.	Sieg.	Ems.	total	%	Sum
2%	2%	7%	4%	8%	2%	15%	22%	40%	131	72%	182
3	2	9	5	10	2	19	29	52			

total — 143   %endemic — 92%

Siegen — Old World

O.	L.—M.Ll.	U.Ll.	Wen.	Lud.	Pri.	Ged.	Sieg.	total	%	Sum
3%	2%	9%	6%	11%	4%	19%	45%	97	62%	157
3	2	9	6	11	4	18	44			

total — 118   %endemic — 82%

Gedinne — Old World

O.	L.—M.Ll.	U.Ll.	Wen.	Lud.	Pri.	Ged.	total	%	Sum
2%	3%	14%	11%	16%	10%	44%	63	47%	135
1	2	9	7	10	6	28			

total — 96   %endemic — 66%

Pridoli — Uralian-Cordilleran

O.	L.—M.Ll.	U.Ll.	Wen.	Lud.	Pri.	total	%	Sum
2%	2%	12%	15%	34%	34%	41	40%	103
1	1	5	6	14	14			

total — 77   %endemic — 53%

Ludlow — Uralian-Cordilleran

O.	L.—M.Ll.	U.Ll.	Wen.	Lud.	total	%	Sum
2%	2%	5%	33%	57%	42	30%	138
1	1	2	14	24			

total — 91   %endemic — 46%

Wenlŏck — Uralian-Cordilleran

O.	L.—M.Ll.	U.Ll.	Wen.	total	%	Sum
6%	6%	10%	77%	31	22%	143
2	2	3	24			

total — 83   %endemic — 37%

		Sum
Upper Llandovery		89

		Sum
L—M Llandovery		78

Sample size

Inspection of Fig.26A—F, Fig.28 and Table III shows that the percentage of cosmopolitan taxa occurring in a biogeographic unit is strongly correlated with the total number of taxa recorded from the biogeographic unit in most cases. The abundance in absolute numbers of specimens of cosmopolitan taxa is high at most localities, possibly indicating something about their broader tolerance to any environment and their ability to make a successful go of things in most places. This higher absolute abundance of specimens belonging to cosmopolitan taxa (*Atrypa*, *Leptaena*, *Eospirifer*, *Howellella*, *Nucleospira*, *Ambocoelia*, etc.) insures that biogeographic units that have been poorly sampled will have a far higher percentage of cosmopolitan taxa represented than those biogeographic units which have been extensively sampled (many more localities that include just about all potential communities). In other words, a poorly sampled biogeographic unit will suggest the presence of a far higher percentage of cosmopolitan taxa than is actually the case. Therefore, one may not accept the raw total numbers of endemic and cosmopolitan taxa known from a biogeographic unit as reflecting the real situation unless one has an excellent appreciation of the sampling. For example, Fig.26A—F lay out the data for the Gedinne through Givet taxa occurring in Nevada (the biogeographic units change through time as this is a biogeographic boundary area). Note that in Fig.26A—F the percentage of cosmopolitan taxa in Nevada is far higher, associated with a small total number of taxa, than in the co-occurring Appohimchi and Old World units, the latter both being characterized by a far higher total number of taxa. This difference reflects two artifacts in the sampling of the Nevada Lower and Middle Devonian: (*1*) a far smaller number of localities have been studied; (*2*) a far smaller number of benthic communities have been sampled. Therefore, it would be wrong to conclude that the Lower and Middle Devonian of the Nevada region is characterized by a far lower total number of taxa, and a far higher percentage of cosmopolitan taxa than either the Appohimchi or Old World units. However, it is reasonable to conclude that the Old World Realm does include more than one unit comparable in size to the Nevada area so that it is unreasonable to expect the Nevada area to originally have included a total number of taxa comparable to that recorded for the Old World unit, i.e. it is unfair to expect the same total number of taxa in a well sampled realm and a well sampled unit of rank lower than a realm. A second example is provided by the Malvinokaffric Realm Devonian. Despite the low total number of taxa (far lower than present in the poorly sampled Nevada Devonian) we know that the Malvinokaffric Devonian has been very well sampled (hundreds of localities scattered over more than two continents), and to be a very good approximation of the real total number of articulate brachiopod taxa (possibly even better than that present in either the Appohimchi or World units!). MacArthur and Wilson (1963) have earlier commented on the same relation when dealing with living materials. It is thus evident, although hardly surprising, that the sampling of a biogeographic unit raises the same questions as does the sampling for total taxic diversity at individual stations (Fisher et al., 1943; Murray, 1968). In other words, an understanding of the biogeographic significance of total number of taxa and percentage of cosmopolitan taxa present in a biogeographic unit depends on a good understanding of the sampling. A raw compilation from the literature may be very misleading! Caveat Emptor!

Fig.26 indicates the essential unity of the non-Malvinokaffric Silurian-Devonian units insofar as their common evolutionary source materials and also the high degree of environmental similarity through time that they possess in common. The exceptionally low diversity and low percentage of cosmopolitan taxa present during both the Silurian and Devonian of the Malvinokaffric Realm is also brought out in this figure and in Table III. This difference emphasizes both the biologic and environmental distinctiveness of the Malvinokaffric Realm.

Changes in provincialism through time

Inspection of Fig.26A—L and Table III shows the profound change in Silurian through Late Devonian provincialism through time. Further research will enable interested students to further refine and subdivide the scheme presented here, but the overall trends cannot be expected to change. The pronounced increase in level of provincialism from the beginning of the Silurian through the Lower Devonian and then down again through the Upper Devonian is a fact of life.[42]

Inspection of Fig.26A—L indicates these changes in provincialism both as pertains to total number of endemic taxa per unit of time and as to percentage of endemic taxa per unit of time. Earlier (section "Total and percentage rates of appearance, evolution and extinction for the Silurian and Devonian") I have considered the waxing and waning of provincialism in terms of worldwide number of new taxa per unit of time and also percentage of new taxa per unit of time. The figures for number of endemic taxa and percentage of endemic taxa per unit of time back up the worldwide figures for new taxa during the same time units. However, the information on endemic taxa per unit of time analyses the same data in a slightly different manner despite the similarity of the conclusions. The thrust of all these approaches is to emphasize the changing production rate of new taxa per unit of time, i.e., changing rate of evolution.

Fig. 27. pp. 161—174. Stratigraphic ranges of the Llandovery through Famenne brachiopods, with Caradoc-Ashgill extent indicated for taxa having pre-Llandovery, reasonably close antecedents or actual extent.

T	N	E
0	0	0
7	4	7
14	4	11
21	8	11
21	6	8
18	9	3
15	5	6
12	3	2
12	3	3
11	5	2
7	6	1
3	1	2
2	2	0
0	0	0

Schuchertella

Bagrasia
Leptalosia
Buxtonia
Irboskites
Eostrophalosia
Oligorachis
Truncalosia
Agramatia
Orbinaria
Praewaagenoconcha
Sinoproductella
Strophoproductus
Chonopectus
Acanthatia
Hamlingella
Steinhagella
Leioproductus
Galeatella
Mesoplica
Manoproductus
Laminatia
Semiproductus
Sentosia

Davidsonia
Orthopleura
Xystostrophia
Devonalosia
Productella
Helaspis
Spinulicosta
Stelckia
Devonoproductus
Whidbornella
Chonopectoides
Steinhagella

"Schuchertella"
Hipparionyx
Iridistrophia
Areostrophia
Aesopomum
Morinorhynchus

Notoparmella
Notanoplia
Callicalyptella
Boucotia

Coolinia

Fardenia

T	N	E			T	N	E				T	N	E
22	19	22			0	0	0				0	0	0
10	5	7			0	0	0				3	0	3
12	6	7			0	0	0				5	0	2
9	3	3			0	0	0				10	4	5
8	3	2			0	0	0				8	3	2
7	1	2			2	0	2				7	3	2
9	6	3			3	2	1				4	1	0
4	2	1			1	0	0				4	0	1
3	0	1			2	0	1				7	4	3
3	1	0			2	1	0				7	2	5
3	1	1			1	0	0				9	6	5
2	0	0			3	1	2				5	4	0
2	1	0			4	1	2				1	1	0
1	1	0			4	4	1				0	0	0

Genera (range bars, left to right):

Metacamerella, Camerella, Porambonites, Syntrophia, Noetlingia, Parastrophinella, Grayina, Anastrophia, Llanoella

Antirhynchonella, Brevilamula, Clorindella, Barrandina, Clorinda, Boucotides, Gypidula, Wyella, Levigatella, Amsdenina, Clorindina, Gypidulina, Sieberella, Gypidulella, Devonogypa, Ivdelinia=Procerulina, Leviconchidiella, Zdimir, Biseptum, Carinagypa, Pentamerella

Stricklandia, Microcardinalia, Ehlersella, Costistricklandia, Aenigmastricklandia, Kulumbella, Plicostricklandia

	T	N	E
	3	0	3
	3	1	0
	2	0	0
	2	2	0
	0	0	0
	0	0	0
	0	0	0
	1	0	1
	20	12	19
	18	14	10
	5	5	1
	6	6	6
	2	2	2
	0	0	0

Retichonetes

Atribonium
Colidium

Eoconchidium
Pseudoconchidium
Callipentamerus
Pentameroides
Pentamerus
Bisulcatella
Sapelnikovia
Supertrilobus
Lissocoelina
Harpidium
Carmanella
Stenopentamerus
Pentamerifera
Eopentamerifera
Capelliniella
Brooksina
Rhipidium
Ectorhipidium
Pinguaella
Kirkidium
Pararhipidium
Khodalevitchia
Bisulcata
Shrockia
Conchidium
Aliconchidium
Cymbidium
Lamelliconchidium
Vagranella
Spondylostrophia
Severella
Spondylopyxis
Subriana
Holorhynchus
Nondia
Borealis
Virgianella
Virgiana
Platymerella

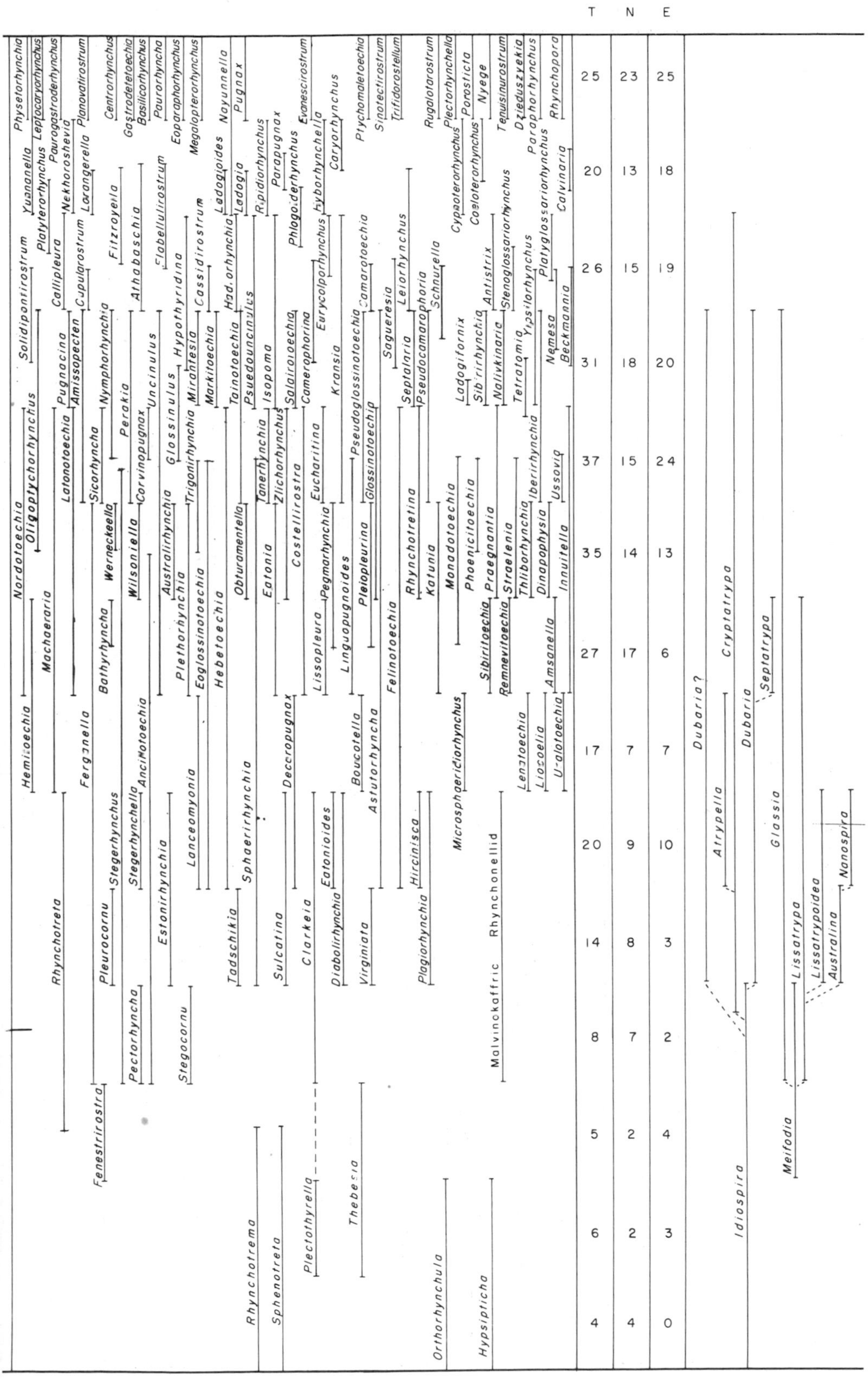
T N E
25 23 25
20 13 18
26 15 19
31 18 20
37 15 24
35 14 13
27 17 6
17 7 7
20 9 10
14 8 3
8 7 2
5 2 4
6 2 3
4 4 0
Physetorhynchia
Yunnanella
Solidipontirostrum
Nordotoechia
Hemitoechia
Leptocaryorhynchus
Platyterorhynchus
Oligoptychorhynchus
Machaeraria
Paurogastroderhynchus
Nekhoroshevia
Callipleura
Pugnacina
Latonotoechia
Rhynchotreta
Planovatirostrum
Lorangerella
Cupularostrum
Amissopecten
Fergonella
Fenestrirostra
Centrorhynchus
Fitzroyella
Nymphorhynchia
Sicorhyncha
Bathyrhyncha
Pleurocornu
Stegerhynchus
Gastrodetoechia
Perakia
Werneckeella
Stegerhynchella
Pectorhyncha
Basilicorhynchus
Athabaschia
Corvinopugnax
Wilsoniella
Ancillotoechia
Paurorhyncha
Flabellulirostrum
Uncinulus
Estonirhynchia
Eoparaphorhynchus
Hypothyridina
Glossinulus
Australirhynchia
Plethorhynchia
Megalopterorhynchus
Cassidirostrum
Mirantesia
Trigonirhynchia
Eoglossinotoechia
Lanceomyonia
Stegocornu
Markitoechia
Hebetoechia
Nayunnella
Ladogioides
Had.orhynchia
Tainotoechia
Obturamentella
Tadschikia
Pugnax
Ladogia
Pseudouncinulus
Sphaerirhynchia
Rhynchotrema
Ripidiorhynchus
Isopoma
Tanerhynchia
Eatonia
Sulcatina
Sphenotreta
Parapugnax
Zlichorhynchus
Decoropugnax
Phlogoiderhynchus
Salairotoechia
Costellirostra
Clarkeia
Plectothyrella
Evanescirostrum
Hyborhynchella
Camerophorina
Eucharitina
Lissopleura
Eatonioides
Diabolirhynchia
Caryorhynchus
Eurycolporhynchus
Kransia
Pegmarhynchia
Linguopugnoides
Virginiata
Thebesia
Ptychomaletoechia
Camarotoechia
Pseudoglossinotoechia
Glossinotoechia
Pleiopleurina
Boucotella
Astutorhyncha
Sinotectirostrum
Saguresia
Trifidorostellum
Leiorhynchus
Septalaria
Felinotoechia
Pseudocamarophoria
Rhynchotretina
Katunia
Hircinisca
Plagiorhynchia
Orthorhynchula
Schnurella
Rugaltarostrum
Plectorhynchella
Cyphoterorhynchus
Porosticta
Coeloterorhynchus
Nyege
Ladogifornix
Sibirirhynchia
Monadotoechia
Phoenicitoechia
Sibiritoechia
Microsphaeridiorhynchus
Malvinokaffric Rhynchonellid
Hypsiptycha
Antistrix
Tenuisinurostrum
Stenoglossariorhynchus
Nalivkinaria
Praegnantia
Remnevitoechia
Dzieduszyckia
Paraphorhynchus
Ypsilorhynchus
Tetratomia
Straelenia
Thliborhynchia
Lenatoechia
Platyglossariorhynchus
Iberirhynchia
Dinapophysia
Liosoelia
Rhynchopora
Calvinaria
Nemesa
Beckmannia
Ussovia
Innuitella
Amsanella
U-alatoechia
Dubaria?
Cryptatrypa
Atrypella
Dubaria
Septatrypa
Glassia
Idiospira
Lissatrypa
Lissatrypoidea
Australina
Nanospira
Meifodia

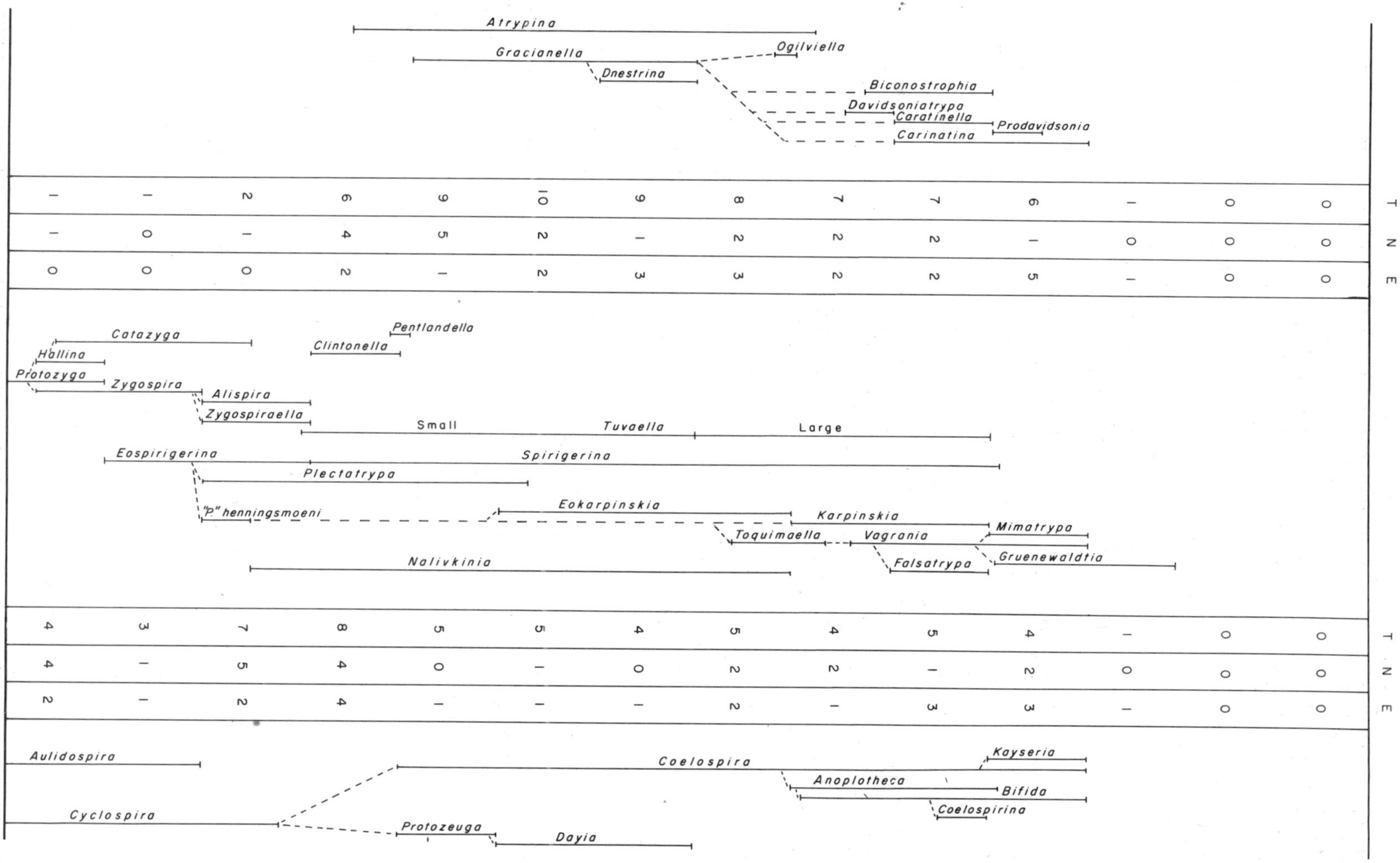

Atrypina
Gracianella
Ogilviella
Dnestrina
Biconostrophia
Davidsoniatrypa
Caratinella
Prodavidsonia
Carinatina
T 1 1 2 6 9 10 9 8 7 7 6 1 0 0
N 1 0 1 4 5 2 1 2 2 2 1 0 0 0
E 0 0 0 2 1 2 3 3 2 2 5 1 0 0
Pentlandella
Catazyga
Clintonella
Hallina
Protozyga
Zygospira
Alispira
Zygospiraella
Small
Tuvaella
Large
Eospirigerina
Spirigerina
Plectatrypa
"P" henningsmoeni
Eokarpinskia
Karpinskia
Mimatrypa
Toquimaella
Vagrania
Nalivkinia
Gruenewaldtia
Falsatrypa
T 4 3 7 8 5 5 4 5 4 5 4 1 0 0
N 4 1 5 4 0 1 0 2 2 1 2 0 0 0
E 2 1 2 4 1 1 1 2 1 3 3 1 0 0
Aulidospira
Coelospira
Kayseria
Anoplotheca
Bifida
Cyclospira
Coelospirina
Protozeuga
Dayia

T N E

Camarophorella

T	N	E
0	0	0
4	0	4
6	0	2
15	7	9
9	2	1
7	2	0
6	1	1
6	1	1
5	2	0
4	3	1
1	1	1
2	1	1
2	0	1
2	2	0

Protatrypa
Punctatrypa
Atrypinella
Spinatrypina
Invertrypa
Spinatrypa
Atrypa
Isospinatrypa
Desquamatia
Variatrypa
Anatrypa
Reticulatrypa
Hollardiella

Dicamaropsis
Camarospira
Dicamara
Charionella
Charionoides
Merista
Camarium
Meristina
Meristella
Meristelloides
Pentagonia
Pradoia
Triathyris
Anathyris
Meristospira
Athyris

Cryptothyrella
Hindella
Hyattidina
Whitfieldella
Didymothyris
Dichozygopleura
Septathyris
Leptathyris
Buchanathyris

Plectothyrella
Protathyris
Glassina
Greenfieldia
Cyrtia
Plicocyrtia
Havlicekia
H.(Hedeina)
Nikiforovaena
Eospirifer
Janius
H.(Macropleura)
Striispirifer
Nucleospira

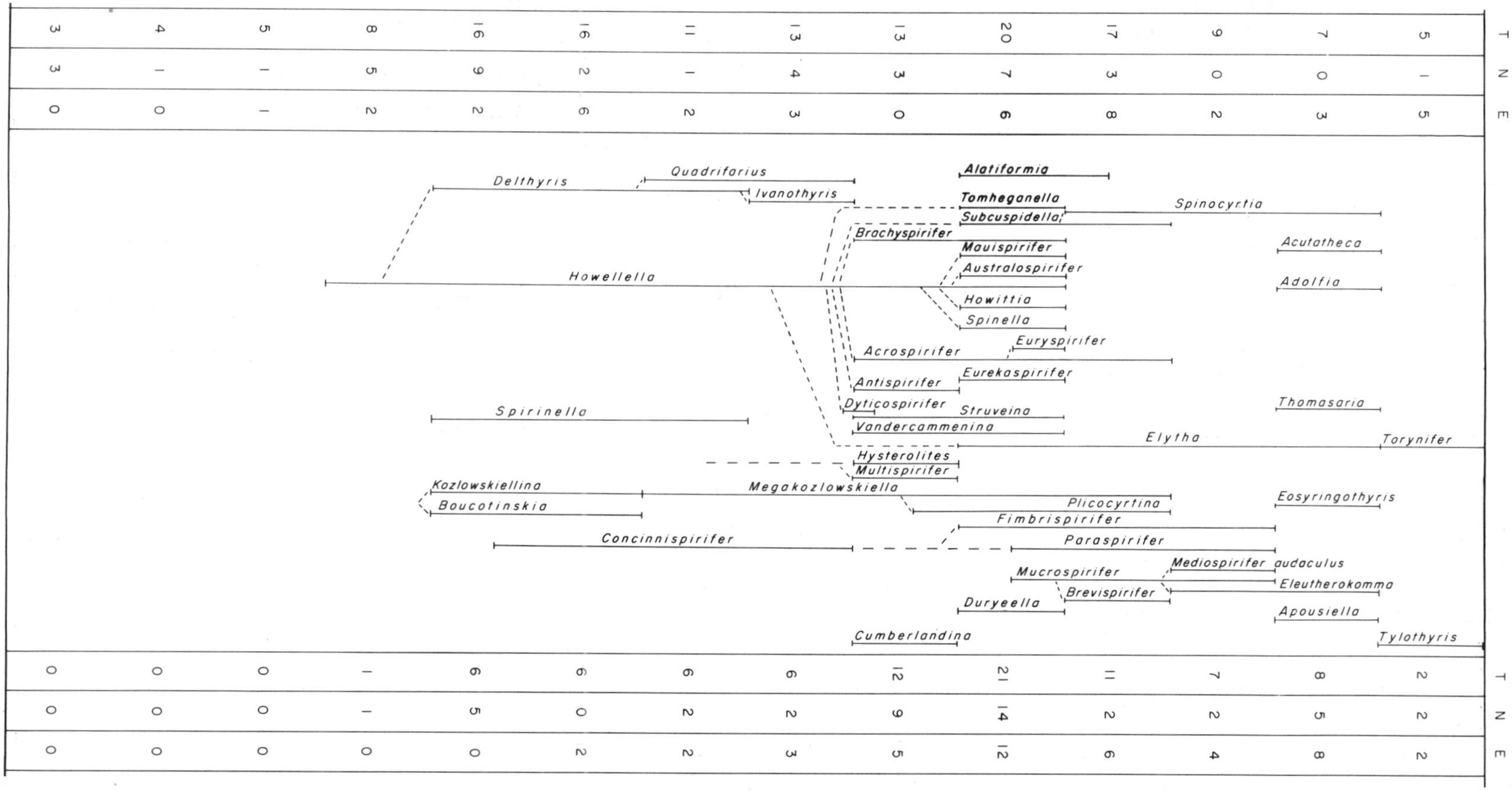
Delthyris
Quadrifarius
Ivanothyris
Alatiformia
Tomheganella
Subcuspidella
Spinocyrtia
Brachyspirifer
Mauispirifer
Australospirifer
Acutatheca
Howellella
Adolfia
Howittia
Spinella
Euryspirifer
Acrospirifer
Eurekaspirifer
Antispirifer
Dyticospirifer
Struveina
Thomasaria
Spirinella
Vandercammenina
Elytha
Torynifer
Hysterolites
Multispirifer
Kozlowskiellina
Megakozlowskiella
Eosyringothyris
Boucotinskia
Plicocyrtina
Fimbrispirifer
Concinnispirifer
Paraspirifer
Mediospirifer audaculus
Mucrospirifer
Eleutherokomma
Brevispirifer
Duryeella
Apousiella
Cumberlandina
Tylothyris

Genera (range bars): Costispirifer, Amoenospirifer, Theodossia, Cyrtospirifer, Syringospira, Indospirifer, Cyrtiopsis, Sphenospira, Lazutkinia, Austrospirifer, Arctospirifer, Undispirifer, Candispirifer, Tenticospirifer, Lirospirifer, Proreticularia, Elythna, Allanella, Spurispirifer, Obesaria, Reticularia, Bojothyris, Quadrithyrina, Quadrithyris, "Quadrithyris" plicate, Warrenella, Najadospirifer, Xenomartina, Reticulariopsis, Prosserella, Cingulodermis, Brachythyris, Litothyris, Strophopleura, Eoreticularia, Pinguispirifer, Palaeospirifer, Parallelora, Tenellodermis, Chimaerothyris, Tenelaspira, Cyrtina, Tecaocyrtina, Acanthospirina, Syringothyris, Cyrtinaella, Moravilla, Timenia, Pustulatia, Ambothyris, Spinoplasia, Plicoplasia, Crurithyris, Ladjia, Emanuella, Metaplasia, Ambocoelia, Echinccoelia, Bisinocoelia

T	N	E		T	N	E
12	9	12		3	2	3
9	5	6		2	0	1
6	3	2		2	1	0
14	7	11		1	0	0
12	8	6		2	0	1
8	3	3		2	0	0
6	1	1		2	2	0
5	1	0		0	0	0
4	2	0		0	0	0
2	2	0		0	0	0
0	0	0		0	0	0
0	0	0		0	0	0
0	0	0		0	0	0
0	0	0				

T	N	E				T	N	E
1	0	1				3	0	3
6	0	5	Macclarenella			4	1	1
9	5	3	Parazyga			17	10	14
5	3	1	Plectospira	Rensselandia	Cryptonella	9	4	2
4	1	2	Retzia	Meganterella · Chascothyris, Conomimus, Subrensselandia, Bornhardtina, Stringocephalus, Acrothyris, Geranocephalus, Enantiosphen, Enantiosphenella, Stringomimus	Cranaena · Cryptonella · Reeftonella, Cimicinella, Cimicinoides	11	9	6
3	0	0	Leptospira, Pseudoparazyga, Trematospira	Meganteris · Malvinokaffric "Cryptonella", Proboscidina		4	1	2
4	3	1				4	3	1
1	0	0	Rhynchospirina			1	0	0
1	0	0				1	0	0
1	1	0	Homoeospirella, Homeospira			3	3	2
0	0	0				0	0	0
0	0	0				0	0	0
0	0	0				0	0	0
0	0	0				0	0	0

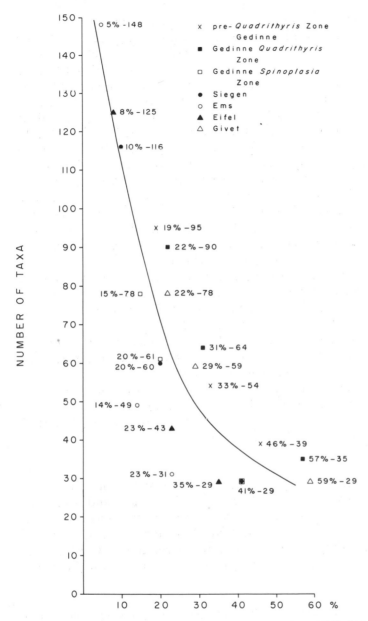

Fig. 28. Plot of number of taxa known from non-Malvinokaffric Devonian biogeographic units against percentage of cosmopolitan taxa. Note the close correlation between number of taxa (regardless of time interval) and percentage of cosmopolitan taxa. This is clearly an important sampling artifact determined by the abundance in actual numbers of specimens of cosmopolitan taxa as contrasted with the greater rarity of endemic taxa in actual numbers of specimens.

Cosmopolitan taxa and rates of evolution and extinction

The following partial listing of cosmopolitan taxa for the Lower Llandovery through Givet time interval (taxa that occur in three or more biogeographic units during the Devonian or two units during the Late Silurian; they also include many of the non-"rare" Llandovery and Frasne taxa) is striking. The bulk of these taxa are slowly evolving. These taxa do not belong to the "index fossil", "guide fossil" group. They have little utility for refined dating and zonation of Silurian-Devonian rocks. In other words, they are slowly evolving taxa. The inescapable correlation here between large population size and slow rate of evolution is evident.

Acrospirifer, Aegiria, Aesopomum, Ambocoelia, Amphistrophia, Ancillotoechia, Antirhynchonella, Atrypa, Atrypina, Brachymimulus, Brachyprion, Brevilamula, Clorinda, Coolinia, Cyrtia, Cyrtina, Dalejina, Dicaelosia, Dicaelosia (long-lobed), *Dolerorthis, Eoplectodonta, Eospirifer, Ferganella, Glassia, Grayina, Gypidula, Hedeina, Hesperorthis, Howellella, Iridistrophia, Isorthis, Janius, Kirkidium, Leangella, Leptaena, Leptaenisca, Leptostrophia, Lissatrypa, Machaeraria, Megakozlowskiella, Mesodouvillina, Merista, Meristella, Meristina, Nucleospira, Pentamerus, Plectatrypa, Plectodonta, Plicoplasia, Proreticularia, Protathyris, Protochonetes, Protomegastrophia, Ptychopleurella, Resserella, Rhynchospirina, Rhynchotreta, Salopina, Schizophoria, "Schuchertella", Sieberella, Skenidioides, Sphaerirhynchia, Spinatrypa, Spirigerina, Stegerhynchus, Striispirifer, Strophonella, Tropidoleptus, Virgiana.*

These are the genera that are easy to find at localities of appropriate age; these are the taxa represented by specimens present in most collections. These are the taxa that force us to search for the "rare" specimens more useful for zonation.

Taxic duration

Table III and Fig.29—32 indicate both the time duration and provincial as contrasted to cosmopolitan distribution of Silurian and Devonian brachiopods. (All of the statistical material in this section has been compiled from work in progress to be entitled "Distribution of Silurian and Devonian Articulate Brachiopods". In this paper references will be made to the occurrences of the various taxa in time and space with the neces-

---

Fig. 29. Compilation of time duration (units 1 through 13 represent any span throughout the interval Caradoc through Frasne) for cosmopolitan and provincial (including "rare") taxa in the interval Lower Llandovery through Frasne. All of the cosmopolitan taxa are cosmopolitan for at least one of the time intervals Lower Llandovery through Frasne, but not necessarily for their entire time of occurrence. Malvinokaffric taxa are not considered in this compilation, although adding them would not affect the overall

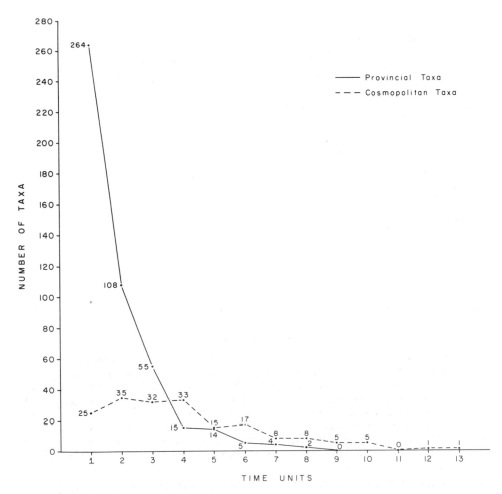

conclusions. Cosmopolitan taxa include taxa that occur as disjunct occurrences and in boundary regions (thus slightly increasing the number of cosmopolitan taxa). The number of time units assigned to a taxon includes ranges through the Caradoc (for Silurian taxa occurring earlier), but does not include post-Frasne time units (Famenne and later) for the few taxa that span the Frasne-Famenne boundary. Note the lower number of cosmopolitan taxa. Note the far greater average time span of cosmopolitan taxa (i.e., far lower rate of evolution). All of the time units Lower Llandovery through Frasne have been assigned unit value (as done in Fig. 22); however, if a figure proportional to their probable relative length were chosen (as is indicated in Fig. 19) the conclusions would not differ significantly as these relative lengths do not depart very greatly from unity (a total range of no more than 0.5 to 1.5). The significance of this compilation regarding the relation between geographic distribution, time duration, size of interbreeding population and rate of evolution should be noted carefully.

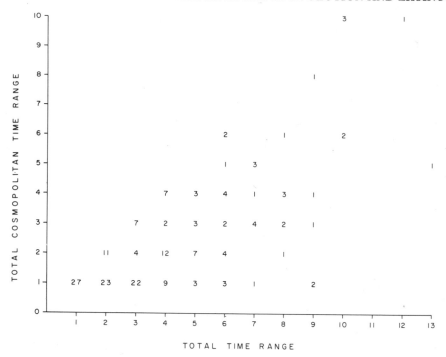

Fig. 30. Plot of total time range (measured in the same units as in Fig. 29) against total cosmopolitan time range (measured in same units) for cosmopolitan taxa (chiefly genera). Note the tendency for taxa having a longer total time range to have a longer total cosmopolitan time range. This is further evidence favoring the concept that cosmopolitan taxa having a long time range have a lower rate of evolution than taxa approaching or having a provincial distribution (that is, taxa being cosmopolitan for only a short part of their time range approach a more provincial behaviour than do those acting in a cosmopolitan manner for a longer part of their total time range).

sary documentation). Inspection of Table III makes the following points clear:

(1) The Llandovery and the Frasne are the times of maximum cosmopolitanism.

(2) Maximum provincialism is reached in the Ems both as regards low number (or percentage) of cosmopolitan taxa and highest percentage (or number) of endemic taxa.

(3) The cosmopolitan taxa include those with the longest time duration by far, many of them persisting from the earlier parts of the Silurian and some even from the Ordovician.

(4) The Old World Realm contains a higher percentage of taxa with a long time duration (largely the "Silurian holdovers" of Boucot et al., 1969).

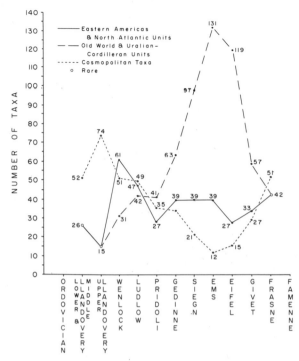

Fig. 31. Changes in number of taxa during the interval Lower Llandovery-Frasne. Note that the time intervals Lower Llandovery-Upper Llandovery and Frasne are recorded as either "Rare" or "Cosmopolitan" whereas the intervening interval is divided into "Cosmopolitan" and "Eastern Americas-North Atlantic" plus "Old World-Uralian-Cordilleran". The height of provincialism is reached in the Ems. The Upper Llandovery peak is interpreted as the result of a large influx from a very restricted source (probably the southeastern Kazakhstan region in large part). The Wenlock anomaly correlates well with a maximum in reef abundance.

(5) The Eastern Americas Realm contains the highest percentage of newer taxa.

(6) It must be understood that the statistics presented here provide an overall trend. The actual numbers of taxa are subject to change both as regards further taxonomic studies and as regards slight changes in correlation. There is no reason, however, to suspect that the *percentages* indicated in this Table will be subject to significant change. The general outlines of the faunal development are clear. Further refinements in both taxonomy and correlation may be expected to further emphasize the trends shown in the Table rather than smoothing them out.

(7) The lengthy time duration of the cosmopolitan taxa, those very taxa that occur in largest abundance in most collections, further empha-

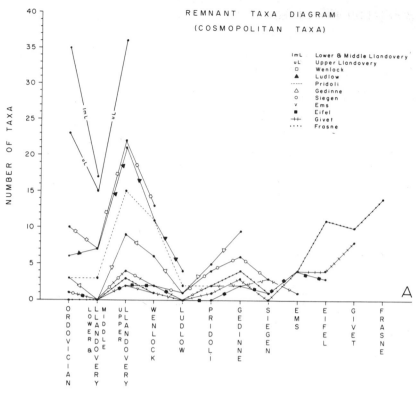

REMNANT TAXA DIAGRAM
(COSMOPOLITAN TAXA)

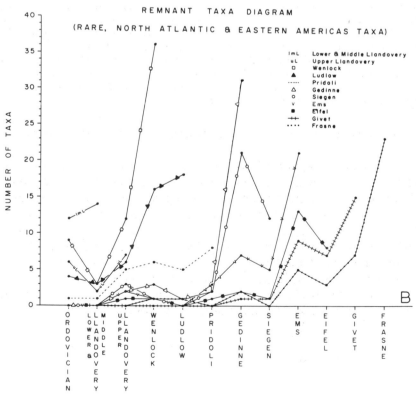

REMNANT TAXA DIAGRAM
(RARE, NORTH ATLANTIC & EASTERN AMERICAS TAXA)

REMNANT   TAXA   DIAGRAM

(RARE,   OLD   WORLD   &   URALIAN – CORDILLERAN   TAXA)

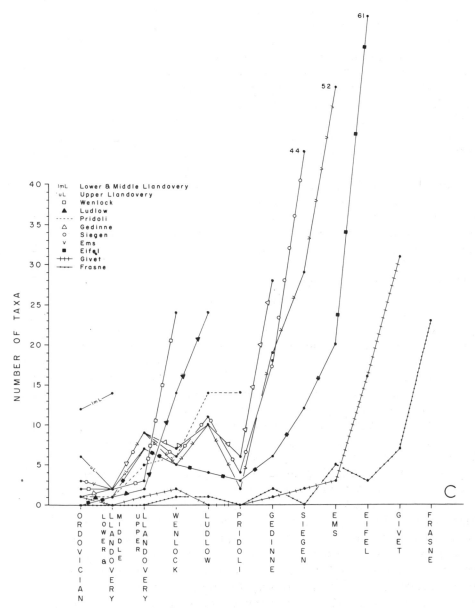

Fig. 32. Plot of data from Table III. Note the radically different pattern shown by Cosmopolitan as contrasted with Provincial distributions. This is partly a result of the decrease in number of cosmopolitan taxa during a time of high provincialism and also the result of longer time duration of individual cosmopolitan taxa. Note the greater time persistence of cosmopolitan taxa on the average. Note the greater abundance, at any specific time, of new taxa in provincial units as contrasted with cosmopolitan taxa during the same time unit.

sizes the strong correlation between large population size and slow rate of evolution, as well as the reverse.

Time duration of cosmopolitan and provincial taxa

Fig.29 indicates beyond any shadow of doubt that cosmopolitan taxa have a far greater time persistence than do provincial taxa. This conclusion strongly reinforces that provided earlier on other grounds that taxa characterized by large populations (including cosmopolitan taxa) evolve far more slowly than do those characterized by small populations (including provincial taxa). The percentage of exceptions, as shown in Fig.29 is small in both instances. There is obviously going to always be a gradation between taxa characterized as provincial and those characterized as cosmopolitan. A measure of this gradation may be viewed by taking the so-called cosmopolitan taxa (see Fig.30) and graphing their total time duration against their cosmopolitan time duration. It is easily observed that there is a strong tendency for longer lived cosmopolitan taxa to also be cosmopolitan for a greater percentage of their total time span than is the case for shorter lived cosmopolitan taxa.

Boundary region biogeography

Working out the biogeographic character of the faunas from a given region is not difficult if that region is located well away from biogeographic boundaries. Disjunct elements derived from other contemporaneous biogeographic entities and relict elements held over from an earlier time are the only real complications that may tend to partially obscure the biogeographic unity of the fauna. However, faunas obtained from a region that turns out to be situated in a boundary area are a very different matter. Initially they appear confusing in that they "mix" elements that one would not expect to occur together (all of these comments assume that one has available a full spectrum of animal communities representing as many environments as were present).

The analysis of the biogeographic story in a boundary region must be done in as fine stratigraphic detail as possible in order to learn whether the apparent "mixtures" of elements one is not accustomed to observing together actually occur in the same stratum. If these elements from the "mixtures" do not occur in the same stratum, or at the same locality, then they may represent evidence for a fluctuating biogeographic boundary rather than for a mixed fauna. If one does not perform the biogeographic analysis stratum by stratum, locality by locality, there is the danger that the boundary region will be interpreted as one in which there is a mixing of faunas with the implication that the environments characterizing the biogeographic entities have also been mixed, and that the faunal barriers have somehow become blurred in this area. This false sense of gradation and blurring will badly obscure any understanding of the reasons responsible for the provincialism.

The statistical data for the Nevada Devonian (including during appropriate time intervals portions of the Old World and Eastern Americas Realms) in Fig.26A—F makes this point clear. Nevada during the Devonian was situated in a biogeographic boundary region adjacent to the limits of the Old World and Eastern Americas Realms. Inspection of Fig.26A—H indicates that beginning in the Gedinne (pre-*Quadrithyris* Zone) the fauna was chiefly Old World plus cosmopolitan taxa, followed in the *Quadrithyris* Zone by a mixture of about two parts Old World taxa to one part Eastern Americas Realm taxa plus, of course, cosmopolitan taxa. During the subsequent *Spinoplasia* Zone (later Gedinne) a definite shift of the Nevada region into the Eastern Americas Realm took place (essentially no endemic Old World Realm taxa recognized in the available samples). With Siegen time (*Trematospira* Zone) there is a beginning of the shift back to Old World Realm affinities (although the Eastern Americas Realm endemic taxa outnumber the Old World Realm taxa by more than 2:1 in the available sample). During Ems time (*Acrospirifer kobehana* and *Eurekaspirifer* zones) the shift back to Old World Realm conditions continues with the presence of a large predominance of Old World Realm endemic taxa over Eastern Americas Realm taxa (although with the introduction of a significant number of taxa endemic to Nevada which form the justification for a Cordilleran Region of the Old World Realm). During Eifel time (*Leptathyris* and *Warrenella kirki* zones) there are no Eastern Americas Realm taxa, a large number of Old World endemic taxa and a large number of taxa endemic to Nevada, all of which are assigned to the Cordilleran Region of the Old World Realm. During Givet time this situation for Eifel time continues in similar manner except that a larger number of cosmopolitan taxa is present, as is true all over the world during this time interval. Inspection of Fig.28 indicates that the small size of the Nevada samples (representing a very limited number of communities and relatively few actual localities) probably will have tended to make the Nevada faunas appear more cosmopolitan than they actually are. The inadequacy of the sample, when compared with that from better known regions such as the Appohimchi Subprovince of the Appalachian Realm and the Rhenish-Bohemian Region of the Old World Realm indicates the problem of using the Nevada percentages and total numbers of taxa for any but broad, outline purpose.

Ecologic correlations

Fig.36 presents an ecologic compilation of the Benthic Assemblage range for the taxa dealt with in Part II, Chapter 6. Although the data presented there is far more limited than might be wished this does not prevent some conclusions from being made. It appears clear that articulate brachiopods, whether provincial or cosmopolitan in their distribution, occur most abundantly in communities having a narrow Benthic Assemblage spread (Fig.36A).

It also appears clear that the majority of the taxa (Fig.36B), whether occurring over a narrow Benthic Assemblage spread or a wide Benthic Assemblage spread occur in the Benthic Assemblages 3-5 region. In other words, the Benthic Assemblages 1 and 2 regions in addition to being populated by fewer taxa are no more favorable to cosmopolitan, long-ranging taxa than are the more offshore Benthic Assemblages. There is, therefore, no indication that the offshore region is less "stable" environmentally than the onshore region *if* environmental stability is a function of the abundance of cosmopolitan, long-ranging taxa (cosmopolitan taxa are, in general, far more long-ranging than provincial taxa).

Sources of error in the tabulations

The chief sources of error in the compilation of the biogeographic statistics derived from Fig.27 are as follows:

(1) Differences of opinion regarding the correct generic assignment of species that will alter the total range of a genus.

(2) Differences of opinion regarding the correlation of localities from which a genus is known.

(3) Differences of opinion regarding the validity of various genera including opinions that hold the genus in question to have been split unmeaningfully or not split enough.

(4) No systematic attempt has been made to indicate on Fig.27 the range of a genus within a time unit where it does not span the entire unit; in most instances it has arbitrarily been shown to span the entire unit. This will, in cases where the genus does not span the entire unit, result in biasing the compilations somewhat. However, the very gross nature of the anomalies shown by the compilations suggests that such errors in plotting will not significantly affect the conclusions at this stage of the study.

(5) The differences in relative time length of the units employed in Fig.27 would slightly alter the magnitude of some of the effects graphed in this section, but not enough that normalization would have a very significant effect.

(6) Williams' (1957) thought-provoking paper on "monographic evolution", using articulate brachiopods as the example, raises the question about the validity of any statistical data based on numbers of genera per time unit or per geographic area. We have no basic reason to conclude that

the relative percentages of taxa known from the Phanerozoic as a whole for each small group will remain constant as taxonomic and geographic information about fossils gradually improves. However, my own impression of Late Ordovician through Devonian articulate brachiopods suggests that changes in these percentages will not be serious enough to invalidate the conclusions arrived at in this book, although the relative percentages may well change somewhat. For example, we know far too little yet to dare compare the numbers of provincial and cosmopolitan taxa present in the various regions recognized for the Devonian; the Caradoc and Ashgill of the Andean region contain rich brachiopod faunas that are virtually unknown. My impressions are subjective in that I have not quantified them but they are based on as thorough a familiarity with the articulates of this time interval as is currently available for the group as a whole (my familiarity in certain sectors is woefully deficient, but overall I feel that I have seen enough to permit the generalizations to have validity). I am certain, for example (or at least as certain as one may be concerning most data), that there are far more Lower Devonian brachiopod taxa recognized and potentially recognizable than there are Lower Silurian taxa; I do not think that further intensive study of Lower Silurian faunas will substantially change this picture. My impression is that the potential uncertainties in our knowledge of Silurian-Devonian articulate brachiopods are very low as compared with our knowledge of Lower-Middle Cambrian and Triassic brachiopods; for the latter two time intervals monographic effects may still be very serious.

Further refinements in both correlation, geographic distribution and taxonomy will, undoubtedly, permit additional and more precise conclusions of many sorts to be drawn from the basic data. These refinements should be made as soon as practical. However, because such refinements are endless in nature I felt justified in summarizing the state of the art at this time. I do not wish to suggest that my conclusions are to be taken as final and not subject to future modification. These are uncharted waters full of hidden rocks on which I have probably scraped bottom more than once. My only prayer is that I am not headed straight for the bottom rather than on a course headed at least in the general direction of a better understanding of basic relations.

Comment

Crude though this account of Silurian-Devonian "Biogeographic statistics" may be, untrammeled by the lumber of elegant statistical devices, it is clear that the data of fossil distributions in time and space have much to tell the interested biologist and paleontologist willing to take the time. Future, more sophisticated analyses of this data should afford a correspondingly more profound understanding of the forces at work that have molded the fossil record. The dry body of faunal lists and illustrations will come alive once again with the story of the ebb and flow of animal evolution on the grand scale played against the physical background characteristic of each time interval of the past.

The Permo-Triassic Change

The marked change in the shallow-water marine invertebrate faunas of the world adjacent to the Permo-Triassic boundary has been a source of chagrin and speculation for almost one-hundred years (Newell, 1956). The prime problems are as follows: (*1*) the extinction in the Late Permian of most of the diverse Paleozoic invertebrates at the generic, family, and superfamily levels plus some in the ordinal level and of a few classes; (*2*) the appearance in the earliest Triassic of a fauna consisting chiefly of new genera, new families, and new superfamilies of invertebrates; (*3*) the geographically widespread but small number of taxa present in the earliest Triassic.

Initial consideration of this dramatic Permo-Triassic change almost supports a non-evolutionary concept of faunal origins as the changes are so dramatic. However, there is a little evidence, for example in Pakistan, for mixtures of Permian and Triassic taxa. The lack of evidence for gradual change in morphology with time is probably, as suggested by Newell (1956), a consequence of very small populations with consequent smaller chance for preservation (particularly when the chances for minor disconformities in this part of the record are considered as pointed out by Newell.

Before discussing the possible mechanisms responsible for these dramatic changes, changes so different in magnitude from the bulk of those recognized elsewhere in the Phanerozoic fossil record, it is necessary to summarize the physical geology known for the Permo-Triassic boundary interval in order to find data essential for an understanding of the biologic facts known to us.

(*1*) The Permo-Triassic boundary interval is one of maximum regression of shallow seas from the continents, exceeding in scope the regressions of the Late Tertiary, Early Cambrian, Late Ordovician, and Late Cretaceous as well as all others known during the Phanerozoic.

(*2*) The climate of the Permo-Triassic boundary interval is concluded (Kummel, 1973) to have been equable worldwide, as there is no evidence for widespread glaciation or extremes of any sort not distributed on a worldwide basis.

(*3*) Reef environments were essentially absent.

Consideration of the above data leads to the following conclusions:

(1) Permo-Triassic boundary faunas of shallow-water type were restricted to *very* narrow ribbons adjacent to the continental margins. This restriction means small populations if area occupied by faunas is thought to have any relation to size of population.

(2) The equable climate rules out the possibility of climatic belts having acted as biogeographic barriers to faunal migration and mixing.

(3) Major extinctions affecting the Permian fauna are easily interpreted as the result of diminished population size, lowering of climatic barriers that earlier acted to permit the existence of provincially differentiated faunas and disappearance of reef environments. A higher number of communities thus were present during the Late Permian than during the Early Triassic.

(4) The widespread, low taxonomic diversity earliest Triassic faunas are consistent with the presence of a limited number of animal communities.

(5) The major taxonomic differences and lack of taxonomic continuity at the generic, familial, superfamilial levels and even some at the ordinal level between the latest Permian and earliest Triassic are evidence for an extremely rapid rate of phyletic evolution possible under conditions of very reduced population size.

(6) The reduction in number of taxa occurring near the boundary, i.e., heavily reduced rate of taxonomic and cladogenetic evolution, is consistent with both the major extinction event at the end of the Permian and the very limited number of taxa present in the earliest Triassic.

(7) The absence on land for any evidence of other than routine evolutionary rates for either plants or tetrapods is consistent with this picture. Fig.33 summarizes the Permo-Triassic boundary problem.

The Late Cretaceous-Early Tertiary Change

The significant faunal changes affecting shallow-water, shelf-region marine invertebrates in the Late Cretaceous-Early Tertiary boundary region are worth considering from the view of population-size effects. There is general agreement about worldwide extinction affecting a number of major groups (not just ammonites and belemnites, although by no means all groups, Bramlette, 1965). There is also general agreement that many groups show good evidence for continuing existence and organic evolution across the boundary, as well as diversification across the boundary. The end of the Cretaceous sees an increasing amount of climatic differentiation over that present in the earlier Cretaceous and Jurassic (which were far more equable intervals worldwide). The end of the Cretaceous also sees

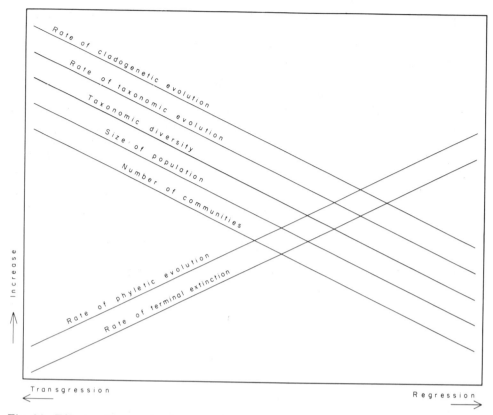

Fig. 33. Effects of regression-transgression under conditions of uniform climate, cosmopolitanism and non-reef regime conditions.

the presence of more extensive tracts of continent above sea level, and also a high level of reef activity. The more extensive areas of land on the continents beginning to appear in the boundary interval makes possible the presence of a more highly provincial shallow-water, shelf fauna due to the subdividing effects of isotherms (associated with the more highly differentiated climate) crossing land masses so as to form a large number of biogeographically separated regions with an appropriately increased number of communities. The climatic changes help to explain an increased extinction rate for some groups, just as the increased possibilities for provincialism permit speeding up of the evolutionary process for other groups due to smaller population sizes and isolation.[43]

Reasons for Differing Reef Abundances in Time

Discussions summarized graphically in Fig.22M, emphasize that the areas of reef and reef-like structures present on modern continents during the Silurian-Devonian have not been constant. Consideration of reef distributions in the Mesozoic and Cenozoic may be obtained from study of the appropriate papers in Hallam (1973). It is clear that the latitudinal distribution of Mesozoic and Cenozoic reefs is highly correlated with the position of the appropriate isotherms, which change back and forth in time. It is reasonable to conclude that such may also have been the case in the Silurian-Devonian. Unfortunately, we have a far poorer concept of isotherm distribution during the Silurian-Devonian than during the later Mesozoic and Cenozoic. The relative absence of Llandovery reef-type structures, as well as evaporites, may indicate a cooler climate present worldwide, even in those regions well away from the Malvinokaffric Realm. The abundance of Late Wenlock-Ludlow reefs, as well as evaporites, may correlate with an increase in temperature worldwide. The Pridoli decrease in reef-type structures may reflect yet another perturbation followed by the Devonian increase correlating with a gradual overall thermal increase. The Famenne absence of reefs is more likely the result of the massive terminal Frasne extinction event (possibly related to a physical event of unknown type) having wiped out all of the reef-forming organisms. It is not until Lower Carboniferous time that a new group of reef-forming organisms appears again, with different antecedents and habits than those existing in the pre-Carboniferous. The inference about reef abundance being tied to worldwide thermal level cannot be proved at this time, but is reasonable.

Deep Sea

The essential absence on the continents of benthic shelly invertebrates from Benthic Assemblage 6 and deeper positions (the Pelagic Community of this paper) suggests that a deeper-water and probably even a deep-sea shelly benthic invertebrate fauna was not present in the Silurian or Devonian. Beds in this position yield abundant fossils of planktic organisms like graptolites, conodonts, acritarchs, chitinozoans, radiolaria and benthic trace fossils. The presence of the low-diversity *Notanoplia* Community and also the Starfish Community adjacent to the boundary between Benthic Assemblages 5 and 6 further reinforces this impression. The deep-sea shelly benthic invertebrates may have been a post-Devonian development. It is unlikely, even after considering the difficulty geologists have in recognizing deep-sea deposits on a purely physical basis, that deep-sea shelly benthic invertebrates of Silurian and Devonian age would have been in existence if they are completely absent in the Benthic Assemblage 6 position. The presence of some rare transported shelly benthic invertebrates in turbidite sequences only serves to reinforce this conclusion.

Rowe's (1971) summary of the density of shelly benthic organisms per area of modern sea bottom from the shelf into the deep ocean shows conclusively that there is a pronounced drop-off. However, the far lower population density of the modern deep seas is still inadequate as an explanation for the apparent absence of shelly benthic fossils in positions much farther offshore than Benthic Assemblage 5 or 6. The collections of deep-sea shelly benthos made by Sanders and Hessler (1969) from the deep sea serve to underline Rowe's (1971) results.

Watkins (1973) has commented that the presence of well-laminated strata of Mississippian age that lack megafossils in the Bragdon Formation of northern California is evidence of a deep-sea (or at least deeper than shelf margin) environment lacking in abundant organisms capable of producing bioturbating effects. Well-laminated strata, with sharp bedding-plane contacts, are also characteristic of the pre-Carboniferous (including the Silurian-Devonian) record in positions well offshore of Benthic Assemblage 6. The lack of shelly benthic megafossils (planktic groups like the graptolites and goniatites are common in many places) in these deposits is common as is the lack of any evidence of bioturbation (i.e., the sharp bedding-plane contacts). Thus we are drawn to the conclusion that not only were shelled benthic invertebrates absent in positions offshore and deeper than Benthic Assemblage 6, but that soft-bodied organisms capable of producing bioturbation were also uncommon except for a few trace-fossil producing organisms. (Hessler and Sanders, 1967, emphasize the large number of annelid worms present in their deep-sea samples).[44]

The marked change in graptolitic faunas at the Ordovician-Silurian boundary, coinciding with similar changes in the benthic, shallow-water invertebrates, may be interpreted to indicate that the graptolites did not belong to the oceanic plankton. One would not expect that the physical changes held responsible for the marked changes in population size that correlate with the extinction event effecting shallow-water benthos should have similarly effected a widespread oceanic planktic group. The high rate of phyletic evolution and lowered rate of taxonomic and cladogenetic evolution shown by the graptolites across the Ordovician-Silurian boundary is similar to that for the brachiopods.

If one accepts Hedgpeth's (1957) division of the marine environment into oceanic and neritic, as approximating the 200-m contour, there is then no evidence for an oceanic Silurian or Devonian shelly benthic fauna. It is also not clear whether or not there was an oceanic planktic fauna, although the Silurian-Devonian plankton (graptolites, acritarchs, etc.) certainly did inhabit the neritic environment.

Explosive Evolution?

Our information concerning the brachiopods of the Silurian and Devonian permits some conclusions about the possibilities for "explosive evolution". If one's hypothesis is that shortly after the "first appearance" of a group it experiences an interval of "explosive evolution" (= high, cladogenetic production rate of taxa) we can consider the hypothesis against the known facts of the Silurian-Devonian brachiopod record. Inspection of the data in this book shows conclusively that there is no necessary correlation between time of "first appearance" and a high ("explosive") rate of evolution. We have, to the contrary, evidence for high rates of phyletic and cladogenetic evolution at varying times in the history of groups of genera and species belonging to single families or subfamilies. The coming in of many groups at the family level during the early part of the Late Llandovery is unaccompanied in these groups by any evidence for an early interval of rapid evolution followed by a subsequent decline in rate of evolution. Rather, the Silurian-Devonian data suggest that groups evolve as small, isolated populations (we commonly lack any information about their rate of evolution during these circumstances, as we commonly are ignorant of their location and must infer it merely from their subsequent, fully-formed appearance) following which they experience an interval of explosively rapid geographic distribution. Following on this interval of explosively rapid distribution their rate of evolution may be either rapid or slow depending on the size of the interbreeding population. For example, we find that the species of *Eocoelia* (Fig.18), widespread geographically but making up a relatively small population, phyletically evolve rapidly as contrasted to the generic-level evolution of the coeval pentamerinid taxa that make up large populations (Fig.18, 27). Subsequently, the pentamerinids experience, during the Late Wenlock-Ludlow, an interval of rapid cladogenetic and taxonomic evolution at the generic level (Fig.27) consequent with the small, isolated populations existing in scattered reef environments. During the Llandovery and Wenlock there does not appear to be any significant difference in rate of phyletic evolution among the various stricklandids (Fig.27) from their first appearance at the base of the Llandovery to their time of extinction near the end of the Wenlock. The chonetids (Fig.17) are widespread during the bulk of the Silurian (although possibly evolving originally from a point source!), but evolve slowly after their first appearance in the Llandovery, following which during the Early Devonian conditions of smaller

population size (correlated with both provincialism and conditions of regression) they cladogenetically and phyletically evolve rapidly, following
which in the later Devonian they again evolve slowly under conditions of
lowered provincialism and transgression. The terebratuloids of the Devonian present still another case. They appear suddenly, presumably from a
small, geographically isolated source (else we would expect to find them
in the fossil record of these earlier times) where evolution had been going
on for some time with rates unknown (possibly large, however, due to the
small size of the inferred populations) until becoming suddenly widespread near the Silurian-Devonian boundary. Some groups (see Fig.20, 21)
then experience an interval of rapid "explosive" phyletic and cladogenetic
evolution, but this interval correlates well with a time of high provincialism and regression, with no evidence on hand that the Silurian-Devonian is
actually the time when the terebratuloids first evolved (one must not
confuse the time of first appearance in the stratigraphic record with the
time when a group first evolves in a biologic sense!). Other groups
(Fig.27, the rensselandids) of terebratuloids evolve very slowly, if at all,
during the Lower Devonian and through the Eifel, until the Givet when
they manifest a sudden burst of rapid cladogenetic evolution that may be
tied in with the presence of the proper type of reef or reef-related environment (they are reef-related organisms in large part, preferring rough-water
environments of the type associated with reefs), following which they
become extinct as a group before the beginning of the Frasne.

   In other words, "explosive evolution" among Silurian-Devonian
brachiopods may occur at any time during the history of a group *once*
conditions conducive to small, isolated interbreeding populations are met.
This information about Silurian-Devonian brachiopods makes one wonder
if the commonly cited examples of "explosive evolution" affecting other
groups of animals early in their evolutionary record are any different. It
would be desirable to test the record of these other groups against its
relation to interbreeding population size in time, and also pay particular
attention to the problem of first appearance in the stratigraphic record
(the widespread, sudden occurrence of fossil traces of the group) against
the actual time when the group may first have evolved far earlier under
conditions of geographically isolated, small populations that may never
have been preserved in the fossil record, or at least are to be located only
with great difficulty owing to their excessive rarity.

Plate Tectonics and Continental Drift

   Despite the resolve indicated in the Preface to avoid current questions
of continental drift and plate tectonics, the recent appearance of the *Atlas*

of Palaeobiogeography (Hallam, 1973) makes a few remarks tempting. The *Atlas* makes clear that Mesozoic and Cenozoic Realm boundaries are approximately parallel to present-day latitudes. This parallelism precludes any large-scale north—south movements of the major land masses with the possible exception of Peninsular India for which data are still inadequate. Realm boundaries in the Permo-Carboniferous appear to also parallel present-day latitudes, although the data presented in the *Atlas* is not as comprehensive on this point as might be wished. The Malvinokaffric Realm boundary during the Silurian and Devonian also approximates present-day latitudes (although the Eastern Americas and Old World boundaries do not). The situation for the Ordovician is less clear in this regard; that for the Cambrian is still very vague as presented in the *Atlas*. All of this information suggests that the tendency for Silurian and young-er Realm boundaries to parallel present-day latitudes militates against extensive north—south relative movement of major land masses *if* Realm boundaries are related at all to isotherms that themselves are parallel to latitudes. Thus one is left with the possibilities for east—west movements completely open insofar as biogeographic data is concerned. Much more work on Permo-Carboniferous biogeography is needed before such a con-clusion may be accepted for the pre-Triassic, not to mention the lack of adequate compilation for the Cambro-Ordovician relative to this question.

Hallam's *Atlas* makes clear that a warm-water Tethyan Realm is present from at least the Jurassic onward, and possibly from the Permian or even the Carboniferous. However, this evidence favoring climatic bipolarity is not present in the pre-Carboniferous.[4][5] On the contrary, during both the Silurian and Devonian the evidence favors a cool or cold Southern Hemi-sphere Malvinokaffric Realm having no Northern Hemisphere counterpart, only a Northern Hemisphere warmer region bounding the Malvinokaffric Realm. The situation for the Cambrian and Ordovician is not clear, al-though the Caradoc-Ashgill climatic picture is probably similar to that for the Silurian and Devonian.

Correlation by Means of Fossils and Precision

Once it is recognized that different groups of fossils are characterized by differing rates of evolution (morphologic, cladogenetic, taxonomic or phyletic) the problem of precision in stratigraphic correlation by means of fossils must be considered. It is clear that fossil groups undergoing rapid evolution will provide more precise tools for correlation than those under-going slow change. The thesis presented here is that historically, groups of fossils representing originally small populations will evolve more rapidly than those representing originally large populations. It is also the thesis of this treatment that the rapidly evolving small populations tend to be segregated from the large populations both as separate communities and as smaller biogeographic entities. If this is indeed the case it may be concluded that correlations within such smaller-population communities and smaller biogeographic entities will be far more precise than within larger-population communities and larger biogeographic entities. Actual experience with Silurian and Devonian fossils shows these conclusions to hold very well. For example, correlations based on the phyletically rapidly evolving species of *Eocoelia* and *Stricklandia* are far more precise than those based on the taxonomically slowly evolving *Pentamerus* and *Pentameroides* (Fig.18). Correlations between taxa belonging to communities occurring in rapidly evolving groups and those belonging to communities containing slowly evolving taxa will be correspondingly poor, except in boundary regions where a certain amount of mixing may occur. The same situation holds for biogeographic entities of differing size.

This information indicates that the fossil zones of one biogeographic region and another as well as between one community and another will commonly be of unequal absolute duration, even after environmental changes affecting time duration of a taxon in one place have been accounted for. Stehli and Wells (1971) have earlier arrived at the same conclusion employing hermatypic coral data. The conclusions presented here differ, however, in extending the work of Stehli and Wells from a contrast and comparison of low-latitude versus high-latitude corals to a broader base in which population size varies inversely with precision in correlation regardless of latitude. In part the conclusion adopted here is in agreement with the old rule of thumb that placed greatest weight for purposes of correlation on rare fossils, although it is here emphasized that "rarity" must be examined on a worldwide basis and not used as a mere local phenomenon controlled in one area by rapidly changing environment that may be totally absent in another region where the taxon in question is actually very common and widespread.

These remarks apply chiefly to benthic organisms. Boucot et al. (1974) consider the separate, although related question of correlation precision based on depth-zoned planktic fossils.

Geographic Speciation

Throughout this treatment runs the implicit thread of the overall importance during the past of geographic speciation as opposed to sympatric speciation. Although this may be, in part, the result of the greater difficulty of working with fossils (absence of soft parts, absence of color patterns on shells, and the like that might be of specific importance if known, leading to a potential lumping together of unlike species into polyphyletic species), the overall pattern inevitably emphasizes the importance of geographic speciation as the major mechanism. The correlation of high cladogenetic evolutionary rate (production of more than one descendent taxon from a single progenitor) with times of widespread provincialism and abundance of reefs (reefs may be thought of as scattered "microprovinces" in poor communication with each other except in very small regions) together with the widely increased number of communities possible under these conditions is strong circumstantial evidence in favor of geographic speciation. This information from the fossil record agrees well with that for the Recent summarized lately by Mayr (1970, and earlier publications).

Eldredge and Gould's (1972) "Punctuated Equilibria" and "Phyletic Gradualism" are merely alternate ways of phrasing the normal situation in which rapid evolution (their "Punctuated Equilibria") occurs in geographically isolated small populations (with consequent far higher probability of gaps in the fossil record) while slow evolution occurs in geographically widespread large populations (with small chance of gaps in the fossil record). They err a bit, in their enthusiastic, correct support of rapid, geographic speciation as a most vital mechanism, by trying to downplay the obvious fact that slow speciation co-occurs with rapid speciation. The term "Punctuated Equilibria" is a bit unfortunate as it may suggest to the unwary that "saltations" and "macroevolution" are supported by them and by the fossil record. There clearly is no need to appeal, in explaining the fossil record, to saltations or macroevolution as the significance of rapid evolution under conditions of isolation and small populations in terms of fossil preservation is entirely adequate as an explanation completely consistent with the mechanisms of change put forward by the biologist.

I have emphasized that the fossil record discussed in this book correlates very strongly with the concepts of geographic speciation. However, it should not be concluded from this that examples do not exist in which an appeal to some type of mechanism capable of producing sympatric speciation of a sort will not better explain the facts in a particular situation. Examples of closely related species occurring in the same deposit may be easily explained in terms of slight differences in niche (snails preferring attached algae as opposed to closely related species foraging on the bottom sediment beneath the algae; taxa reproducing or producing spat that set at slightly different times of the year, etc.). It is, of course, a moot point as to whether one should consider examples of this type to be strictly speaking sympatric. In addition, there are some examples which may be easily explained as resulting from the development of closely related species in isolation and then a later changing of distributional barriers in a manner permitting the taxa to occupy the same area where various ecological situations (activities of predators for example) may permit them to co-exist despite the fact that they utilize the same resources.

However, the type of sympatric evolution discussed by Kohn (1958, 1959), which also is similar to that occurring when level-bottom organisms evolve peripherally into reef organisms, which involves distinct, ecologically different environments present in the same locale probably has occurred many times during the past. In a way, this type of sympatric speciation may be considered as geographic but on a very small scale. Sympatric speciation in the sense of a single form dwelling in a specific environment giving rise to two or more taxa dwelling in that precise environment down to the smallest, significant detail is unsupported by the fossil record as of any importance.

Geographic speciation, completeness of the fossil record and rates of evolution

Since 1859 one of the most vexing properties of the fossil record has been its obvious imperfection. For the evolutionist this imperfection is most frustrating as it precludes any real possibility for mapping out the path of organic evolution owing to an infinity of "missing links". The fossil record is replete with evidence favoring organic evolution provided by short sequences of species with overlapping morphologies arranged in a clinal manner with time; the same is true for many sequences of genera and even for a fairish number of families. However, once above the family level it becomes very difficult in most instances to find any solid paleontological evidence for morphological intergrades between one suprafamilial taxon and another. This lack has been taken advantage of classically by the opponents of organic evolution as a major defect of the theory. In other words, the inability of the fossil record to produce the "missing links" has been taken as solid evidence for disbelieving the theory. Proponents of the theory have emphasized the gradational nature of the many specific, generic and familial samples in rebuttal, and appealed to the incompleteness of the fossil record. The problem of Precambrian "missing links" between phyla, classes and lower ranks has been dealt with by Rhodes (1966), who rightly points out that any developments occurring prior to the first occurrence of hard parts in a group will be reflected as gaps in the record with the majority of the major gaps occurring prior to the Cambrian.

Consideration of geographic speciation, the seemingly dominant mechanism by which most new forms have arisen, the likelihood that population sizes in rapidly evolving taxa undergoing speciation will be small, and the additional problems of the probabilities of preserving any organism in the fossil record provides one with another argument favoring the concept of organic evolution. The fossil record is strongly biased in favor of organisms characterized by large, geographically widespread populations. In other words, we should predict a good representation in the fossil record of slowly evolving forms but a very poor representation of rapidly evolving forms (as earlier pointed out with force by Mayr, 1963; see also Rhodes, 1966, p.33). This relation virtually guarantees that any group characterized during its history by any taxa subject to rapid evolution under conditions of endemism coupled with a later or earlier interval of wider distribution of antecedent or descendent slowly evolving related taxa will be also characterized by "missing links". In other words, the presence of significant gaps in the phylogenetic record implies a time interval of rapid evolution under conditions of small, geographically restricted population size. This is, admittedly, a conclusion based on negative evidence. But, such an argument is not an admission of defeat in the search for missing links, that others have concluded to have been nonexistent, but merely a recognition that the discovery of these transitional forms will have a far lower probability than will the discovery of the corresponding widely distributed, slowly evolving forms. This problem also points up the necessity of trying to obtain more complete geographic coverage in our study of fossils. The concentration on fossils collected in classic regions rich in well preserved specimens is understandable, and very necessary if we are to better understand problems of morphology, but the filling out of the phyletic record is in no small part accomplished as well by studying poorly preserved specimens obtained from outlandish regions of the world. In all candor it must be admitted that the discovery of new fossil occurrences in outlandish regions goes on at a slow, irregular pace with the scrappy, inadequate collections subjected in many cases to initially inadequate attention. Most such discoveries are made by field geologists concerned little, if at all, with the paleontologic significance of their work. Compounding the field man's attitude is that of the stratigraphic paleontologist assigned to provide routine "age assignments" to the field geologist (consider as well the "holier-than-thou" attitude taken by many paleobiologists towards the stratigraphic paleontologist doing this "routine" work on poorly preserved, inadequate collections and one has all the ingredients necessary for maintaining very slow increase in our knowledge of organic evolution). The study of the inadequate collections of poorly preserved fossils obtained from outlandish regions should be considered a rare opportunity by the paleobiologist rather than as an onorous chore better delegated to a stratigraphic paleontologist. There is nothing routine about the study of these fossils collected from outlandish regions as each collection commonly presents challenges morphologic, taxonomic, phylogenetic, ecologic and biogeographic more than adequate to test the mettle of the most experienced, sophisticated paleobiologist. Any consideration of current density of information concerning life of the past immediately shows that for most of the world removed too far from the shores of the North Atlantic our knowledge is very uneven (not that the circum-North Atlantic regions have reached a condition of perfect knowledge; the truth is far from that).

In summary, the presence of gaps in the fossil record is a necessity for the concept of organic evolution if we are willing to agree that the record of life is characterized by organisms possessed of widely differing rates of evolution closely related to size of interbreeding population, distributional area and a world in which the chances for preservation of organisms as fossils has always been low.

Living examples of the types we would predict to have occurred in the "gaps in the record" do occur (small populations, small chances for preservation as fossils, etc.). Examples are such organisms as *Sphenodon, Latimeria* and *Neopilina*. The absence of any fossil record for the interval between the Middle Devonian and the present for *Neopilina* and its ancestors, and correspondingly appropriate absences for *Sphenodon* and *Latimeria*, make the point emphatically that it is "cricket" to infer an interval of "cryptic" evolution to have taken place during the gaps in the fossil record of the known fossil organisms. Whether or not this type of inference may be applied to the Precambrian animals is still very uncertain as other mechanisms may be involved.

Must all small populations evolve rapidly?

Biologists have produced evidence indicating that in most instances evolution probably proceeds most rapidly under conditions of small population size. The conclusions reached by the biologist are supported by the circumstantial evidence of the fossil record. However, does it necessarily follow that *all* small populations must evolve rapidly? There are a certain number of situations where it is easy to infer relatively small population sizes and also nothing that one would be inclined to call very dramatic or rapid evolutionary changes. Fig. 29 suggests that some endemic brachiopod taxa have indeed persisted for as lengthy time intervals as have some of the longer-lived cosmopolitan taxa. This situation suggests the possibility that if the endemic taxa in question do represent smaller populations in time than the corresponding cosmopolitan taxa there must also be a possibility for a small percentage of the endemic taxa to evolve slowly despite the fact that the great majority evolve very rapidly. Of course, it may be that such cases of endemic taxa evolving slowly may really be correlated with differences in population density that cancel out the area effect. There is, as always, the possibility that the endemic taxa concluded to have been subject to slow evolution may in actual fact have been poorly analysed taxonomically; that in reality they represent far more than one taxon, that they do represent rapidly evolving taxa. However, consideration of the presence of "living fossils" and the probability that some highly endemic taxa (some of which are also "living fossils") do indeed persist for great time intervals relatively unchanged (Simpson's, 1944, bradytelic forms) indicates that although population size is adequate to explain the bulk of the evolutionary rate control problem and is considered here to be the first-order control, there must be other, lower-level controls that are very important in the control of evolutionary rates.

Van Valen (1973), employing a number of secondary sources including the *Treatise of Invertebrate Paleontology*, has summarized the time-duration (survivorship curves) data for a large number of animal groups as well as a few plant groups. His summaries show, as has been shown by earlier workers (Simpson, 1944) as well, that most taxa have a relatively short time duration, that a smaller number have a moderate time duration, and a few have a long time duration. Summary of the Silurian-Devonian brachiopod survivorship data (Fig. 29) indicates the same conclusion. However, analysis of the Silurian-Devonian brachiopod data shows conclusively that the great bulk of the taxa having a long time survivorship are cosmopolitan, with the great bulk of the short time-duration taxa being endemic. Consideration of Van Valen's summaries makes it very clear that the same interpretation may be applied to his data. For example, the overall shorter time duration of vertebrates, and especially of mammals, the shorter time duration of rudistid bivalves as contrasted with the bulk of the bivalves which are level-bottom types, correlates very well with endemic groups being characterized by shorter time durations than cosmopolitan groups. The very small number of long time-duration endemic types do, however, pose special questions that are very difficult to answer. Possibly some of Van Valen's hypotheses, although not applying as a first-order control to the bulk of the long-term duration taxa, may apply as explanations for the small groups of endemic, long time-duration taxa. In other words, rate of evolution correlates very well with population size for the great majority of organisms, but there is a small residue where this explanation may not apply. The effort to obtain a better understanding of the controls governing this small residue's long time duration is important.

Van Valen's (1973) graphical survivorship curves probably afford a very realistic summary of the real situation as regards longevity of relatively

short and medium-duration taxa. However, for the taxa that are characterized by long time duration, a relatively small percentage and number, the
situation is complicated by the presence in the geological record of intervals of major extinction. The major intervals of extinction (Permo-Triassic, terminal Cretaceous, Rhaetic, terminal Frasne, terminal Ashgill, etc.
events) would not affect the statistics of the short- and medium-duration
taxa significantly, but the case with the long-duration taxa may well be
different. In addition, Van Valen's approach assumes that waxing and
waning intervals of provincialism, transgression—regression changes,
changes in reef abundance (or actual absence), changes in food supply,
etc., all of which should have a profound effect on rates of cladogenetic
and phyletic evolution (particularly marked at the specific and generic
levels, progressively less marked in appropriate higher categories), have no
effect on rates of evolution (and extinction) in a statistical sense. So... it
would appear that Van Valen is summarizing overall trends that probably
mask a number of underlying rate changes and forces that may well modify the absolute duration of individual taxa.

Thus, Van Valen has concluded that the extinction (and evolutionary)
process goes on unaffected, as far as rate is concerned, by wide swings in
selection pressure, or at least that wide changes in selection pressure will
affect each group of organisms in the same manner. Van Valen also implies that various groups of organisms are characterized by extinction rates
of their own.

The effect of percentage of rare and endemic taxa as compared to
cosmopolitan and abundant taxa must be considered when trying to interpret the survivorship curves of the type employed by Van Valen (1973).
Time intervals of high cosmopolitanism (compare Table III, Llandovery
with Wenlock-Eifel units), relatively less common in the Phanerozoic record than are times of moderate to high provincialism, will provide a higher
percentage of cosmopolitan, slowly evolving taxa than of rare or endemic,
chiefly rapidly evolving taxa. Therefore, during a time interval of cosmopolitanism the survivorship curves plotted on a log scale as done by Van
Valen will depart very markedly from linearity. During a time of moderate
to high provincialism there will be a far higher number of rapidly evolving
taxa as compared with slowly evolving taxa. During both cosmopolitan
and provincial time intervals there will be a small number of very slowly
(bradytelic) evolving taxa. As cosmopolitan time intervals are in the
minority during the Phanerozoic it is clear that the "linear" feature of
Van Valen's compilations is in largest part a reflection of this fact. In
other words, if one sums up a limited number of normal distributions with
a large number of exponential distributions the resultant curve will have
an essentially exponential form. An additional component that significant-

ly increases the percentage of short time-duration taxa is the fact that there are far more taxa present during times of high provincialism (with most being short-lived!) than during times of high cosmopolitanism (with many being long-lived!). The cards are stacked when one summarizes the Phanerozoic record for the time duration of genera belonging to any group of organisms in favor of a very large number of short-lived genera. Until the survivorship curves are analysed for second- and lower-order components it will not be obvious that the entire record is not character-ized during the entire Phanerozoic by an exponential curve for survivor-ship.

PART II.
SUPPORTING PALEOECOLOGIC AND
BIOGEOGRAPHIC DATA

CHAPTER 5

Silurian—Devonian Community and Biogeographic History

Late Ordovician, Ashgill Background

Sheehan (1973a, p.152) has raised the logical question concerning whether or not there was a reorganization of community associations and contents prior to the Silurian. The climactic Late Ordovician, Late Ashgill or earliest Llandovery extinction event discussed at some length here further underlines the importance of Sheehan's question. I have not carefully studied, and am not very familiar with Ashgill communities. However, the general impression I possess is as follows. The Early Llandovery fauna is in largest part made up of taxa that persisted from the Ashgill (Boucot, 1968a). As far as the brachiopods are concerned the bulk of the taxa persisting into the earliest Llandovery appear to be closely related at the specific level to Ashgill, Old World Realm, European to Central Asian taxa. The associations of the Old World Realm Ashgill appear to occur in communities having much in common with those occurring in the earliest Llandovery. For example, the virgianid communities of the Ashgill (*Holorhynchus* and *Eoconchidium* Communities) differ in no regard from the virgianid communities of the Llandovery (*Virgiana* and *Platymerella* Communities) as they belong to the same Community Group. The *Dicaelosia-Skenidioides* Community of the Ashgill (for example, that reported by Potter and Boucot, 1971) appears to differ in no significant regard, except for the absence due to extinction in the equivalent Silurian unit of typical Ordovician genera like *Christiania* and *Anoptambonites*, from the same unit in the Llandovery. Admittedly, the Llandovery *Cryptothyrella* Community has not been recognized in the Ashgill. However, the overall difficulty in distinguishing Ashgill age Old World Realm faunas from Early Llandovery North Silurian Realm faunas in the absence of "key" genera like *Stricklandia* underlines the community similarities. It is no accident, from a community point of view, that stratigraphic units like the Ellis Bay Formation of Anticosti Island, Quebec pose difficult questions of age as the Ashgill fauna of the Old World Realm has so much in common with that of the Early Llandovery North Silurian Realm fauna.

Siluro-Devonian Communities

Silurian

North Silurian Realm, North Atlantic Late Silurian Region and Uralian-Cordilleran Late Silurian Region

Quiet-water communities. The Silurian quiet-water, high-diversity (10-20 brachiopod species) benthic marine communities (Fig.3-5) have been recognized primarily because of the common occurrence of articulated shells, chiefly brachiopods, associated with a fine-grained, marly matrix, that include a large proportion of very small specimens. Rarely are these shells in growth position. Units such as the Henryhouse Formation, Rochester Shale, Slite Group marls, Mulde Marl, Laurel Limestone, Hemse Group marls, Upper and Lower Visby Group marls, Elton Shales, Wenlock Shale (Buildwas Beds, for example), Waldron Shale, and Wenlock Limestone (level-bottom facies only) are so characterized. All of these units may be assigned to one of the taxonomically diverse, quiet-water communities shown on Fig.4. All of these units contain many minor faunal subdivisions that may eventually be termed community units of some type.

The inferred intertidal *Salopina* Community (Fig.4) shares many taxa with the inferred subtidal *Striispirifer* Community, although occurring landward of it. Because of this the *Salopina* Community is assigned a quiet-water position. The shells of the *Salopina* Community are commonly disarticulated and may occur in sand-size matrix as contrasted to the *Striispirifer* Community, but the greater chances for intertidal exposure to rough-water in Benthic Assemblage 2 after death makes this situation reasonable.

There will, of course, be both rough- and quiet-water environments on rocky bottoms, in the reef regions, and in hypersaline regions, although the great bulk of the quiet-water environments discussed in this book occur in normal salinity regions bottomed by fine-grained sediment. Ultimately it will be possible to discuss a climax concept for which the first shells to occupy a fine-grained sediment bottom will provide a substrate suitable for later arrivals unable to survive on a fine-grained sediment bottom.

A gradient of increasing taxonomic diversity is evident from the *Salopina* Community out to the *Dicaelosia-Skenidioides* Community (Fig.4) and then to equivalents within the undivided communities of Benthic Assemblage 5. Eventually it will be practical to provide a diagram showing the Benthic Assemblage spread or range of many of the taxa within these quiet-water communities. Watkins and Boucot (1973) present a table showing the Benthic Assemblage range of brachiopod genera in the Upper Llandovery of the Welsh Borderland. For example, *Salopina* is restricted to Benthic Assemblage 2, but the associated *Howellella* certainly extends into the *Dicaelosia-Skenidioides* Community, i.e., a Benthic Assemblage 4 position. *Dicaelosia* and *Skenidioides* occur in Benthic Assemblages 4 and 5 quiet-water positions. *Clorinda* does not occur in abundance above Benthic Assemblage 5. A diagram of the data in Part II, Chapter 6 would give an approximation of the Benthic Assemblage spread referred to above.

A second type of quiet-water benthic marine community (Fig.3, 4) is represented by the *Protathyris, Didymothyris, Atrypella, Dayia, Gracia-nella,* and *Dubaria* communities. These widespread, low-diversity, single brachiopod-genus communities of numerically abundant spire-bearers commonly occur as articulated shells in a shaly matrix. *Atrypella* Community has been observed in life position (Smith, 1972). These communities commonly show a size-frequency distribution rich in small specimens with few large specimens, further emphasizing their quiet-water habitat. There clearly must be an ecologic factor or factors correlating strongly with quiet-water, to account for the low diversity of such communities, but at present we have no good understanding of just what is involved. They are assigned to Benthic Assemblages, despite the absence of "key" genera in the same matter as described above for the "bracketing" of the *Striispirifer* Community that also lacks "key" genera.

One possibility (Boucot, 1963b) is an environment in which a dense growth of marine plants acted to exclude most taxa of brachiopod except for the few forms able to attach to the plants themselves (*Terebratulina septentrionalis* of the present attaches to *Zostera*). The importance of bottom-attached algae in the quiet-water environment is uncertain.

Paine (1971) describes modern examples in the intertidal region where dense growths of algae effectively exclude benthic organisms (removal permitted adjoining benthic shells to rapidly invade the area formerly occupied by the algae). Shells living attached to such algae would, after death, probably make up a randomly oriented group of articulated shells. Shells dwelling on the bottom below such algae would be more likely to form a death assemblage preserved with its umbones directed downward. Smith (1972) describes examples of *Atrypella* Community (low diversity, quiet water) with the umbones directed downward in those examples that retain a high ratio of articulated shells. It is possible that some of the patchy distribution of communities observed is a result of patchy occur-rences of algae that caused the occurrence of low-diversity communities or the absence of shells. It is clear that dense growths of algae tend to provide a restrictive environment, and to lower the normal faunal diversity expected in a Benthic Assemblage. Such an explanation pertains, however, only in the photic zone.

The trace fossil communities of the Silurian appear to agree in all regards with the information outlined by Seilacher (1967). For example, Greiner (1973) cites the presence of *Costistricklandia* Community brachiopods associated with *Zoophycos* Community in northern New Brunswick, i.e., a Benthic Assemblage 4 position suggested for *Zoophycos*.

Rough-water communities. The rough-water marine benthic communities (Fig.3-5) are recognized chiefly within Benthic Assemblages 3 and 4. Their absence from Benthic Assemblage 5 may be due to water depth being effectively below the rough-water environment. Failure to recognize rough-water environment communities in Benthic Assemblage 2 probably results from our present inability to decipher the significance of the *Eocoelia* and *Cryptothyrella* communities (Fig.3, 4). Certain rough-water communities contain heavy-shelled pentameroids, others contain heavy-shelled trimerelloids, and taxonomically have low diversity. Many of the rough-water Benthic Assemblage 3 communities occur both adjacent to reefs (Ingels, 1963) in a rough-water environment and in rough-water environments quite separate from reefs.

The reconstructed illustrations of Ziegler et al. (1968, fig.4, 6, 8) for the *Eocoelia*, *Pentamerus*, and *Stricklandia* "communities" in living position are at variance with the concept of quiet-water and rough-water discussed above (they show genera "living together" that I conclude did not live together). However, it must be recalled that these reconstructions are based almost entirely on Welsh Borderland material, where as I have pointed out previously (Boucot, 1970) the steep bottom slope and close spacing of community boundaries permits far more mixing of adjacent communities than would occur in a region of low slope, as well as the possibility for small-scale patchy original distributions. It is important to note the non-correspondence between the reconstructions of Ziegler et al.'s (1968) *Pentamerus* and *Stricklandia* "communities" with the photographs of Ziegler et al. (1966) showing actual *Pentamerus* and *Stricklandia* communities in life position. Ziegler et al.'s (1968) reconstructions are mixtures of taxa that did not live together in the same biotope (for example, we have no specimens of *Pentamerus* or *Stricklandia* in life position associated with abundant shells of other genera), although they may have been members of immediately adjacent or successive communities (see Fig.15).

Rocky-bottom communities ("hard" substrate communities). Rubel (1970, p.78) has introduced the term *Linoporella* Community (Fig.3) for a Llandovery fauna whose physical situation strongly suggests that the community lived in both a rocky-bottom and level-bottom location.

Eurypterid communities. In the Silurian the Hughmilleriidae-Stylonuridae Community of Kjellesvig-Waering (1961) and Caster and Kjellesvig-Waering (1964, fig.4 and 5) occurs landward of Benthic Assemblage 1. The Hughmilleriidae-Stylonuridae Community (of specific eurypterid genera listed by Kjellesvig-Waering, 1961, and Caster and Kjellesvig-Waering, 1964) inhabited brackish water rather than the nonmarine environment. Although acritarchs consistent with brackish water are present (dominant- ly Sphaeromorphitae), it is clear from the absence of articulate and inar- ticulate brachiopods, trilobites, corals, echinoderms, and other marine forms that this community does not reflect an open marine environment. It is well developed in the Silurian part of the British Downtonian, in the Wenlock-Ludlow part of the Oslo "Downtonian", in the Shawangunk For- mation of eastern New York, in the Wills Creek Formation of West Vir- ginia (this locality is near the boundary with a Benthic Assemblage 1 fauna), and has recently turned up in the Bolivian Late Silurian (Kjellesvig- Waering, 1973). Cyathaspid and anaspid vertebrates occur as far landward as this eurypterid community, as well as seaward into at least Benthic Assemblage 3.

Kjellesvig-Waering's (1961) Eurypteridae Community occurring in platy limestone ("waterlime") can best be interpreted (Fig.4, 5, this book) as representing a quiet-water hypersaline environment (North American oc- currences are related to evaporites). The community occurs in depths ranging from Benthic Assemblage 1 through 2 and probably into 3. Not- able occurrences are in New York and Ontario (Bertie Group), Estonia (Rootsikula and Paadla Horizons), and Podolia (Ustya Horizon).

The Carsinosomidae-Pterygotidae Community of Kjellesvig-Waering (1961) occurs in the normal marine level-bottom communities from Benthic Assemblage 1 through 3 with ordinary benthic marine shells and plankton. Therefore this last eurypterid community is not used as a sepa- rate entity for large-scale mapping purposes despite its recognition by Kjellesvig-Waering as one of the three major Silurian eurypterid associa- tions. The Carsinosomidae-Pterygotidae Community eurypterids, occur- ring together with benthos belonging to a variety of communities may represent high trophic-level carnivores associated with or even interacting with a variety of benthic communities in a manner similar to that com- monly observed in the modern marine environment for nektic predators (some cephalopods and fishes).

Malvinokaffric Realm

Little study has been devoted to Malvinokaffric Realm Silurian communities. The *Clarkeia, Heterorthella* and *Amosina* Communities (Fig.6) appear to occur in about Benthic Assemblage 2 position. The absence of monograptids from the *Clarkeia* and *Amosina* Communities is consistent with most Benthic Assemblage 2 occurrences. The *Australina* Community (Fig.6) is taxonomically more diverse than the *Clarkeia, Harringtonina* or *Amosina* Communities; the presence of monograptids interlayered with the *Australina* Community is consistent with a Benthic Assemblage 2-3 position (*"Eocoelia" paraguayensis*, an Early Llandovery possible precursor of *Harringtonina* occurs with climacograptids). The overall limited diversity (never as many as 10 brachiopod taxa, commonly 1, 2, or 3) of Malvinokaffric Silurian shelly faunas may also be related to their inferred cold-water environment that in turn may be responsible for the absence of reef environments as well as limestones in general (Berry and Boucot, 1972a). The Hughmilleridae-Stylonuridae Community has just been recognized in Bolivia (Kjellesvig-Waering, 1973).

Mongolo-Okhotsk Region. Little is known ecologically about the *Tuvaella* Community (Fig.4) of the Mongolo-Okhotsk Region except that it is low-diversity (1—5 brachiopod taxa) and probably occurs in Benthic Assemblage 2 position.

Devonian

Eastern Americas Realm Early Devonian level-bottom communities
Fig.7 is an attempt to subdivide the Appohimchi Subprovince Early Devonian communities (discussion of the communities present in the Amazon-Colombian Subprovince of the Eastern Americas Realm is not attempted owing to lack of information; however, there is no evidence that they differ significantly from coeval Appohimchi Subprovince communities known in North America). Many of the stratigraphic subdivisions within the Helderberg appear to contain faunas belonging to more than one community. Unfortunately, the available faunal lists have probably lumped the members of different communities in a manner that obscures and overgeneralizes the real relations. Within Benthic Assemblage 1 no subdivisions are recognized, with the possible exception of the *Globithyris* Community. The *Globithyris* Community may belong to a very shallow part of Benthic Assemblage 2 and/or to a deep portion of Benthic Assemblage 1. It represents a very specialized environment of uncertain nature (Boucot, 1963b). The presence of *Salopina* in both the *Mutationella* I and *Chonostrophiella* Communities, as well as their position relative to other communities, indicates a Benthic Assemblage 2 position. The *Globithyris* Community is a low-diversity, quiet-water entity, the *Mutationella* I and

Chonostrophiella Communities are high-diversity, probably quiet-water. The Tentaculitid Community contains *Howellella* and *Tentaculites* in abundance; the absence of *Howellella* from Benthic Assemblage 1 any-where in the world indicates that the low-diversity, quiet-water Tentacu-litid Community also can best be placed in Benthic Assemblage 2. Follow-ing Anderson (1971), it represents a very specialized environment, consis-tent with its relatively low diversity. The *Hipparionyx* Community (Big Shell Community of Boucot and Johnson, 1967a) has medium diversity, and is interpreted to have occurred in rough-water, based on the disarticu-lated and relatively large, thick-shelled fossils. The *Hipparionyx* Commu-nity is also placed in Benthic Assemblage 2 for paleogeographic reasons, i.e., deeper-offshore and shallower nearer-shore units adjoin in one region or another. The Gypidulinid Communities are similar in all their characteris-tics to the same units in the Silurian, and for the same reasons.

The deeper-water Appohimchi Subprovince Early Devonian communi-ties (Fig.7) present the following problems. When the Gypidulinid Com-munity elements sometimes lumped with it are dissociated from the *Cyr-tina* Community it is a high-diversity unit sandwiched between the Benthic Assemblage 2 *Tentaculites* Community, and the deeper-water *Dicaelosia*-bearing *Dicaelosia-Hedeina* Community. The *Cyrtina* Commu-nity is interpreted as a quiet-water unit. The *Dicaelosia-Hedeina* Commu-nity probably includes several subdivisions, but present information is inadequate for further splitting. It is considered here to be a high-diversity, quiet-water entity. The inclusion of gypidulinids in many places with *Dicaelosia-Hedeina* Community faunal lists may indicate that the range of the Gypidulinid Communities overlaps somewhat, or more likely that col-lecting has not been sufficiently thorough in the past to permit their discrimination. The *Beachia*, *Amphigenia* and *Costellirostra* communities are interpreted as quiet-water, high-diversity units. The inability to subdi-vide them reflects a greater depth range for their taxa than is characteristic of many other communities. I have not observed significant taxic segrega-tions within the area of these three communities. The trace fossil *Taonu-rus* (*Zoophycos*) Community is a quiet-water Benthic Assemblage 3 through 5 unit, lacking shells, that must represent specialized conditions not understood at present (see Chapter 6, and Seilacher, 1967).

The assignment of the *Beachia* Community (Fig.7) to a Benthic Assem-blage 3 through 5 span is based on both geologic and ecologic reasons. First, the *Beachia* Community of Oriskany age in Maine, and the underly-ing *Dicaelosia-Hedeina* Community of the Beck Pond Limestone, inter-tongue with the relatively barren, deep-water, flysch-type strata of the Seboomook Formation (Boucot and Johnson, 1967a). Flysch-type sedi-ments found to date within the Silurian coincide in most cases with the

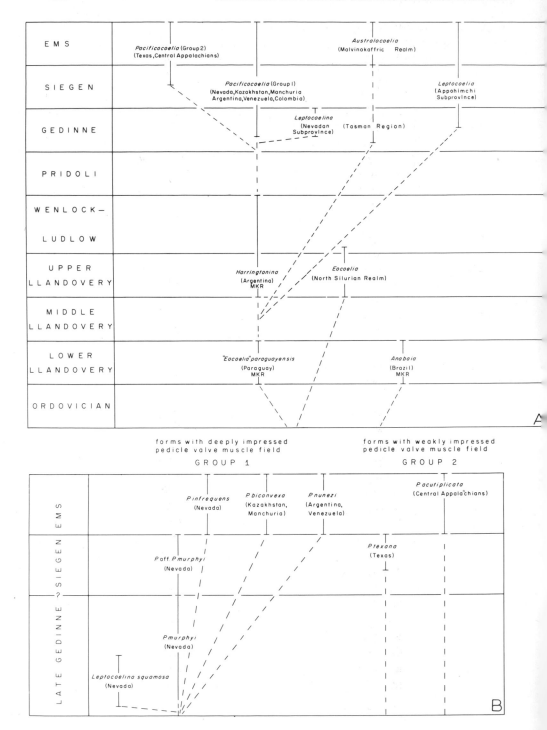

Benthic Assemblage 5-6 position or deeper; there is no reason to assign the flysch facies in the Devonian to a shallower position. Second, the partly coeval *Dicaelosia-Hedeina* Community fauna of the Beck Pond Limestone occurring adjacent to flysch-type strata belongs somewhere in the Benthic Assemblage 4-5 position (Fig.7). Third, the presence of Starfish Community beds adjacent to the *Beachia* Community (Boucot and Yochelson, 1966) in one area indicates a deep lower range for the *Beachia* Community. The alternative of placing the *Beachia* Community entirely in Benthic Assemblage 2 is not justified by the geologic or ecologic evidence presently available. If the *Beachia* Community were assigned entirely to Benthic Assemblage 2 the resulting absence of shelly megafossils from Benthic Assemblages 3 through 5 could not be explained. This question eventually might be further resolved by means of planktic depth-zoned microfossils like conodonts or acritarchs. A similar problem is faced in both the Malvinokaffric Realm *Australospirifer* Community (Fig.9) and the deeper-water Tasman Region shelly communities (Fig.10) as well as the "Normal" Community of the Rhenish Complex of Communities (Fig.8) which lack much continuity from the Silurian. I conclude that the ancestral Silurian leptocoelid genus *Harringtonina* belongs to a community in Benthic Assemblage 2-3 as shown by the associated climacograptids in Paraguay with "*Eocoelia*" *paraguayensis*. However, the time elapsed between the Early Llandovery-Ludlow interval and the Gedinne makes extrapolation of these depth data to the range of any Devonian leptocoelid uncertain (*Leptocoelia* of the Appohimchi Subprovince; *Pacificocoelia* of the Eastern Americas Realm and a few regions outside; *Australocoelia* of the Malvinokaffric Realm). *Leptocoelia*, characteristic and abundant in the *Beachia* Community, is concluded to range from Benthic Assemblage 2 through 5, *Pacificocoelia* to occur in Benthic Assemblages 3 through 5, and *Australocoelia* from Benthic Assemblage 2 through 5. The more distantly related (Fig.34A) Silurian genus *Eocoelia*, occurring in abundance in Benthic Assemblage 2 (Fig.4) has no bearing on the question.

Fig. 34A. Family tree for the genera of the Leptocoelidae (Brachiopoda). Note the far greater proliferation of genera/unit of geologic time during the Early Devonian (smaller areas occupied by all during this provincial time interval) as contrasted to the entire Silurian *(Eocoelia* very widespread during Late Llandovery - Early Wenlock — with a rapid rate of speciation due to small size of interbreeding population). The size of the areas occupied by both *Harringtonina* and *Anabaia* are poorly known, but the relatively low number of Silurian genera (3) contrasts markedly with that for the Early Devonian alone (5 major, generic level groups, one still un-named). *Pacificocoelia* (Group 2) upon further study may turn out to contain at least two generic level taxa. Note the significant gaps in both time and space in their distribution that imply possible migration routes in both time and space.

B. Family tree for the Early Devonian leptocoelid species. Compare the species proliferation during this time of provincialism with that for *Eocoelia* (Fig. 34A) during a time of cosmopolitanism.

The significance of *Zoophycos* in settling this problem is high as this trace fossil is well documented from the Benthic Assemblages more offshore than 2 in the Silurian (Greiner, 1973, for example) which makes it unlikely that we should restrict the Devonian range. Therefore, I conclude that the *Beachia* and the derived *Amphigenia* Communities belong in the Benthic Assemblage 3-5 span.

The *Plicoplasia* Community appears to be merely a quiet-water, high-diversity, slightly deeper-water extension of the *Hipparionyx* Community characterized by smaller shells. The Eurypteridae Community is a quiet-water, low-diversity, hypersaline unit occurring in Benthic Assemblage 2 of the Michigan Basin.

Old World Realm Early Devonian level-bottom communities

Rhenish-Bohemian Region. In the Early Devonian rocks of the Appohimchi Subprovince there is no correlation between the brachiopod fauna and a particular rock type (e.g., carbonate versus terrigenous). Boucot et al. (1969, p.5) therefore concluded that the common association of Rhenish faunas with terrigenous matrix and Bohemian faunas with carbonate matrix could not be explained by the lithofacies differences (the classical Hercynian Problem) but was depth-related (the Bohemian faunas having lived in deeper water farther from shore). The occurrence of *Meganteris*, *Euryspirifer*, and other Late Ems Rhenish genera in carbonate rocks in Czechoslovakia, northern Spain and North Africa is evidence that Rhenish faunas are not completely restricted to terrigenous strata (Boucot et al., 1969). It is now concluded that the controlling factor in the distribution of Rhenish and Bohemian faunas was not depth overall but the distribution of quiet level-bottom and quiet reef or reef-related environments operating over bottoms with differing physical properties unrelated to mineralogy (texture, sorting).

The Early Devonian of the circum-Mediterranean region (Iberian Peninsula, Istanbul region, North Africa) contains both Rhenish and Bohemian Complexes of Communities (Fig.8) (Boucot et al., 1969). These distinct complexes represent differing environments rather than the presence of faunal barriers leading to provincialism. An additional example of Rhenish fauna occurring in both calcareous and terrigenous matrix is the Benthic Assemblage 2 quiet-water *Quadrifarius* Community (Fig.5) in part of the calcareous Skala Horizon of Podolia, and in terrigenous units of the Pridoli (Stonehouse Formation of Nova Scotia, Pembroke Formation of southern Maine and New Brunswick) and the Gedinne (Gres de Gdoumont of the Stavelot region in the eastern Ardenne and Eifel). The *Quadrifarius* Community is concluded to have lived in quiet water (Fig.8).

Boucot et al. (1969) referred to a "Rhenish Community" and a "Bohemian Community" in discussing the complex and rich faunas of both the Schiefergebirge and the Prague region. Boucot (1970, p.605) modified this nomenclature by referring to a "Normal" [46] Community for the Rhenish faunas, thought to occur in what is here termed Benthic Assemblage 3 through 5 position, while recognizing that it contained more than one community in the common sense. He noted that the Bohemian Community of Boucot et al. (1969) was probably a complex of reef and reef-related faunas, but did not coin a separate term for the complex. These confusing usages are here replaced by "Rhenish Complex of Communities" and "Bohemian Complex of Communities".

Benthic Assemblages 1 and 2 are represented by several communities within the Rhenish-Bohemian Region. Present data are inadequate to untangle them. Perhaps the Rhenish Complex of Communities as presently understood is adapted to a muddy bottom (clay and silt-size material), as opposed to the Bohemian Complex of Communities adapted to a coarser, cleaner bottom (sand-size material), but this point is unclear, and I now advocate a quiet-water level-bottom as opposed to a quiet-water reef or reef-related environment because of the shell articulation data. The reef-related nature of some Bohemian Complex of community occurrences (that at Zlaty Kun near Prague in the Konieprus Limestone, for example) may afford quiet-water habitats preventing post-mortem shell disarticulation. Many of the Bohemian faunules contain a high percentage of articulated brachiopods as contrasted with the uniformly disarticulated condition of most Rhenish Complex of Communities brachiopods. Fig.8 indicates the inferred Community relations within the Rhenish-Bohemian Region for the Early Devonian. The Starfish Community of Benthic Assemblage 6 is well represented in the Rhineland (Fig.8) by the famous Bundenbach starfish-bearing slates of the Hunsruck Region.

Fuchs (1971; written communication, 1972) has provided a remarkable analysis of both sedimentary petrography and animal community data for the Late Siegen and Early Ems of the eastern Eifel. He concludes that in addition to the importance of rough and quiet water one must also consider the effect of fine-grained detritus as opposed to coarse-grained detritus as well as the effects of differing depth. He has been able to show the presence of distinctive associations of pelecypods as well as brachiopods and other groups in his communities. Fuchs (1971, fig.9) has summarized the community information departing from the shoreline as follows:

(1) a region containing red beds including abundant plant remains plus vertebrate remains and rare mutationellinid brachiopods (shoreward extension of the *Mutationella* II Community);

(2) rare red beds and a fauna characterized by abundant mutationellinid brachiopods, ostracodes and plant remains plus certain pelecypods corresponding to the *Mutationella* II Community;

(3) red beds absent with a fauna including abundant chonetids, *Trigonirhynchia* and *Subcuspidella* plus tentaculitids and pelecypods that corresponds to the Chonetid Community;

(4) first appearance of acrospiriferids with abundant homolanotids, gastropods, tentaculitids, pelecypods, and *Platyorthis* plus chonetids, *Trigonirhynchia* and *Rhenorensselaeria*, which may correspond to a mixture of the *Rhenorensselaeria* Community, Chonetid Community and the inward margin of the Rhenish Complex of Communities;

(5) maximum abundance of acrospiriferids, chonetids, stropheodontids and some pelecypods that probably belongs to the Rhenish Complex of Communities;

(6) finally, a fauna with rarer brachiopods, varied pelecypods, and first orthoceroids that probably corresponds to the outer margins of the Rhenish Complex of Communities.

Fuch's work is clearly the first directed at a careful splitting up and understanding of Rhenish benthic communities; as it progresses we may expect a far better understanding of the real relationships. The *Tropidoleptus* Community probably occurs in Benthic Assemblage 2 here as it is interpreted to occur elsewhere (Malvinokaffric and Eastern Americas Realms).

The Rhenish-Bohemian Region is unique during the Lower Devonian for the wide extent of its black-shale facies Pelagic Community rich in graptolites. This development is known from Thuringia on the north—southerly into Algeria. This situation makes clear that the planktic larvae of the shelly facies organisms were able to cross wide expanses of water lacking suitable bottoms in order to explain the many patchy occurrences of Benthic Assemblages 1 through 5 that occur in North Africa and Europe surrounded on all sides by Pelagic Community bearing rocks.

Uralian Region. Within the Uralian Region of the Old World Realm there is a community subdivision for the Early Devonian parallel to that within the Rhenish-Bohemian Region (it should be similar to that shown in Fig. 8). The carbonate rocks of the Carnic Alps, the Urals, various parts of both Chinese and Russian Turkestan and the Altai-Sayan contain faunas that contrast markedly with and virtually surround those of the coeval terrigenous and volcanogenic rocks of southeastern Kazakhstan and portions of the Altai. The similarity to the Rhenish-Bohemian complexes of communities is very marked among the spiriferids; strongly plicate forms occur within the terrigenous rocks and relatively smooth forms are abundant in the carbonate rocks in both regions. The ammonoids and trilobites show parallel occurrences. These distinctions are formalized here by recognizing a Uralian Complex of Communities and a Dzhungaro-Balk-hash Complex of Communities for the Early Devonian of the Uralian Region (Fig.2). The Dzhungaro-Balkhash Complex of Communities appears to be equivalent in sense to the Dzhungaro-Balkhash Province as used, for example, by Maksimova et al., 1972; it is clear that Soviet paleontologists conceive of this faunal unit as a separate biogeographic entity, whereas I conceive of it as merely an ecologic entity belonging to the earlier named Uralian Region established originally to include both the Uralian Complex of Communities and the Dzhungaro-Balkhash Complex of Communities. Within each of these complexes of communities, comparable in definition and scope with the Rhenish and Bohemian Complexes of Communities, numerous associations of fossils could properly be termed communities in the sense of Berry and Boucot (1971, 1972b). For example, within the Uralian Complex of Communities a quiet-water, low-diversity *Karpinskia* Community can be recognized complete with abundant articulated shells and abundant small specimens. As with the Rhenish and Bohemian Complexes of Communities it is concluded that the Uralian Region complexes of communities faunal differences are based on quiet-water level-bottom as opposed to quiet-water reef or reef-related conditions. The Uralian Complex of Communities brachiopods commonly are present in an articulated condition as opposed to the almost invariably disarticulated condition of the Dzhungaro-Balkhash Complex of Communities shells. The presence of reefs from place to place in both the Bohemian and Uralian Complexes of Communities environs favors relatively shallow-water conditions, i.e., the cause for the differences with the Rhenish and Dzhungaro-Balkhash Complexes of Communities probably is not depth. The Dzhungaro-Balkhash Complex of Communities contains certain disjunct Eastern Americas Realm taxa, but a high percentage of provincial taxa will ultimately be described (this statement is based on my inspection of extensive collections).

Tasman Region. Information about benthic level-bottom communities belonging to the Tasman Region (Fig.10) is still very preliminary. It was earlier suggested (Boucot, 1970, p.610) that the *Notoconchidium* Community in the Tasman Region Early Devonian is a Benthic Assemblage 2 community. The disarticulated condition of most collections from clean-washed sandstones (Florence Quartzite, for example, with some pebbles) suggests a turbulent depositional environment for these large aggregations of *Notoconchidium*. Similar occurrences in New Zealand make a mere coincidence with a mature sand source area less likely. The "Normal"

Community of Boucot (1970, p.610) is subdivided into four communities (Fig.10). The first, termed the *Spinella-Buchanathyris* Community, is characterized by a taxonomically low-diversity fauna occurring commonly in fine-grained sediment as disarticulated shells for the most part. Because of its commonly disarticulated condition this first community is assigned a Benthic Assemblage 2 position. The second, termed the *Quadrithyris* Community, is characterized by a far higher diversity, and commonly includes abundant articulated shells consistent with a quiet-water habitat. The third, termed the *Maoristrophia* Community (it does not, however, include the *Maoristrophia* occurrences of the New Zealand Region Reefton Beds), is characterized by high diversity. The fourth, termed the *Notanoplia* Community, is characterized by a low-diversity assemblage. The *Notanoplia* Community commonly occurs in fine-grained mudstones, shells are commonly articulated, and the paleogeographic location is commonly seaward of the *Maoristrophia* Community; i.e., the *Notanoplia* Community probably represents deeper and quieter-water conditions. Savage's (1970, pp.656-657; oral communication, 1972) report of *Notanoplia* Community in the lower part of the Maradana Shale, overlain by *Maoristrophia* Community in the upper part of the Maradana Shale, overlain by *Quadrithyris* Community in the lower limestone of the Mandagery Park Formation, overlain by possible shallow-water or even non-marine, coarse, cross-bedded sandstone in the unfossiliferous upper part of the Mandagery Park Formation (the *Notoconchidium* Community is missing) is consistent with the above interpretation of an upward shallowing, quiet-water environment. The presence of *Skenidioides* and *Dicaelosia* in the *Maoristrophia* Community is consistent with a Benthic Assemblage 4-5 assignment, as is the absence of these genera from the *Quadrithyris* Community. Interbedded Starfish and *Maoristrophia* Communities in the Clonbinane District, Victoria (J.A.Talent, written communication, 1973) argue for this placement. The deeper parts of this Region are characterized by graptolites (Pelagic Community) and also the Starfish Community of Benthic Assemblage 6.

New Zealand Region. The New Zealand Region is based on fossils derived from the structurally complex Reefton-area Reefton Group. A number of collections have been obtained from structurally isolated exposures and also from loose blocks. Consideration of the faunas permits dividing up the unlike faunas into communities. The condition of the shells and the associated lithologies, plus information about the behaviour of similar shells in other biogeographic units permits a number of conclusions to be reached about environment of growth and environment of deposition. *Mutationella* Community III is represented by abundant *Pleurothyrella* occurring in an articulated condition associated, however, with tetracorals in limestone. The association of *Pleurothyrella* with corals is distinctly different from the occurrences of the genus in the Malvinokaffric Realm. Numerous *Pleurothyrella* in a disarticulated condition are also found in a coarse-grained quartzite associated with a much smaller percentage of disarticulated *Howellella* and "*Cryptonella*", in a death assemblage that has undergone a certain amount of transportation. The New Zealand Region *Pleurothyrella* occur in *Mutationella* Community III, but in a somewhat different environment favorable to tetracorals. The occurrence of disarticulated *Pleurothyrella* in a coarse-grained sand is consistent with a shallow-water, Benthic Assemblage 2 position. The varied brachiopod fauna characterizing the other Reefton Group collections is by no means randomly distributed. However, since the number of studied samples is still too small to permit dividing these other collections into definite communities they are lumped for the time being into an undifferentiated quiet-water level-bottom Reefton Complex of Communities. The abundant large spiriferids, chonetids, dalmanellids, stropheodontids, terebratuloids and rhynchonellids form a very diverse assemblage, occurring for the most part in mudstone.

Community assignments for Lower Devonian brachiopods in Nevada: Nevadan Subprovince of the Eastern Americas Realm and associated Old World Realm

Information in Johnson (1970a) on the Early Devonian brachiopods of central Nevada allows some tentative community assignments. It should be recognized that for those fossil zones for which collections were obtained from a wide area, as opposed to those from which collections were obtained from only a restricted area, there is the probability that the combined faunal lists for the fossil zones (lumping all localities together) contain elements present in more than one community and elements restricted to only certain communities, with the faunal lists themselves giving no inkling of the ecologic relations. [47] In addition, Table IV lists over thirty genera in both the *Quadrithyris* and *Eurekaspirifer pinyonensis* zones, a situation strongly suggesting that taxa belonging to more than one community are present (within the Silurian-Devonian, diverse brachiopod communities seldom contain over twenty genera). The faunas of *Quadrithyris* through *Eurekaspirifer pinyonensis* zones (Table IV, data taken from Johnson, 1970a, table 9) show an overall general similarity. The interlayering of faunal entities representing differing biogeographic entities (Old World Realm undifferentiated, Nevadan Subprovince of the Eastern Americas Realm, Cordilleran Region of the Old World Realm, Fig.25) makes community diagnosis more tenuous than might be wished,

TABLE IV

Occurrence of brachiopod genera in Johnson's (1970) *Quadrithyris, Spinoplasia, Trematospira,* "*Spirifer*" *kobehana* and *Eurekaspirifer* zones. The first two zones are Gedinne, the third Siegen, the last two Ems. Data taken from Johnson (1970a, table 9). * Indicates correction or change.

	Q	S	T	K	P
Acrospirifer		x	x	x	
Aesopomum	x				
Ambocoelia	x				x
"*Anastrophia*" = *Grayina*	x				
Ancillotoechia			x		
Anoplia		x	x	x	x
Astutorhyncha					x
Atrypa	x	x		x	x
Atrypina	x				
Bifida					x
"*Brachyprion*"	x				
Brachyspirifer					x
"*Camarotoechia*"	x				
"*Camerella*" = *Llanoella*	?	x			
"*Chonetes*"				x	x
Coelospira	x	x	x	x	x
Cortezorthis					x
Corvinopugnax					x
Costispirifer			x		
Cymostrophia					x
Cyrtina	x	x	x	x	x
Cyrtinaella	x				
Dalejina	x	x	x	x	
Discomyorthis			x		
Dolerorthis	x				
Dubaria	x				
Dyticospirifer		*	x		
Elytha			x		
Elythina					x
Eurekaspirifer					x
Gypidula	x			x	x
Howellella	x	x			x
Hysterolites					x
Katunia				x	
"*Leiorhynchus*"					x
Leptaena	x	x	x	x	x
Leptaenisca	x	x			
"*Leptocoelia*" = *Pacificocoelia*		x	x	x	x
Leptocoelina		x			
Leptostrophia	x	x	x	x	x

TABLE IV (continued)

	Q	S	T	K	P
Levenea	x	x	x	?	x
Lissatrypa	x				
Machaeraria	x				
McLearnites				x	x
Megakozlowskiella	x	x	x	x	
Megastrophia		x	x	x	x
Meristella		x	x	x	
Meristina	x				x
Mesodouvillina	x				
Metaplasia			x	x	
Muriferella					x
Mutationella?					x
Nucleospira	x	x	x		x
Orthostrophella		x			
Parachonetes				x	x
Pegmarhynchia		x			
Pholidostrophia		?	?	x	
Phragmostrophia				?	x
Pleiopleurina		x	x		
Plicoplasia		x			
Protocortezorthis	x				
Pseudoparazyga		x	x		
Quadrithyris	x				
Rensselaeria			*	x	
Rensselaerina?		x			
Reticulariopsis			x	cf	
Reticulariopsis?	x				
Rhipidomella					x
"Salopina"				x	
Schizophoria	x	x			x
"Schuchertella"	x	x	x	x	x
Sieberella	x	cf	x		
Spinatrypa	x				
Spinella					x
Spinoplasia		x			
Spinulicosta?					x
Spirigerina	x				
Stropheodonta		x	x	x	x
"Strophochonetes"		x	x		x
Strophonella	x	?	x	x	x
Toquimaella	x				
Trematospira		x	x		
Trigonirhynchia					x

but the overall similarities allow some conclusions. The general aspect of the *Quadrithyris* Zone fauna is like that of Savage's (1968, 1969, 1970, 1971, and Part II, Chapter 6) *Quadrithyris* Community (see Fig.7, 10) and is also close to that of the Appohimchi Subprovince Benthic Assemblage 3 *Cyrtina* Community despite the many biogeographically induced taxonomic differences between their faunas. Possibly the gypidulids of the *Quadrithyris* Zone fauna are derived from a nearby Benthic Assemblage 3 Gypidulinid Community; the *Dubaria* fauna belongs to the quiet-water *Dubaria* Community. Physical mixing and transportation may have resulted in the amalgamation of at least several communities, but the few *Quadrithyris* Zone faunas show a parallelism with the *Cyrtina* Community (compare Table IV and Part II, Chapter 6 for the common genera), plus *Dubaria* and possibly *Gypidula* Communities, and are assigned to Savage's *Quadrithyris* Community.

The *Spinoplasia* Zone fauna is a generalized "Helderberg" fauna resembling the *Cyrtina* Community. In many ways it is similar to the *Plicoplasia* Community. In any event, a correlation with Benthic Assemblage 3 appears reasonable. The large collections from this zone are from localities that have undergone sorting and disarticulation with the possibility that more than one community is included within the *Spinoplasia* Zone due to transportation and mixing. The notable absence of dicaelosids in this community as well as in the preceding *Quadrithyris* Zone suggests a position shallower than Benthic Assemblage 4 for the bulk of the material described to date.

The *Trematospira* Zone, including both of its subzones, is another Nevadan Subprovince unit that is not easy to categorize. Elements of the Gypidulinid Communities may be present, although transported from their growth site. The general aspect of the fauna has as much in common with the *Plicoplasia* Community as with any other within the Appohimchi Subprovince. The absence of both dicaelosids and chonostrophids is consistent with this interpretation, as is the absence of *Hipparionyx*, rensselaerinids, and eurythyrinids in general, i.e., many of the characteristic genera of the *Hipparionyx* Community. The *Trematospira* Zone overall is easiest to assign to Benthic Assemblage 3.

Despite their more provincial nature, both the *Acrospirifer kobehana* and *Eurekaspirifer pinyonensis* Zone faunas in general appear best correlated with the *Gypidula* Community plus the *Plicoplasia* Community (i.e., varying mixtures of both), with Benthic Assemblage 3 indicated.

Underlying the *Quadrithyris* Zone are additional zones of the Gedinne, including the *Gypidula pelagica* Zone (Gypidulinid Community) and intervening un-named zones that include *Dicaelosia* plus a reasonably rich shelly fauna of *Dicaelosia-Skenidioides* and *Dicaelosia-Hedeina* Communities aspect. The shells are taxonomically very different from those in the Appohimchi Subprovince units of the Early Devonian although indicating a Benthic Assemblage 3 and 4 or possibly slightly deeper assignments.

The apparent absence in the *Quadrithyris* through *Eurekaspirifer pinyonensis* zones of any but Benthic Assemblage 3 and deeper units in central Nevada and a few other localities along strike to the southwest is notable. However, the bulk of the collections have been obtained from a limited area and only a small number of really rich localities occur for the *Quadrithyris* and *Spinoplasia* zones, and to a lesser extent for the *Trematospira* and *Acrospirifer kobehana* zones. It may be predicted, as rich localities are found west of those previously studied, they will supply more information about deeper water communities. A case in point is the recent description of a Notanoplid Community at Carlin, Nevada (Boucot and Johnson, 1972) in a position consistent with Benthic Assemblage 6, well away from and to the west of any of the typical coeval brachiopod faunas belonging to shallower environments. The situation eastward is unpromising because of increasing dolomitization in that direction; even worse is the presence of a Middle Devonian (pre-*Leptathyris* zone) disconformity representing removal of the Early Devonian beds (this unconformity may be the platform expression of the important metamorphic and possible orogenic event affecting the Central Metamorphic Belt of the Klamath Mountains to the west, and the potential unconformity between the Late Eifel Kennett Formation and associated units with the Early Devonian units in Siskiyou County, California).

The section at Coal Canyon in the Simpson Park Range, Nevada (from which the bulk of Johnson's, 1970a, *Quadrithyris* Zone collections have been obtained) is ecologically instructive. The Wenlock part of the Roberts Mountains Formation belongs in Benthic Assemblage 5 or deeper as inferred from the relatively rich assemblage of graptolites (see Fig.16 for deeper occurrence of taxonomically rich graptolite assemblages) and to the relative absence of shelly faunas as contrasted with sections rich in Benthic Assemblage 3 to possibly 4 shells in the same formation nearby in the Roberts Mountains. The absence of shells in the Ludlow and Pridoli parts of the formation in the Coal Canyon section, as opposed to the rich faunas present in the same formation nearby in the Roberts Mountains, also is consistent with a Pelagic Zone position. The Gedinne units above the Silurian in the Coal Canyon section consist of scattered limestone lenses containing mainly disarticulated and somewhat fragmented shells. The fragmented shells are entirely consistent with extensive transport from a nearby topographically higher source and final deposition in a graptolitic, pelagic zone site. The overlying *Quadrithyris* Zone consists of graptolitic beds with a single "breccia" bed from which shelly, *Quadrithyris*

Community and some gypidulid faunas have been obtained. The breccia bed is entirely consistent with transport from a topographically higher Benthic Assemblage 3 site to a pelagic zone depositional site. The limestone blocks in the "breccia" bed are very heterogeneous lithologically, some are well-rounded, and differing faunules occur in blocks of different lithology; information consistent with a diverse, upslope source area. The occurrence in Coal Canyon (Boucek, 1968, p.1278) of dacryoconarid tentaculitids in the beds just above the Windmill Limestone breccia bed is consistent with Pelagic Community fossils immediately above the breccia bed, i.e., with the breccia representing a shallower environment set of materials transported and deposited in deeper water downslope from the growth site or sites. The overlying younger Early Devonian section at Coal Canyon becomes increasingly calcareous and is consistent with shallowing of the section up to a point. The *Spinoplasia* Zone is represented by very scattered fossils, chiefly the leptocoelid genus *Pacificocoelia* among the brachiopods that actually may have lived where they were found rather than having been transported from an original growth site. Leptocoelid brachiopods in the Devonian have far greater depth ranges than they do in the Silurian (*Leptocoelia* in the Appohimchi Subprovince ranges from the *Hipparionyx* Community, through the *Plicoplasia*, *Mutationella* and *Beachia* Communities to be the only shells present in the area where the facies changes into unfossiliferous flysch; *Australocoelia* in the Malvino-kaffric Realm is present from the boundary with the mutationellinid communities containing *Pleurothyrella* or *Scaphiocoelia* through the limits of the *Notiochonetes* and *Australospirifer* Communities). It may be more than coincidence that the only starfish found in many years collecting in Somerset County, Maine (Boucot and Yochelson, 1966) occurred at a locality where *Leptocoelia*-bearing beds grade laterally into unfossiliferous flysch-type strata, i.e., a Starfish Community occurrence comparable to the Starfish Community of other regions. Therefore, the Coal Canyon *Pacificocoelia* occurrence is consistent with a position in Benthic Assemblage 5. The Coal Canyon *Spinoplasia* Zone rocks are similar to the platy limestone forming the bulk of the Roberts Mountains Formation, despite their being assigned to another rock unit, and contrast markedly with the more massive shelly limestones of the type *Spinoplasia* Zone Rabbit Hill Formation. In any event the extensive transportation attributed to the Coal Canyon *Quadrithyris* Zone fossils permits their having originally lived within Benthic Assemblage 3.

In summary, the central Nevada Early Devonian brachiopod sequence probably represents a fairly uniform environmental setting with both low-diversity, rough-water *Gypidula* Community faunas and high-diversity, quiet-water *Quadrithyris*, *Cyrtina*, *Plicoplasia*, or *Dicaelosia-Hedeina* type communities ranging within Benthic Assemblage 3 and slightly deeper for some of the older beds. Deepening to the west is indicated (*Callicalyptella* of the *Carlin* area belongs to the Benthic Assemblage 6 Notanoplid Community).

Malvinokaffric Realm Devonian level-bottom communities

Boucot (1971) defined a set of communities (Fig.9) for the Realm, showed their distribution in South Africa, and discussed their occurrence elsewhere. The community characteristics present in Benthic Assemblage 1 are similar to those encountered in other biogeographic and temporal entities in the same shoreline-related position. Benthic Assemblage 2 position is assigned the mutationellinid communities (those with abundant *Scaphiocoelia* or *Pleurothyrella*) and the *Globithyris* Community (Isaacson, 1973a), the latter possibly also extending into Benthic Assemblage 1. Seaward occur the *Notiochonetes* and *Australospirifer* Communities, both representing conditions more favorable to high diversity than the mutationellinid communities. The presence of abundant, disarticulated plicate spiriferids in high-diversity Early Devonian communities inferred to reflect quiet-water level-bottom conditions in parts of the Dzhungaro-Balkhash Complex of Communities, the Rhenish Complex of Communities, the *Plicoplasia, Mutationella* I and *Beachia* as well as the *Australospirifer* Community is significant environmentally.[48] The deepest community identified within the Malvinokaffric Devonian is the Starfish Community (occurrences in Parana and South Africa). The Ambocoelid Community is also present in this Province. The *Zoophycos* Community is present in South Africa (Plumstead, 1969, p. 32) and in Bolivia (P.E. Isaacson, oral communication, 1973).

Middle Devonian communities

Although there are scattered comments in the literature concerning communities, not enough information has accumulated that a worldwide summary can be written. For example, in the Appohimchi Subprovince, neither the Onondaga nor the Hamilton Groups have been subjected to the necessary scrutiny, although a mass of information allows certain major categories to be roughly defined, particularly for the Hamilton. These have been well-known for some time and may approximate Benthic Assemblages, or even actual communities.

The situation in most other parts of the world Middle Devonian is similar to that in the Appohimchi Subprovince, with a few useful exceptions (Struve, 1963; Copper, 1967; Lecompte, 1968; Winter, 1971, for example).

During the Middle Devonian we have excellent examples of heavy-shelled, rough-water, low-diversity communities similar in these respects to those of the Silurian. These include the *Stringocephalus* Community (includes *Stringocephalus* as well as the rarer taxa belonging to the Stringocephalidae) and the *Zdmir* Community (a Middle Devonian gypidulinid community).

A certain amount of community information is provided by Ehlers and Kesling (1970) for the Middle Devonian of Michigan. The Middle Devonian of this region belongs to the Eastern Americas Realm. The absence of *Tropidoleptus* Community and also *Globithyris* Community shells, of Benthic Assemblage 2, as shown by the absence of the *Tropidoleptus* Community and the globithyrinid genus *Rhipidothyris*, suggests that Benthic Assemblage 3 and deeper level-bottom communities only are present in the Michigan Basin Middle Devonian as contrasted with the Central Appalachian region. This conclusion is supported by the physical nature of the Middle Devonian sediments in this region that lack any evidence for the presence of shoreline or very nearshore, shallow-water features such as are abundant near the shoreline region to the east in eastern New York, Pennsylvania and Virginia. The abundance of *Gypidulinid* Communities in the Michigan Basin Devonian makes clear the presence of Benthic Assemblage 3 and possibly deeper communities. A number of reef and reef-related communities are to be met with in the Appohimchi Subprovince Devonian, both in the Central Appalachians and the Michigan Basin. The black shale styliolinid associations clearly belong to the Pelagic Community, a conclusion agreeing with their offshore position.

Community Evolution

Introduction

Community evolution incorporates the consequences of both organic evolution and biogeographic evolution (the development of biogeographic units in time and space, as well as their extinction), but is independent of both. In other words, the evolution of communities is not a carbon copy in time and space of either organic evolution or biogeographic evolution although community evolution employs the materials provided by organic evolution in major geographic divisions dictated by biogeographic evolution. Watkins and Boucot (1973) have presented an example of community evolution and considered the problems that arise from the treatment of actual materials.

Community evolution may be defined in a number of ways. If the evolving taxa present in a community were derived from antecedents present in the same community and gave rise to descendents present in the same community we could find that community evolution and organic evolution run strictly parallel. Such is not, however, always the case. Taxa whose antecedents were present in unrelated communities, may occur together by adaptation of their environmental requirements. The reverse situation is equally true as shown by Arkell's classical Jurassic-Recent example (Arkell, 1956, pp.615-616) in which three commonly associated shallow-water Jurassic bivalve taxa (*Trigonia*, *Astarte*, *Pholadomya*) are today present in the boreal region (*Astarte*), shallow warm water around Australia (*Neotrigonia*) and the abyssal region (*Pholadomya*). A widespread community may, in geologic time, be divided by a barrier to migration into homologous communities that become progressively different in their taxonomic contents, although functionally similar. As new taxa arise they may usurp the position of old taxa and be responsible for the extinction of old taxa, partition existing niches, or simply replace extinct taxa functionally. All of these differing phenomena are a part of what may be termed community evolution. Lastly, there is the problem of new environments becoming habitable by newly evolved organisms due either to changes in physical conditions, or to changes in the biota that create new biologically conditioned environments. Community extinctions may occur for the same reasons in reverse.

With these constraints in mind it is clear that "community evolution" involves a number of distinct phenomena. The community evolution, for example, to be found in Tertiary coral-reef environments populated with typical Tertiary organisms may be compared with that to be found in Permian reefs in functional terms but the largely different origins of the taxa involved counsels caution. The controlling physical and/or biological variables present in these two reef environments may be sufficiently different to make even functional comparisons misleading.

I propose, therefore, to employ the terms analogous community evolution and homologous community evolution in considering the development of Silurian-Devonian communities. Analogous communities will be defined as those that occupy functionally similar positions but which contain few if any taxa with a common phylogenetic origin. They may be of widely differing age. Homologous communities will be defined as those containing a reasonable (a subjective quantity based on experience) percentage of taxa with a common phylogenetic origin, thus raising the probability of their sharing a higher percentage of common environmental requirements *if* the amount of geologic time separating them is not too large. However, it is important to keep in mind that these concepts of analogous and homologous communities are generalizations. There will certainly be cases where the "homologous" descendents of a particular community will have become modified for very different environmental requirements.

It is also important to understand that both analogous and homologous communities may be either contemporaneous or non-contemporaneous. Non-contemporaneous homologous communities, for example, may be represented by a time sequence of species communities belonging to *Eocoelia* (one may speak of the *Eocoelia hemisphaerica* Community, succeeded in time by the *Eocoelia intermedia* Community, succeeded in time by the *Eocoelia curtisi* Community, succeeded in time by the *Eocoelia sulcata* Community, succeeded in time by the *Eocoelia angelini* Community). Contemporaneous homologous communities are conceived as having been derived from common taxa, or a common taxon, occurring in older beds, but containing derivative taxa developed under conditions of isolation (sympatric species would occur together as fossils). The various late Lower Devonian mutationellinid communities may be spoken of as contemporaneous homologous communities (*Cloudella* Community, *Mutationella* Community II, and *Mutationella* Community III). Non-contemporaneous analogous communities are defined as taxonomically unrelated taxa, or a single taxon, performing the same function at different times (at the same place or different places). Contemporaneous analogous communities are defined as associations of taxonomically unrelated taxa, or unrelated single taxa, performing the same function during the same time interval, but at different places; they are kept separate and not permitted to compete with each other by isolating mechanisms.

I am assuming here that, in general, closely related contemporaneous and descendent taxa tend to have similar environmental requirements rather than the reverse (G.G. Simpson, written communication, 1972, points out, however, that "It is generally assumed and has often been substantiated that closely related sympatric taxa usually have different environmental controls."). Ultimately each case of this sort must be judged on its own merits. Experience with Silurian and Devonian brachiopods viewed against their physical background (characteristics of the strata, etc.) suggests that closely related taxa tend to have more in common environmentally than not, although as Simpson has pointed out, it would be a mistake to assume that closely related taxa have identical requirements, particularly if they are sympatric. Fig.35 suggests that most of the Silurian-Devonian brachiopod taxa having close relations to each other in either an ancestor-descendent or common ancestor manner do have fairly close environmental requirements. There are, however, a number of obviously conspicuous exceptions to this generalization; all are discussed under the section on Siluro-Devonian community evolution that follows. Ideally I would like to plot biogeographic boundaries in three dimensions, as in Fig.25, and then superimpose the family trees present within each biogeographic entity in such a manner that applicable community boun-

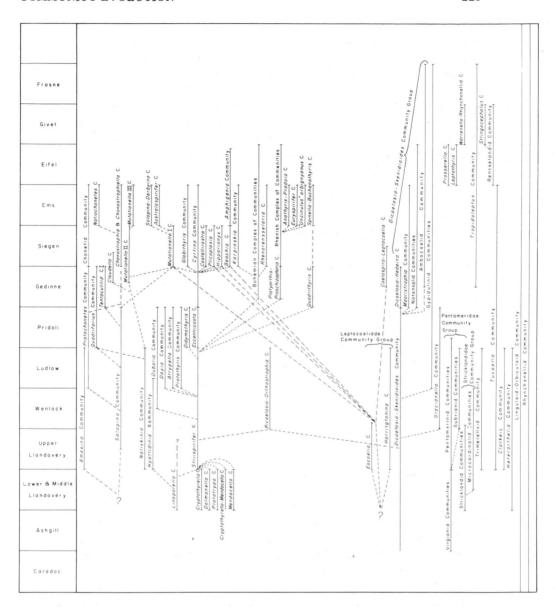

Fig. 35. Stratigraphic range and inferred relations of Llandovery-Ems level-bottom brachiopod communities. Dashed lines indicate taxonomic continuity, and inferred community continuity. The Siegen-Ems age *Vagrania-Skenidioides* Community is considered to be derived from the *Dicaelosia-Skenidioides* Community of the Silurian through an as yet unspecified Gedinne intermediate.

daries in time could be drawn to include the appropriate taxa. Failure to follow such a procedure leads to the type of confusion present in Bretsky's (1969) concept of community evolution (he actually discusses succession of communities!) for Paleozoic invertebrates for which analogous and homologous units are thoroughly mixed. We must carefully analyse the community situation before we have any hope of understanding its development; premature generalization will not aid in attaining understanding. Bretsky (1969, fig.1) for example, indicates the derivation of North Silurian Realm communities, containing no taxa with phylogenetic origins in the Late Ordovician North American Realm from Late Ordovician communities (*Eocoelia* from *Sowerbyella-Onniella* and *Pentamerid-Atrypa* from *Zygospira-Hebertella*). The "evolution" of these Silurian communities from the named Ordovician precursors is unsupported by shared taxa, or taxa with common phylogenetic origin. The best that might be hoped is that Bretsky's taxonomically unrelated communities might be analogous communities (with or without homeomorphic forms) in time, although no evidence for such a conclusion is presented by Bretsky except that both the Ordovician and Silurian communities occur in relatively shallow water. The recognition that communities separated from each other by time and space are analogous is far more difficult than recognizing those which are homologous because of their large percentage of closely related taxa. Recognition that communities are indeed analogous depends on mapping out common boundaries with adjoining communities that may be related to each other in time and space, by analysis of population structure and filling out of analogous ecological niches, and also by careful evaluation of any available physical criteria present in the associated stratified rocks or reconstructions of the regional paleogeography, lithofacies and the like.

Berry (1972, p.76) is essentially thinking along the same lines when he states that "The Toquima-Table Head community type is apparently essentially analogous in its position relative to the *Clorinda* Community of the Silurian..." He has implicitly recognized that the Ordovician taxa of the Toquima-Table Head community have little if any taxonomic continuity with those of the Silurian *Clorinda* Community although both are interpreted to occupy what I term the same Benthic Assemblage position.

Finally, homologous communities in most instances will have short time spans, corresponding to the taxa involved, whereas some analogous communities may be recognized throughout the Phanerozoic. Analogous communities may be very discontinuous in their time occurrence, depending on the recurrence of similar conditions that permit unrelated taxa to perform the same functions (fill the same or similar ecologic niches). Raup and Stanley (1971) employ the term "ecologic replacement" to describe the occurrence in time of analogous communities.

Siluro-Devonian community evolution

Fig.35 diagrams the inferred relations of the Silurian and Devonian-level bottom brachiopod communities. Homologous relations are indicated by dashed lines or by continuity of lines. Observe that certain communities have contents derived from more than one preceding community, which indicates that evolution in the direction of altered niche breadth for some taxa has occurred or that migration has played a role. A number of trends may be seen after inspection of Fig.35 and consideration of Part II, Chapter 6. High-diversity communities may yield both high- and low-diversity homologs. Low-diversity communities may yield descendents to high-diversity communities. Low-diversity communities may continue for a considerable span of time essentially unmodified, and the same is true for certain high-diversity communities. Homologous relations are determined from environmental position combined with knowledge of the taxonomic relations of the common taxa.

It is clear that during the course of evolution some taxa will have descendents with markedly narrower environmental requirements whereas others will have far broader requirements. Such developments complicate the problem of discussing community evolution, but must be comprehended into any acceptable scheme. The important point to be worked out is the degree to which closely related taxa have evolved away from each other in terms of their environmental requirements. It is clear from the data on Silurian-Devonian brachiopods that there are few cases of "air-breathing, land-perambulating fish or swimming, water-breathing tetrapods". It is probable that all, or nearly all, of the closely related taxa have evolved and changed at least a small amount in terms of their environmental requirements. The observation that the majority of taxa and their descendents occupy the same Benthic Assemblage position, and commonly occur in homologous communities, strongly suggests that the environmental divergence of closely related taxa has been small during the time interval considered. Therefore, it is reasonably safe, and actually more conservative in the absence of other evidence, to conclude that closely related taxa of Silurian-Devonian brachiopods will have very similar environmental requirements than to conclude the reverse.

One of the most prominent changes shown on Fig.35 is the widening during the Early Devonian of the Benthic Assemblage spread of several communities (*Australospirifer* Community of the Malvinokaffric Realm, *Beachia* Community of the Appohimchi Subprovince). The *Australospirifer* Community is derived from Benthic Assemblage 2 *Mutationella* I Community and occupies a Benthic Assemblage 3-5 spread. The *Beachia* Community is derived from the *Cyrtina* Community (a Benthic Assemblage 3 entity) and occupies a Benthic Assemblage 3-5 spread as does the *Amphigenia* Community derived from it.[49]

Community groups

Inspection of Fig.35 shows the presence of several groups of communities that are concluded to have homologous relations, plus other groups that are either unrelated or for which we do not yet understand the possible relations.

The presence of these community groups, containing a large number of taxa with common antecedents and descendents argues in favor of most related taxa having similar ecologic tolerances during the time interval diagrammed, although there are some conspicuous exceptions. In the following discussion the term "Community Group" will refer to a number of communities that have homologous relations, although they may incorporate taxa from unrelated communities through the process of changing niche breadth, and also give off taxa to unrelated communities through the process of changing niche breadth. We are dealing with a dynamic, ever changing situation that does not lend itself to easy description. The important thing to follow is the ecologic path taken by individual taxa and their descendents in time by comparing Fig.35 with the information contained in Chapter 6. The complex web that results gives some idea of the gradually changing environmental positions and tolerances of various taxa in the various communities.

Fig.35 and the conclusions made from it about the relative constancy of environmental requirements present in most descendent taxa is very important. At least for the groups of Silurian and Devonian brachiopods involved this is good evidence that there is a strong tendency for descendent taxa to have similar environmental requirements on the average with their antecedents. There is no evidence here that evolution has effected the environmental requirements of this group in a random manner. It would be going too far to say that Benthic Assemblage 5 taxa are always restricted in time to that position, but it is clear that this is the overall tendency.

The noticeable restriction of rough-water shells through time, as evidenced by the Pentamerinid, Gypidulinid, Stricklandid, Microcardinalid and Trimerellid Groups with all of their various taxa, to a very restricted Benthic Assemblage position has been noted. Taxa assigned to these groups show no evidence of diverging environmentally into either other Benthic Assemblages, into other than single-taxon communities, or into quiet-water environments. This environmental behaviour suggests that all of these thick-shelled groups (and possibly similar groups like the stringocephalids and rensselandids) may have become so modified during the process of natural selection as not to have been able to have moved from the positions into which they found themselves. It is notable that the cosmopolitan, high-diversity community taxa that were present during the same time intervals as these taxa, including such shells as *Atrypa*, *Leptaena* and *Nucleospira* never co-occur with them. The rapid cladogenetic evolution of the Pentamerinids during the reef-rich Wenlock-Ludlow interval suggests no lack of inherent capability for further evolution associated with the environmental restrictiveness. The steady phyletic evolution of the Gypidulinids, Stricklandids and Microcardinalids likewise suggests no inherent incapability for further evolution. In view of this data it is puzzling that the linguloid and orbiculoid brachiopods that are so largely, although not entirely, restricted to Benthic Assemblage 1 in abundance from the Cambrian to the present should show such an apparent low level of either phyletic or cladogenetic evolution. Therefore, I must come back again and re-emphasize the lack of attention to morphology and taxonomy that has been the plight of these two inarticulate groups. The museums of the world contain very limited collections of linguloids and orbiculoids, publications dealing with their morphology are similarly limited, they tend to be disregarded in the field when collecting more "important" shells, all of these factors indicating to me that we are not in possession of all the necessary information regarding these two groups that is necessary to the making of a reliable estimate of their characteristics for rate of evolution purposes.

Pentameridae Community Group

Fig.35 relates the Virgianid, Pentamerinid and Subrianid Communities to each other. These varied communities tend to form single-taxon, rough-water communities present in one place or another from the Ashgill into the Pridoli. Endemism and continuing organic evolution are responsible for the various taxa not occurring together in a single community. All of the taxa are characterized by heavy deposits of shell material in the umbonal regions that permitted them to keep their anterior margins upright, out of the sediment. Their environmental tolerances do not appear to have changed significantly during their total range. Some of the individual communities have very restricted occurrences geographically and consist of taxa that evolved rapidly, whereas others had a very wide distribution and appear to have evolved very slowly. The Pentameridae Community Group is an excellent example of a group of homologous communities that persisted for a relatively long time interval without much change in environmental requirements.

Stricklandidae Community Group

The Stricklandid and Microcardinalid Communities have much in common in addition to their common taxonomic ancestry. Both groups of communities occur in the Benthic Assemblage 4 position. The microcardinalids are commonly associated in the Late Llandovery of the North American Platform with *Pentameroides* but it is not clear whether this association is primary or secondary, as both earlier Llandovery and Wenlock pure microcardinalid communities occur unmixed with abundant specimens of other taxa. The Stricklandidae Community Group is a second Silurian example of a group of closely related organisms that occur in a low-diversity community and do not appear in time to change their environmental requirements.

Leptocoelidae Community Group

The characteristics of the Leptocoelidae Community Group are different in time from those of the Pentameridae and Stricklandidae Community Groups. The Silurian leptocoelid communities occur as low-diversity units occurring in a relatively narrow environmental position as do the Pentameridae and Stricklandidae Community Groups. But, with the Devonian we see that the leptocoelids no longer form low-diversity communities restricted to a narrow environmental range. Rather they form parts of a number of other communities with a range far wider than that possessed by any of their known Silurian antecedents. The Leptocoelidae Community Group forms an excellent example of a major change in environmental range with time.

The Leptocoelid communities have no certain antecedents in the pre-Late Llandovery. They may be derived from the Malvinokaffric Realm Silurian *"Eocoelia" paraguayensis* of the Early Llandovery about whose ecology we know very little.

Dicaelosia-Skenidioides Community Group

The *Dicaelosia-Skenidioides* Community Group forms an easily recognized group of Benthic Assemblages 4 and 5 communities commonly characterized by high diversity and smaller shell size than is present landward. The smaller shell size holds for most of the taxa, including those with other forms belonging to the same genus that occur landward and those that do not have close relatives landward. But, low-diversity communities belonging to this Group are recognized, which suggests that very restrictive conditions may be encountered in this position. Endemism may occur here just as in all other Benthic Assemblage positions and Community Groups. One of the notable derivatives of the *Dicaelosia-Skenidioides* Community Group is the Benthic Assemblage 3 Gypidulinidae Community Group appearing in Wenlock time. The development of a nearer-shore, rough-water, low-diversity group from a more offshore, quiet-water, high-diversity group is notable.

Gypidulinidae Community Group

The Gypidulinidae Community Group is well established by Ludlow time. It includes a large number of endemic communities each containing a single gypidulinid taxon. The environment of these communities is interpreted as rough-water. The Community Group persists from the Ludlow through the Frasne. The taxa making up the Community Group change continually, under varying conditions, as evolution progresses but the homologous nature of the resulting communities is easy to ascertain. The coincidence in time of extinction for the Pentameridae Community Group and arising of the Gypidulinidae Community Group is a good example of ecologic replacement as both groups appear to have occupied the same position and performed the same functions. Wenlock and Ludlow gypidulinids behave both as members of Gypidulinid Communities and as members of high diversity, quiet water Benthic Assemblages 4 and 5 communities such as the *Dicaelosia-Skenidioides* and *Dicaelosia-Orthostrophella* Communities. The Ashgill-Llandovery genus *Brevilamula*, which is antecedent to the gypidulids, occurs only in the quiet-water Benthic Assemblage 5 high-diversity type communities. The Gypidulinidae Community Group taxa form another excellent example of organisms that made a major change in their environmental requirements with time. They differ, however, in this regard from the leptocoelids. The leptocoelids began as a group with narrow tolerances and developed forms with broad tolerances, whereas the gypidulinids began as a group with moderately broad tolerances (Benthic Assemblage 4-5, quiet-water, high-diversity communities) and gave rise to groups with narrow tolerances (B.A. 3, rough-water, low-diversity).

Striispirifer Community Group

The *Striispirifer* Community Group is by far the most complex unit shown on Fig.35. Evidence is suggested on Fig.35 for the addition of material to the *Striispirifer* Community Group from the Leptocoelidae Community Group in the Devonian. Evidence also suggests that the low-diversity, quiet-water, endemic *Nalivkinia* Community of Central Asia gave rise towards the end of the Silurian to the mutationellinids occurring in the low-diversity, quiet-water, Benthic Assemblage 2 units (*Mutationella* II, *Cloudella* Communities). It is also suggested that the low-diversity *Cryptothyrella* Community of the pre-Late Llandovery was the source of the meristinids present in the *Striispirifer* Community Group. The number of taxa shared with the earlier occurring *Linoporella* Community suggests that this may have been the source for the bulk of the taxa present in the *Striispirifer* Community Group. Thus, the *Striispirifer* Community Group may be viewed as a high-diversity, quiet-water, Benthic Assemblage 3 entity that acquired some additions from other sources, but chiefly evolved in place, subject however, to the changes imposed by conditions of provincialism present during part of the interval of its existence.

The chief characteristic of the *Striispirifer* Community Group is its subdivision during the highly provincial Lower Devonian into a number of parallel entities with a common origin from the main Silurian stem. These common entities have faunal characteristics very similar to those present in the Silurian. In addition, taxonomic evidence suggests that from time to time specialized, low-diversity communities developed from *Striispirifer* Community Group antecedents. It is remarkable to note how little exchange there appears to have been between the Benthic Assemblage 4 and 5 *Dicaelosia-Skenidioides* Community Group taxa and those of the *Striispirifer* Community Group, although they both do share some taxa that overlap.

Unassigned groups

In addition to the units discussed here there are many other individual communities shown on Fig.35 that cannot be now assigned to a Community Group. Some comments, however, may be made about them.

The Trimerelloid Community, in a general way, is homologous throughout the Late Llandovery-Ludlow interval. However, the various trimerelloid taxa, when studied from an ecological viewpoint may show that things are not as simple as the present classification indicates.

The *Tuvaella* Community is an analogous unit that does not give rise to others. The *Clarkeia* and *Heterorthella* Communities are similar in this regard.

The brackish-water Hughmilleriidae-Stylonuridae Community of the Late Silurian has homologous community continuations within the post-Silurian. Kjellesvig-Waering (1961) points out that most of the post-Silurian eurypterids are descendents of taxa present within this community and occupy either a brackish or a non-marine habitat. The Eurypteridae and Carsinosomidae-Pterygotidae Communities become extinct after the end of the earliest Devonian.

Also of great importance for an understanding of Silurian community evolution is the sudden appearance in the Late Wenlock-Ludlow interval of a large number of reef and reef-related communities. The Llandovery, earlier Wenlock and Pridoli are times of relatively low abundance of reef and reef-like bodies. For reasons that remain unclear the Late Wenlock-

Ludlow was a time propitious for the development of reefs and reef-like bodies with a consequent increase in related numbers of reef and reef-related communities, as well as overall increase in number of taxa.

Community succession

The previous paragraphs have discussed some aspects of level-bottom brachiopod community evolution during the Silurian and Lower Devonian. Earlier I have tried to make the distinction between community evolution and the appearance of successive analogous communities in what Raup and Stanley (1971) would have termed community replacement. An additional type of small scale vertical change in community is met with in which either changing physical conditions over a short time span or biologically conditioned changes over a short time span create conditions leading to a changing set of communities. The coral and stromatoporoid plus algal reef environment of the Silurian-Devonian, as discussed by Lecompte in a series of papers (see Lecompte, 1961, and bibliography in same), is an example of a biologically conditioned environment that progressively extends from a relatively deep, quiet water, muddy bottom up into a less deep, more turbulent environment that finally culminates in the reef environment with a consequently different set of communities as differing depth and turbulency zones are encountered with each set of organisms making possible shallower conditions for the next set of organisms.[50] In other words, we have here an example of a community succession in which one community paves the way for another. This type of community succession has some things in common with the climax concept of the modern day ecologist concerned particularly with land plants, but it is not an exact parallel. An additional example of a community succession, for which there is no evidence that biologic conditioning results in widely differing depth conditions, is encountered in many Silurian-Devonian quiet-water successions for which we have alternating shaly or marly and shelly layers. Examples include the Waldron Shale, the Slite Marls, and the quiet-water facies of the Wenlock Limestone.[51] All of these units show a regularly alternating succession, on a scale of a few centimeters in many instances, of soft, marly beds with hard shelly beds. The marly and shelly beds contain different faunas of shells. The hard beds tend to contain a large amount of calcarenite in which pelmatozoan debris forms a large part in addition to shells in which brachiopods normally form the largest volume. This regular alternation may be explained as a consequence of intermittant turbulence having spread out muddy material in which the shaly fauna thrives. The shaly faunal communities then accumulate enough hard substrate[52] material through their own activity to make possible the setting of larvae requiring hard bottoms or at least non-shaly bottoms. The hard substrate organisms then thrive until wiped out by the next influx of mud. This process is evidently repeated many, many times in some areas. This alternation too may be termed community succession and held to be distinct from community replacement and community evolution.[53] In a sense community succession, as the term is used here, involves the biologic conditioning of the bottom by one group of organisms in order that other groups may subsequently succede on the same ground. In this sense community succession will normally involve very short intervals of time such as parts of a year or a few years, but certainly not intervals of millions of years such as one would normally predict to be the case for most examples of community replacement (Raup and Stanley's, 1971, ecologic replacement).

Lower Paleozoic gastropod diversity, abundances and potential predator: prey relations

Lower Paleozoic gastropods occur in relatively high abundance in Benthic Assemblage 1. High abundances are also met with in the upper parts of Benthic Assemblage 2 and in the brackish environment. Seaward of these Benthic Assemblage positions it is most unusual, in the pre-Middle Devonian to find high concentrations of gastropods relative to other taxa. In the Benthic Assemblage 3—5 span it is very uncommon to find more then 1% of gastropod specimens in the total count of benthic shells. This abundance relation changes radically during the Upper Devonian, until by Carboniferous time gastropods are widely distributed in moderate to even high abundance locally in well offshore positions. The mechanism responsible for this radical change in offshore gastropod abundance is not understood at present.

Taxic diversity for Lower Paleozoic gastropods does not parallel their abundance. In the Benthic Assemblage 1 and 2 positions, as well as in the brackish environment, taxic diversity is very low. In the Benthic Assemblage 3—5 span, however, taxic diversity is high. For example (Boucot and Yochelson, 1966, table 1) ten and eleven gastropod taxa are reported from two units of Early Devonian age in the Appalachians associated with seventeen and twenty articulate brachiopod taxa respectively. The lower number of gastropod taxa probably reflects in part their relative rarity (less than one gastropod to one hundred brachiopods) which may have biased the sampling seriously.

There are few exceptions to this situation regarding taxic diversity and abundance for Lower Paleozoic gastropods. In the Silurian both oriostomatids and euomphalopterids attain a moderate abundance in many widespread level-bottom, high-diversity communities occurring in the

Benthic Assemblage 3—5 span. Platyceratid gastropods, of the Silurian-Devonian attain a moderate abundance in many Benthic Assemblage 3—5 span high-diversity level-bottom communities, but this abundance may be correlated with the coprophagous relation they have with attached echinoderms.

If one deduces that a reasonable number of the Benthic Assemblage 3—5 span Lower Paleozoic gastropods were carnivorous then the low abundance and high taxic diversity may be interpreted as a predator—prey relation of the type outlined for the Recent by Paine (1966,1971). Clearly it will be difficult to test this hypothesis except by compiling data regarding gastropod diversity and abundance relations systematically relative to other benthic taxa occurring together in Lower Paleozoic strata. There undoubtedly were a host of other Lower Paleozoic predators feeding on the benthos such as, starfish and nautiloids to mention just two, which might also be considered in the same manner to test this possible relation between herbivore, or filter- and suspension-feeder diversity and predator-diversity (or absence!).

CHAPTER 6

Abundant Taxa Characteristic of Siluro-Devonian Level-Bottom Communities

Despite the title of this Chapter the subject matter is restricted largely to level-bottom, brachiopod-rich communities. Level-bottom communities rich in corals (both tabulates and tetracorals) and stromatoporoids are very important on the Silurian-Devonian platforms but have been given short shrift owing to my ignorance and lack of experience with them. What attention these important coral and stromatoporoid communities receive is derived largely from scattered literature; it should not be considered as based on authoritative, comprehensive acquaintance with their problems. Reefs composed largely of algal, stromatoporoidal and coralline material are prominent in the Silurian-Devonian platform carbonate record (although they form a relatively small volume of that record). These reefs appear to be largely restricted to Benthic Assemblage 3. Ignorance of the myriad communities involved in Silurian-Devonian reefs has precluded any attempt to handle these communities on a worldwide basis. Lowenstam (1950) and Manten (1971) have discussed and given some insight into the Silurian reefs of Illinois and Gotland, respectively. Their work indicates some of the inherent possibilities. Enough is known to indicate that the "systematics" of Silurian-Devonian reef communities will be ultimately worked out on a worldwide basis.

With the exception of the communities found in Benthic Assemblage 1, most level-bottom benthic communities not dominated by corals and stromatoporoids are dominated by brachiopods. Bivalves and gastropods plus trilobites, corals and other megafossils seldom make up as much as five percent of the actual number of fossils encountered until the Upper Devonian. Beginning with the Upper Devonian it is common to encounter benthic communities containing both abundant brachiopods and molluscs.

Epifauna and infauna should not be included in the same community. The reasoning is similar to that for separating out nektic and planktic communities from the underlying benthic communities. The controls that govern the distribution of epifauna and infauna are partly similar and partly unlike. It is reasonable to assume that suspension feeders and filter

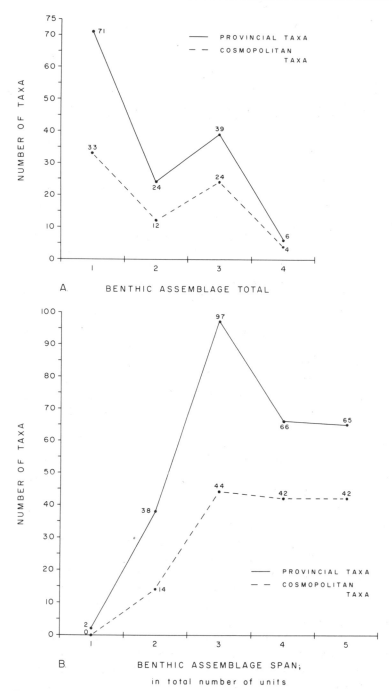

Fig. 36. Benthic Assemblage distribution of articulate brachiopod taxa dealt with in this Chapter.

feeders of different types, whether epifaunal or infaunal, will be similar in their distribution insofar as food supply is concerned, but there may be other important variables concerned with the very real difference in life style that are not similarly distributed. For example, deposit feeders actively working the intertidal zone below the surface of the sediment will not be subjected to nearly as rigorous conditions because of exposure as will epifauna, although the conditions may be somewhat different, due to removal of some water, than in the subtidal environment. For these reasons we suggest that infaunal taxa be at least analyzed separately from epifaunal taxa in order to try and check on the similarities and differences. It is easy to conceive of variables like sediment viscosity, sediment compactibility, pore-water chemistry and the like that may heavily influence the environment for infaunal organisms yet be unrelated to epifaunal distributions.

The communities discussed in this treatment are listed here alphabetically. Only those communities dealt with by myself or discussed with colleagues who have had experience with them are listed (there are undoubtedly far more community units of the same rank and importance as those listed here, but due to ignorance I do not attempt to discuss or define them). Neither trace fossils associated with these communities nor communities consisting chiefly of trace fossils (except for the *Taonurus* and *Skolithos* Communities) are discussed here (see Seilacher, 1967, for a discussion of their depth distribution). In a general fashion Seilacher's (1967, fig.2) trace fossil communities may be tied into the Benthic Assemblages employed here. He outlines, moving away from land, the following: *Scoyenia* facies (associated with red beds); *Skolithos* and *Glossifungites* facies; *Cruziana* facies; *Zoophycos* (= *Taonurus*) facies; *Nereites* facies. The *Scoyenia* and *Skolithos* plus *Glossifungites* facies appear best correlated with the brackish-water or non-marine environment for the first facies and Benthic Assemblage 1 for the second facies. The *Cruziana* facies fits best with Benthic Assemblage 2 and possibly the deep part of 1. The *Zoophycos* facies correlates best with Benthic Assemblage 3 and deeper positions to the 4 or possibly 5 position. The *Nereites* facies appears best correlated with some part of Benthic Assemblage 5 and deeper positions.

The level-bottom communities dominated by corals and stromatoporoids are not dealt with here, although a few entities are briefly considered in order to provide some notion of their relation to associated brachiopod-dominated units.

Fig.22 provides a tabulation of the level-bottom articulate brachiopod-dominated Silurian and Devonian communities. The solid line is a summary of the communities listed in this Chapter. The dashed line represents an estimate of the total number of brachiopod-dominated communities that might be expected to be recognized when more adequate work is done. It represents, for the Devonian, an estimate of the communities to be expected from the various Old World Regions after they have been considered in as much detail as the Eastern Americas and Malvinokaffric Realms.

Watkins et al. (1973) have discussed the differences between Petersen Animal Communities (PAC) based on recurring combinations of species, Parallel Communities (groups of PACs related by common ancestry of contained taxa that are characterized by similar or the same genera and families), and Benthic Assemblages (discussed by them as groups of PACs occurring in the same broad-scale environmental region). In this book, and in this section in particular, no attempt has been made to carefully discriminate between PACs and Parallel Communities, although Benthic Assemblages have been differentiated in all instances from the first two categories. Ultimately it is desirable that Silurian-Devonian communities should be defined and classified both as to PACs and Parallel Communities. However, this work of classification will take considerable time and effort, region by region, stratum by stratum. After the work is completed we will have a far more reliable basis for considering community evolution than is now possible. In terms of homologous and analogous communities it is possible to have one PAC give rise to a succeeding PAC in time and to a contemporaneous PAC laterally with both the lateral and the vertical PACs being termed Parallel Communities. Analogous Communities will not be termed Parallel Communities as their evolutionary origins will be separate.

Ambocoelid Community: Devonian (The Simonson Dolomite contains typical examples)

Abundant individuals of any of the many Devonian ambocoelid genera such as *Ambocoelia, Emanuella, Ladjia, Metaplasia, Echinocoelia, Crurithyris* (but not two or more taxa together!). This community probably corresponds to the "*Emanuella*-Pelecypod Community" of Noble and Ferguson (1971) reported from the Middle Devonian of western Canada. It appears to occur in a relatively deeper water position, seaward to reef-like bodies, in a Benthic Assemblage 4 to possibly 5 position. It occurs in all Realms. The ambocoelid taxa *Spinoplasia* and *Plicoplasia* have not been observed to form communities of this type (the *Plicoplasia* Community is a high-diversity community). However, the presence in the *Coelospira-Leptocoelia* Community (Benthic Assemblage 4) of a few specimens of *Spinoplasia* or *Plicoplasia* is suggestive of a deep-water (B.A. 4 or 5) origin during Gedinne time of the Ambocoelid Communities. Ambocoelids of the Silurian (from the Roberts Mountains Formation) are known in the

Dicaelosia-Skenidioides Community, further evidence of a deep-water origin from either a high- or moderate-diversity community for the Ambocoelid Communities.

Amosina Community: Late Llandovery-Ludlow (Catavi Sandstone and unnamed unit in southern Peru contain typical examples)

Amosina.

Amphigenia Community: Ems-Eifel (Bois Blanc Formation and Camden Chert are typical examples)

Discomyorthis, Platyorthis, Pentamerella, Leptaena, Strophonella, Protoleptostrophia, Anoplia, Plicanoplia, Eodevonaria, Costellirostra, Atrypa, Coelospira, Pentagonia, Leptocoelia, Meristina, Nucleospira, Acrospirifer, Mucrospirifer, Megakozlowskiella, Fimbrispirifer, Elytha, Ambocoelia, Plicoplasia, Cyrtina, Prionothyris, Cloudothyris, Centronella, Amphigenia, Stropheodonta.

Anathyris-Pradoia Community: Ems (upper La Vid Shales contain a typical example; see Wallace, 1972, p.125-126)

Anathyris, Pradoia collettei, "Uncinulus" pila, other brachiopods. The abundance of articulated shells contained in a shaly matrix is consistent with relatively quiet water conditions. A position in Benthic Assemblage 5 is suggested by its relationships to other communities (see Wallace, 1972, fig.1). The absence of this community in the position where it would be expected (the Schiefergebirge, for example) suggests that some other controls in addition to temperature correlated with depth plus a shaly matrix must be involved in its presence or absence.

Arthrophycus Community: Ordovician-Silurian (Tuscarora Sandstone is a typical example)

Arthrophycus. A. Seilacher (written communication, 1973) reports *Arthrophycus* associated with *Cruziana* in Ordovician and Silurian sandstones in Jordan and the Sahara. On this basis, as well as on the stratigraphic sequence and physical character of the Tuscarora, the *Arthrophycus* Community is placed in Benthic Assemblage 1. Seilacher feels that *Arthrophycus* "...is definitely caused by a large worm who browsed through the sediment by a backfill action in search of food". Massa and Jaeger (1971) describe a typical *Cruziana* Community trace-fossil association in a Benthic Assemblage 1, nearshore position from the Silurian of North Africa, that also includes *Arthrophycus.*

Atrypa Community: Upper Llandovery-Frasne (typical examples are wide-spread in the world).

Atrypa "reticularis". The cosmopolitan taxon *A. "reticularis"* commonly occurs in high-diversity communities with a Benthic Assemblage 3—5 span, but it also may occur in an articulated condition in quiet-water argillaceous beds indicating that at least a portion of its environmental range does not overlap with other taxa. The Benthic Assemblage position of this single taxon type occurrence of *A. "reticularis"* includes at least Benthic Assemblage 3.

Atrypella Community: Wenlock-Pridoli (Read Bay Formation is a typical example)

Atrypella (of the *A. prunum*, or *A. scheii*, or *A. phoca* type).

Australina Community: Late Llandovery-Ludlow (Silurian units in the Pre-Cordillera de San Juan)

Australina, Castellaroina, and additional Malvinokaffric taxa.

Australospirifer Community: Ems (Bokkeveld Sandstone contains examples)

Australospirifer, Australocoelia, "Schuchertella", Ambocoelia, Plicoplasia.

Beachia Community: Siegen (Tarratine Formation is a typical example)

Beachia, Dawsonelloides, Protoleptostrophia, Leptostrophia, Meristella, Acrospirifer, Leptocoelia, Platyorthis.

Bohemian Complex of Communities: Late Siegen-Early Ems (Konieprus Limestone is a typical example)

Eospirifer, Janius, Merista, Ivanothyris, Fascicostella, Quadrithyris, rhynchonellids, *Parachonetes, Aesopomum, Radiomena, Talaeoshaleria, Devonaria, Ivdelinia.*

Borealis Community: Late Early Llandovery-Middle Llandovery (*Borealis* Bank of Estonia is a typical example)

Borealis.

Coelospira-Leptocoelia Community (Cap Bon Ami Community of Boucot and Johnson, 1967a): Gedinne (Cap Bon Ami Formation is the typical example)

Coelospira, Leptocoelia, Protochonetes.

Carsinosomidae-Pterygotidae Community: Silurian

Genera of the Carsinosomidae and Pterygotidae eurypterids listed by Kjellesvig-Waering (1961).

Chonetid Community: Siegen-Ems (common in many Ardennes and Rhineland sections)

Plebejochonetes or *Chonetes* sensu strictu.

Chonostrophiella-Chonostrophia Community: Gedinne-Ems (York River Sandstone contains typical examples)

Similar to *Mutationella* Community I but lacks *Mutationella.*

Clarkeia Community: Late Llandovery-Ludlow (Tarabuco Sandstone is a typical example)

Clarkeia.

Clorinda Community: see "Undivided Communities"

Cloudella Community: Upper Gedinne-Siegen (Matagamon Sandstone yields typical examples)

Cloudella.

Conchidium Community: see Subrianinid Communities

Costellirostra Community: Siegen (Upper Grande Greve Limestone is a typical example)

Etymothyris, Costellirostra, Plicanoplia.

Cruziana Community: Silurian-Devonian

Cruziana. Seilacher (1967) points out the position of the *Cruziana* Community between the seaward *Zoophycos* Community and the landward *Skolithos* Community. Silurian data indicates that occurrences of *Cruziana* occur in the Benthic Assemblage 1 position, so, based on Seilacher's information, we conclude that the *Skolithos* Community, also occurring in Benthic Assemblage 1 represents a slightly shallower intertidal position than does the *Cruziana* Community. Seilacher (written communication, 1973) reports the *Cruziana* Community from Silurian sandstones of Jordan and the Sahara. Massa and Jaeger (1971) describe a typical *Cruziana* Community trace-fossil fauna from a Benthic Assemblage 1 position in the Silurian of Libya that also includes abundant *Arthrophycus* (sic *Harlania*). Thus the *Cruziana* Community is concluded to range from Benthic Assemblage 1 through 2.

Cryptothyrella Community: Lower and Middle Llandovery; possibly Ashgill of Baltic region (lower Brassfield Formation and some localities within the Beechhill Cove Formation yield typical examples)

    Cryptothyrella.

Cryptothyrella-Mendacella Community: Lower and Middle Llandovery (Caparo Formation of Venezuela is a typical example)

    Cryptothyrella, Mendacella, Dalmanella, Eostropheodonta, Protatrypa, rhynchonellids.These taxa may, in some instances (*Cryptothyrella* Community, for example) occur in single-taxon communities or in varying mixtures indicating that they all have an environmental range. The *Cryptothyrella-Mendacella* Community is a *Salopina* Community analog but gives rise to taxa that occur in the *Striispirifer* Community in the Late Llandovery.

Cyrtina Community: (Coeymans Community of Boucot and Johnson, 1967a): Gedinne-Ems (Coeymans Limestone contains typical examples)

    Cyrtina, Craniops, Dalejina, Schizophoria, Plethorhynchia, Atrypa "reticularis", Rhynchospirina, Mesodouvillina, Meristina, Howellella, Plicoplasia, Nanothyris, Podolella.

Dayia Community: Ludlow-Pridoli (Mocktree Shale is a typical example)

    Dayia.

Dicaelosia-Hedeina Community (Kalkberg-New Scotland Community of Boucot and Johnson, 1967a): Gedinne (New Scotland and Kalkberg Limestones contain typical examples)

    Hedeina, Dicaelosia, Skenidium, Dalejina or *Discomyorthis, Acrospirifer* or *Howellella, Platyorthis, Orthostrophia, Megakozlowskiella, Levenea, Atrypa "reticularis", Leptaena "rhomboidalis", Cyrtina, Meristella, Isorthis, Eatonia, Costellirostra, Spinoplasia, Coelospira, Trematospira, Leptostrophia, Strophonella, Iridostrophia, "Schuchertella", Anastrophia, Leptaenisca, Nucleospira, Atrypina, Plectodonta.*

Dicaelosia-Orthostrophella Community (termed the Henryhouse Community in Boucot, 1970): Wenlock-Ludlow (Henryhouse Formation is a typical example)

Resserella, Dalejina, Isorthis, Dicaelosia, Dolerorthis, Ptychopleurella, Dictyonella, Amsdenina, Grayina, Strophonella, Amphistrophia, Leptaenisca, Coolinia, Leptaena, rhynchonellids, *Atrypa, Lissatrypa, Nucleospira, Coelospira, Delthyris, Howellella, Meristina,* rhynchospirids, *Orthostrophella.*

Dicaelosia-Skenidioides Community: Ashgill-Pridoli (Elton Beds are a typical example)

Skenidioides, Atrypa "reticularis", Dalejina, Coolinia, Plectodonta, Sphaerirhynchia, Isorthis, Dicaelosia, Dolerorthis, Ptychopleurella, Dictyonella, Grayina, Strophonella, Leptaena "rhomboidalis", Lissatrypa, Nucleospira, Coelospira, Delthyris, Howellella, Merista, Meristina, Homeospira, Resserella.

Didymothyris Community: Pridoli (Kaugatuma Stage contains typical examples)

Didymothyris.

Dubaria Community: Wenlock-Pridoli (Pridoli Limestone contains typical examples)

Dubaria.

Eccentricosta Community: Pridoli (Keyser Formation contains typical examples)

Eccentricosta, Protathyris, Leptostrophia, Rhynchospirina, Nucleospira, Isorthis, Atrypa "reticularis", Howellella, Delthyris, Cupularostrum, Leptaena "rhomboidalis", Machaeraria, "Schuchertella", Coelospira.

Eocoelia Community: Late Llandovery-Early Wenlock (Sodus Shale is a typical example)

Eocoelia.

Eoconchidium Community: see Virgianinid Communities

Eurypteridae Community: Silurian

Genera of the Eurypteridae listed by Kjellesvig-Waering (1961).

Euryspirifer paradoxus Community: Ems (typical examples encountered in upper part of Santa Lucia Limestone and upper La Vid Shales; see Wallace, 1972, p.126)

Euryspirifer, "Dalmanella" fascicularis, Athyris, Anathyris. The commonly articulated condition in which many of the shells are found plus the shaly nature of the matrix suggest a relatively quiet-water environment. The stratigraphic position of this community (Wallace, 1972, p.126) suggests a Benthic Assemblage 4 to 5 position.

Favosites-Massive Stromatoporoid Community: Silurian-Devonian (Unit B of Santa Lucia Limestone contains typical example; see Wallace, 1972)

Cerioid corals (mainly favositids) and massive stromatoporoids. Wallace (1972, p.124) suggests that this community is equivalent to part of Lecompte's (1958, 1961) "Zone turbulente" and Struve's (1963) "Stromatoporoiden Bankriff". The massive nature of both the corals and stromatoporoids plus the well-washed nature of the matrix suggest to Wallace (1972) that this community lived in a very turbulent environment. This community occurs as an unbedded, massive unit. A position near the boundary between Benthic Assemblage 2 and 3 is suggested by its relationship to other communities (see Wallace, 1972, fig.1).

Kaljo (1972) mentions occurrences in the Silurian of Esthonia that may belong to this community. Rickard (1963) presents data suggesting that this community is present within the Lower Devonian of the Helderberg Group in New York (his Biostromes of the Manlius, for example) in a position consistent with a B.A. 2 or 3 assignment (see his fig.3).

The "Massive-Stromatoporoid-Coral Community" of Noble and Ferguson (1971) probably includes the *Favosites*-Massive Stromatoporoid Community, plus some other material, from beds of Middle Devonian age in western Canada. Mohanti's (1972) Facies I (bii) of the Cantabrian Devonian probably belongs here.

Globithyris Community: Siegen-Ems (Tomhegan Formation contains typical examples)

Globithyris. Flood (in Talent et al., 1972) has found articulated specimens of *Globithyris* associated with spiriferoids, rhynchonelloids and chonetoids in carbonate rocks occurring in New South Wales. This occurrence is radically different (high-diversity) from the low-diversity *Globithyris* Community of the Appohimchi Subprovince and Malvinokaffric Realm. This raises the possibility that the Australian *Globithyris* evolved independently, although Flood does not suggest that it is morphologically any different from the *Globithyris* found elsewhere, or that it diverged in its ecologic requirements from those found elsewhere. In any event it does not appear reasonable to assign the Australian *Globithyris* to the same community as those known elsewhere.

Gracianella Community: Wenlock-Pridoli (Roberts Mountains Limestone contains typical examples)

Gracianella. The *Gracianella* Community co-occurs with the *Dubaria* Community in the Carnic Alps at the Valentinertorl, but elsewhere is separate and probably represents a quiet-water environment.

Gypidulinid Communities: Ludlow-Frasne (Becraft Limestone contains typical examples)

Any one of the many species and genera of gypidulid occurring in abundance by itself. The abundant *Zdmir* of the Old World Realm are an excellent example in the Givet.[54]

Harringtonina Community: Wenlock-Ludlow (Silurian units in the Precordillera de San Juan)

Harringtonina, Castellaroina, Australina, Protochonetes.

Heterorthella Community: Llandovery-Ludlow (Catavi Sandstone contains typical examples)

Heterorthella.

Hipparionyx Community (Big Shell Community of Boucot and Johnson, 1967a): Siegen (Oriskany Sandstone is a typical example)

Large specimens of the following: *Acrospirifer, Costispirifer, Hipparionyx, Rensselaeria, Leptocoelia, Beachia, Meristella, Leptostrophia, Platyorthis.*

Holorhynchus Community: see Virgianinid Communities

Homolanotid-Plectonotus Community: Silurian-Givet (Chapman Sandstone contains typical examples)

Species of the trilobite genus *Homolanotus* and the bellerophontid genus *Plectonotus.*

Hughmilleriidae-Stylonuridae Community: Wenlock-Devonian

Genera of the Hughmilleriidae and Stylonuridae eurypterids listed by Kjellesvig-Waering (1961).

Hyattidina Community: Late Llandovery-Wenlock (Reynales Limestone contains typical examples)

Hyattidina.

Karpinskia Community: Siegen-Ems (massive limestones in both the Urals and Altai)

Karpinskia (quiet-water community as evidenced by many specimens still hinged together, and by presence of many small specimens). Present in Uralian Region.

Kirkidium Community: see Pentamerinid Communities

Lamellar Stromatoporoid Community: Silurian-Devonian (Unit B of the Santa Lucia Limestone contains typical example; see Wallace, 1972)

Lamellar growth forms of both corals and stromatoporoids. Wallace (1972, pp.124-125) suggests that this community is equivalent to Lecompte's (1958, 1961) "Zone subturbulent". She suggests that this communities position and greater amount of terrigenous material in the matrix suggest slightly deeper and less turbulent water than the *Favosites*-Massive Stromatoporoid Community (conclusions in agreement with those reached earlier by Lecompte). A position within Benthic Assemblage 3 is suggested by its relationship to other communities (see Wallace, 1972, fig.1). Kaljo (1972) mentions occurrences in the Silurian of Esthonia that may belong to this community. This unit is probably represented in the Lower Devonian of the Appalachians by Boucot et al.'s (1959, 1966) Members 1 and 2 of the Beck Pond Limestone. Noble and Ferguson (1971) mention the presence of this community in the Middle Devonian of western Canada.

Leptathyris Community: Eifel (Eifel age carbonate rocks in Nevada contain the typical examples)

Leptathyris circula (other taxa occur in some number but the abundance of specimens is low; see Johnson, 1970b). The Benthic Assemblage position of the *Leptathyris* Community is not certain, but its similarities in taxonomic contents (see Johnson, 1971b) to the better placed *Warrenella* Community suggest that a Benthic Assemblage 4 or possibly 5 position is reasonable.

Linoporella Community: Lower Llandovery of Esthonia and possibly Late Llandovery of Welsh Borderland. Lower Llandovery of Esthonia (Rubel, 1970) is typical

Platystrophia, Clintonella, Leptaena, Ptychopleurella, Furcitella, Stegerhynchus, Parastrophinella, Atrypopsis, Dictyonella, Savageina, Triplesia, Linoporella, Spirigerina, Cryptothyrella, Hesperorthis, Fardenia, dalmanellids.

Maoristrophia Community: Gedinne-Ems. (Upper Mount Ida Formation and Maradana Shale are typical)

Maoristrophia, Leptaena, orthotetaceans, stropheodontids, rhynchospirid, sphaerirhynchid, *Atrypa, Eospirifer, Howellella, Notoleptaena, Lissatrypa, Dolerorthis, Skenidioides, Dicaelosia, Isorthis, Strophonella, Resserella, Dalejina, Megakozlowskiella, Plectodonta.*

Microcardinalid Communities: Llandovery-Wenlock (Blackgum Formation is a typical example)

Microcardinalia or *Plicostricklandia* (but not both). The intimate association in northern Michigan (Cordell Formation) and eastern Wisconsin (Baileys Harbor), as well as in the Laketown Formation of the southwest, of *Microcardinalia* and *Pentameroides* raises the possibility that the Microcardinalid Communities may overlap the boundary between Benthic Assemblages 3 and 4 in the Late Llandovery. *Plicostricklandia* (the Wenlockage microcardinalid) is not commonly associated with *Pentameroides* or its relatives. An additional possibility is that the Michigan, Wisconsin and southwestern localities represent places oscillating near the boundary between Benthic Assemblages 3 and 4.

Mutationella Community I: Siegen (Tarratine Formation is a typical example)

Mutationella, Chonostrophiella, Leptostrophia or *Protoleptostrophia, Meristella, Acrospirifer, Leptocoelia, Salopina, Platyorthis.*

Mutationella Community II: Gedinne-Siegen (Chapman Sandstone is a typical example)

Mutationella or *Mendathyris.*

Mutationella Community III: Ems (basal Icla Sandstone is a typical example)

Scaphiocoelia or *Pleurothyrella.*

Nalivkinia Community: Silurian (Kuznetsk Basin Silurian yields typical examples)

   Nalivkinia.

Nereites Community: Silurian-Devonian (Waterville Formation has typical examples)

   Many taxa of nereitic trace fossil. The nereitic group of trace fossils is commonly associated with flysch-type sequences occurring seaward of Benthic Assemblage 5, commonly with Pelagic Community fossils like graptolites. The *Nereites* Community provides good evidence for the presence of soft-bodied metazoans.

Notanoplid Community: Gedinne-Eifel (many units in the Melbourne region, Victoria, and in the Orange region, New South Wales)

   Notanoplia or *Boucotia* or *Callicalyptella.*

Notiochonetes Community: Ems (Bokkeveld Sandstone yields typical examples)

   Notiochonetes, Australocoelia, "Schuchertella", Meristelloides.

Notoconchidium Community: Gedinne-Siegen (Florence Quartzite is a typical example)

   Notoconchidium, rhynchospirinid.

Orbiculoid-Linguloid Community: Silurian-Devonian

   Various taxa of orbiculoid and linguloid brachiopods.[55]

Orthoceras Limestone Community: Wenlock-Devonian (this community is widespread in Europe and North Africa; see Berry and Boucot, 1973d)

   Orthoceroid cephalopods, "Bohemian" bivalves, *Hercynella*, crinoidal debris (abundant *Scyphocrinus* in Pridoli age beds). This community is intimately associated with the Pelagic Community as discussed by Berry and Boucot (1973d). The *Orthoceras* Limestone Community of the Silurian-Devonian is not environmentally similar to the communities of the Early Ordovician dominated by nautiloid cephalopods and gastropods discussed by Berry (1972, p.73), although it is possible that some of the younger taxa have antecedents among the earlier taxa.

Pelagic Community: Silurian-Devonian (Road River Formation yields typical examples)

   Graptolites, acritarchs, dacryoconarid tentaculites, conodonts, chitinozoans, ammonoids, radiolaria, probably no in situ benthic shells. Boucot

(1970, pp.606-608) discussed the term Pelagic Community, and it was employed by A.M. Ziegler and A.J. Boucot (in Berry and Boucot, 1970). Earlier included with it was what is here termed the *Orthoceras* Limestone Community. Essentially the Pelagic Community is meant to include fossil assemblages made up of planktic and nektic organisms deposited on the sea floor at depths below those for which benthic shelly organisms may be expected to occur. In other words, depths greater than those for the Silurian-Devonian, in which the Starfish Community of Benthic Assemblage 6 or the *Orthoceras* Limestone Community of Benthic Assemblage 4 or 5 may be expected to occur. The term Pelagic Community is not meant to refer to associations of plankton or nekton deposited in relatively shallow-water conditions, either with or without (in some special lagoonal environments where benthos is absent) benthos. Shallow-water deposits of pure plankton and nekton may commonly be distinguished on the basis of the planktic depth zonation of the fauna. For example, the lower member of the Llandovery-age Ross Brook Formation in Nova Scotia contains virtually no fossils except for a few graptolites (both climacograptids and monograptids) but the composition of the graptolite fauna clearly indicates that in depth terms it belongs in about the Benthic Assemblage 2-3 depth, and is not labeled as a Pelagic Community for this reason. Once a planktic depth zone faunal scheme has been worked out for a group of organisms during a particular time interval it should be easy, if a complete faunal sample is available, to work out its relative depth position leading to the decision about whether or not it belongs to the Pelagic Community.

After fossil planktic communities have been more effectively defined and analysed what is here termed Pelagic Community will probably be assigned a specific term. The shallower planktic communities will be assigned other specific terms. In other words, the Pelagic Community, as used here, is a relatively deep-water community of planktic organisms that contrasts in its taxic composition with other, shallower-water planktic associations.

Boucot (1970) earlier pointed out that Cambrian-Early Ordovician agnostid faunas probably belong in the Pelagic Community position (as is concluded on more concrete grounds by Robison, 1972), and P. Wallace (written communication, 1972) suggests including the Devonian *Buchiola*-Ammonoid associations here as well (Boucot, 1970, earlier suggested that the Devonian Wissenbacher Schiefer type assemblages with ammonoids belong here).

Noble and Ferguson (1971) report typical Pelagic Community fossils and environment from the Middle Devonian of western Canada. The presence in their Pelagic Community of abundant linguloids emphasized the fact that the simple occurrence of linguloids does *not* prove the presence of Benthic Assemblage 1, as it is necessary to have knowledge of the associated fauna and also the nature of the physical geology and lithofacies picture before arriving at this conclusion. Linguloids have a wide depth span both today and in the past.

The black, styliolinid-rich shales of the Eastern Americas Realm Devonian belong to the Pelagic Community as does the very similar Chattanooga Shale and its equivalents elsewhere in North America (Woodford Shale, etc.).

Pelecypod Communities: Silurian-Devonian (Arisaig Group yields typical examples)

Infaunal pelecypods. The term Pelecypod Communities as employed here is meant to apply *only* to bivalve-dominated, infaunal communities in the Benthic Assemblage 1 and 2 position. There are other bivalve-dominated communities in the Silurian-Devonian that are not considered here (the reef-associated *Megalomus* Community, a quiet-water entity as shown by the commonly articulated valves, of the North American Silurian is a good example).

Pentamerinid Communities: Late Llandovery-Pridoli (Salamonie Formation is a typical example)

Pentamerus, Pentameroides, Kirkidium, or any other genus of the Pentamerinae but none of them together.

Pentamerus Community: see Pentamerinid Communities

Pentamerus and *Pentameroides* do not occur together except for the few occurrences like the Schoolcraft Dolomite where both types intergrade as members of a single species; the point of divergence of the two genera.

Pentlandella Community: Late Llandovery. (Beds in the Haswell Burn, Pentland Hills Scotland, conatin typical material)

Pentlandella. The *Pentlandella* Community consists chiefly of specimens of *Pentlandella* occurring in fine-grained matrix, either articulated or disarticulated.

Normally they occur almost entirely as a single taxon community, but may be associated with *Atrypa* "*reticularis*" and a few other taxa. The distribution of *Pentlandella* from the Baltic region to Britain, where it is never a really common shell, is reminiscent in area of the Later Silurian distribution pattern of the more abundant genus *Dayia. Pentlandella* probably is a Benthic Assemblage 2 or very shallow Benthic Assemblage 3 taxon (it may range through 2 into 3) as judged both by its associations and its stratigraphic relations.

Phaceloid Coral-Ramose Alveolites Community: Silurian-Devonian. (Unit A of the Santa Lucia Limestone contains typical examples; see Wallace, 1972)

Phaceloid and ramose corals. Wallace (1972, p.125) points out that this community is associated with a higher percentage of terrigenous material, that the corals are commonly found in life position, and that the absence of any signs of current activity point to a non-turbulent environment. A position either within Benthic Assemblage 3 or 4 is suggested by its relationships to other communities (see Wallace, 1972, fig.1).

Occurrences of this type are common in the Silurian of the Northern Hemisphere. The Community is well represented in many of the coral occurrences in both the Lower and Middle Devonian of the Eastern Americas Realm, as well as in Europe.

The Phaceloid Coral-Ramose Alveolites Community is probably almost equivalent to the "Branching Tabulate-Coral Community" reported by Noble and Ferguson (1971) from the Middle Devonian of western Canada.

The *Spinatrypina-Thamnopora* Community of Johnson and Flory (1972) from the Eifel-age Telegraph Canyon Formation in Nevada characterized by abundant *Cryptatrypa, Spinatrypina, Emanuella*, and *Thamnopora* appears to belong in the Phaceloid Coral-Ramose *Alveolites* category. The presence of a high percentage of articulated brachiopods, some possibly in growth positions on corals, is consistent with this view, as is the microspar nature of the matrix (presumably a lithified mud), all evidence pointing to a quiet-water environment. A Benthic Assemblage 3 or even 4 position is reasonable for the *Spinatrypina-Thamnopora* Community. Mohanti's (1972) Facies IV of the Cantabrian Devonian belongs to this Community.

Platymerella Community: see Virgianinid Communities

Platyorthis-Proschizophoria Community: Gedinne (Schistes de Mondrepuits are a typical unit)

Platyorthis, Proschizophoria, Mesodouvillina, "Schuchertella", Howellella, Podolella.

Platyschisma Community: Pridoli (*Platyschisma* Beds of the Welsh Borderland are typical examples)

Platyschisma. This low-spired gastropod is found in great abundance in beds of Downton age in the Welsh Borderland, but has also been noted in coastal Maine in large aggregations. The *Platyschisma* Community has not been noted except in Pridoli-age beds, but this may reflect lack of attention paid to gastropod-bearing beds plus the smaller chances of preserving brackish environments.

Plicoplasia Community (Glenerie Community of Boucot and Johnson, 1967a): Siegen (Glenerie Limestone is the typical example)

Medium to small specimens of the following: *Acrospirifer, Costispirifer, Hipparionyx, Leptocoelia, Beachia, Meristella, Leptostrophia, Platyorthis, Plicoplasia, Pegmarhynchia.*

Prosserella Community: Eifel (Detroit River Formation is a typical example)

Prosserella. A hypersaline, low diversity, quiet-water (shells commonly articulated) Community.

Protathyris Community: Wenlock-Pridoli (Hardwood Mountain Formation has typical examples)

Protathyris.

Protatrypa Community: Lower Llandovery (6a beds on Malmoya in the Oslo region are a typical example); a single taxon community that probably belongs in about the Benthic Assemblage 3 position.

Protatrypa. Shells belonging to this community may be commonly articulated and occur in a quiet-water, muddy environment, or become disarticulated as in occurrences in Nova Scotia (Glencoe Brook Formation) or in northern Newfoundland. *Protatrypa* may also occur in higher-diversity communities including such taxa as *Mendacella, Eostropheodonta mullochensis* and *Cryptothyrella* of the *Cryptothyrella-Mendacella* Community which indicates something about the environmental span of *Protatrypa.*

Quadrifarius Community: Pridoli-Early Gedinne (typified by the Stonehouse Formation)

Quadrifarius, Salopina, Protocortezorthis, Rhynchospira, Protochonetes.

Quadrithyris Community: Gedinne (Basal limestone of the Mandagery

Park Formation is typical; Savage, 1968, 1969, 1970, 1971, oral communication, 1972)

Savage is the first to recognize this community, and also note its occurrence within the Windmill Limestone breccia of Nevada. *Cyrtina, Dolerorthis, Quadrithyris, Meristella,* rhynchonellids, several types of *Atrypa, Isorthis, Howellella, Plicoplasia, Grayina, Proreticularia.*

Rensselandid Communities: Late Eifel-Givet (Massenkalke has typical examples)

Rensselandia, Subrensselandia, or any other related taxon in the subfamily, but not more than one together.

Rhenish Complex of Communities: Gedinne-Ems (many units in both the Ardennes and Rhineland)

Tropidoleptus, Eodevonaria, Anoplotheca, Septathyris, Anathyris, Acrospirifer, Platyorthis, Rhenostrophia, Meganteris, Schizophoria, Bifida, Hysterolites, "Leptostrophia", Stropheodonta, Athyris.

Rhenorensselaeria Community: Siegen-Ems (York River Formation of the Mount Albert region contains typical examples)

Rhenorensselaeria.

Rhynchonellid Community: Silurian-Devonian (Clemville Formation contains typical examples)

Various species included under the portmanteau term *"Camarotoechia"* (in Britain many of these forms have been identified as *"Camarotoechia" decemiplicata*).

Salopina Community: Late Llandovery-Pridoli (Whitcliffe Beds are typical example)

Salopina, Isorthis, Protochonetes, Howellella, "Camarotoechia", leptostrophid. The occurrences of *Salopina* in the Lower Devonian of Nevada and adjacent regions do not belong to the *Salopina* or related Benthic Assemblage 2 communities. These western North American occurrences in deeper-water positions (Benthic Assemblage 3-5 positions) belong to species apparently modified for different conditions than those found in the Silurian or in the Devonian of the Appohimchi Subprovince and Malvinokaffric Realm. (These western North American forms are almost septate as in *S. submurifera,* and should be included with the ecologically similar *S. crassiformis* of Podolia.)

Salopina-Derbyina Community: Ems (highest marine beds in Icla Devonian)

    Salopina, Derbyina.

Scolithus Community: Paleozoic

Seilacher (1967), in outlining the trace fossil communities, has included the *Scolithus* Community in a very nearshore, shallow-water position. In our scheme it belongs in Benthic Assemblage 1, where it is seldom accompanied by shelly fossils. Scattered examples of this community are recorded from many parts of the Paleozoic, including the Silurian and Devonian.

Spinatrypina-Thamnopora Community: see Phaceloid Coral-Ramose *Alveolites* Community

Spinella-Buchanathyris Community: Ems (Buchan Caves Limestone provides typical examples)

    Spinella, Buchanathyris.

Starfish Community: Silurian-Devonian (Bundenbach Slate is a typical example)

    Starfish.

Stricklandid Communities: Llandovery-Early Wenlock (La Vieille Formation is a typical example)

    Stricklandia or *Costistricklandia* (but not both).

Striispirifer Community (formerly — Boucot, 1970 — termed the Rochester Shale Community): Late Llandovery-Ludlow (Rochester Shale is typical example)

    Atrypa "*reticularis*", *Amphistrophia, Coolinia, Dalejina, Howellella, Leptaena* "*rhomboidalis*", *Mesodouvillina, Meristina, Resserella, Plectodonta, Protomegastrophia, Diabolirhynchia, Striispirifer, Eospirifer, Strophonella, Trematospira, Sphaerirhynchia, Whitfieldella, Ancillotoechia, Rhynchotreta.*

Stringocephalus Community: Givet (The German Massenkalke is a typical example)

    Stringocephalus or any other genus of the Stringocephalinae (but not more than one at a locality).This community of heavy-shelled brachiopods may be present either adjacent to or far removed from reef associations. This community occurs in about the Benthic Assemblage 3 position, and is indicative of rough-water conditions as shown by the commonly disarticulated condition of the shells, the commonly sorted condition of the shells, and the coarse nature of the matrix (sand- and gravel-size material) in most cases.

Stromatolite Community: Silurian-Devonian (Wallace, 1972, p.125, reports this community from the Santa Lucia Limestone)

Stromatolites. Wallace (1972, p.125) considers that the Stromatolite Community represents a nearshore, intertidal environment, subject to periodic desiccation. A bird's-eye limestone texture and suncracks are commonly developed. This community is concluded to be present in about the Benthic Assemblage 1, or slightly landward (brackish) position.

Kaljo (1972) mentions occurrences in the Silurian of Esthonia that may belong to this community. Laporte (1963) reports this unit in the earliest Devonian of the Appalachians

Subrianinid Communities: Wenlock-Ludlow (Bowsprings Limestone is a typical example)

Conchidium, Cymbidium, Subriana, Severella or *Aliconchidium* but none of these together.

Tentaculitid Community I: Gedinne (Manlius Formation yields typical examples)

Howellella, leperditids, tentaculitid tentaculites (see Ludvigsen, 1972).

Trimerelloid Community: Llandovery-Ludlow (Guelph Dolomite yields typical examples)

Trimerella or *Dinobolus* or *Rhynobolus* (but none of these together).

Tropidoleptus Community: Gedinne-Frasne (Hamilton Group yields typical examples)

Tropidoleptus.

Tuvaella Community: Llandovery-Ems (units in the Silurian of Tuva)

Tuvaella, Eospirifer.

"Uncinulus" orbignyanus Community: Ems-Eifel. (Typical example in upper part of Santa Lucia Limestone; see Wallace, 1972)

"Uncinulus" orbignyanus. The stratigraphic position of this community (Wallace, 1972) suggests a position in Benthic Assemblage 3. The low taxonomic diversity implies some form of environmental restriction.

Undivided Communities (for Benthic Assemblage 5, Fig.3, 4, 5): Silurian

Cyrtia, Brevilamula in the Llandovery, *Clorinda* in the Late Silurian, *Coolinia, Atrypa* "reticularis", *Glassia, Resserella, Visbyella, Skenidioides, Dalejina, Leptaena* "rhomboidalis", *Dicaelosia, Plectatrypa, Dolerorthis, Leangella, Pentlandina, Plectodonta.*

Vagrania-Skenidioides Community: Siegen-Ems (Typical examples in the Prongs Creek Formation)

"Dolerorthis", Skenidioides, Salopina (crassiformis type), *Muriferella, Schizophoria, Dalejina, Cortezorthis, Isorthis, Kayserella, Leptaena, Brachyprion, Cymostrophia, Mesodouvillina, Phragmostrophia,* "Schuch-tertella", *Aesopomum, Grayina, Gypidula* sp. 1 (of Lenz), *Latonotoechia, Thliborhynchia, Cupularostrum, Nymphorhynchia, Werneckeella, Vagra-nia, Davidsoniatrypa, Atrypa, Desquamatia, Spinatrypina, Spinatrypa, Protatrypa, Nucleospira, Ambocoelia, Plicoplasia, Cyrtina, Reticulariopsis, Atrypina, Howellella, Parachonetes.*

The *Vagrania-Skenidioides* Community (see Ludvigsen, 1970; Lenz and Pedder, 1972 for faunal lists) is a high-diversity community, belonging to the *Dicaelosia-Skenidioides* Community Group (see Fig. 35), that occurs in the Benthic Assemblage 4—5 position. Occurrences are widespread in Arctic and Northern Canada (the Arctic Islands, Yukon, northern British Columbia, Alaska Boundary region), and the Community is also represent-ed in the *pinyonensis* zone of Nevada in the Cortez Range (the more easterly occurrences of the *pinyonensis* zone, following Niebuhr's, 1973, studies, belong to different communities occurring in the Benthic Assem-blage 3 position). The *Vagrania-Skenidioides* Community may be the source, at least in largest part, of the *Warrenella*-rhynchonellid Communi-ty of the Middle Devonian. The *Vagrania-Skenidioides* Community is probably present in the Lower Devonian of Novaya Zemlya (Boucot et al., 1969) and elsewhere in Siberia and Central Asia, and probably it repre-sents quiet water.

Virgiana Community: see Virgianinid Communities

Virgianinid Communities: Ashgill through Middle Llandovery (Mayville Dolomite is a typical example)

Eoconchidium, Holorhynchus, Virgiana, Platymerella, Borealis, Nondia, Pseudoconchidium or *Virgianella* but none of these together.

Warrenella-Rhynchonellid Community: Givet (Common in Givet age beds of western Canada and western U.S., especially in Nevada)

Noble and Ferguson (1971) define this community and it is also discussed by Johnson (1971b, pp.306-307, upper fauna of Table 2). It is in a position that corresponds to Benthic Assemblage 4 and possibly 5 (shallower in position than the Pelagic Community).

Warrenella, Leiorhynchus, other rhynchonellids. Enough taxa are present to define it as a moderately high-diversity community, although not high-diversity.

Zdmir Community: see Gypidulinid Communities

Zoophycos (Taonurus) Community: Devonian (Esopus Formation yields typical example)

Taonurus caudagalli. The occurrence of the low-diversity (essentially no shelly fossils; only the trace-fossil genus *Taonurus*) *Taonurus* Community between the *Amphigenia* Community and its equivalent with *Etymothyris* in the Green Pond Outlier, N.Y. (Boucot et al., 1970) indicates a Benthic Assemblage 3 to 5 span.[56] The preservation of the *Taonurus* structures suggests no deep reworking of the sediment surrounding the burrow. The absence of shells must indicate conditions inimical for shelly growth or preservation. This Community is fairly widespread in the Middle Paleozoic of the world, and occurs elsewhere in the geologic column. P. Wallace (written communication, 1972) writes in reference to *Zoophycos* (sic. *Taonurus*): "One pecularity is that the beds *never* have any body fossils in, though beds immediately above and below may be very shelly. One suggestion is that "*Zoophycos*" (whatever it is) ate the shells and that CO_3 was subsequently removed by solution. Seems pretty logical. Certainly *Zoophycos* is always very abundant where it does occur." Seilacher (1967) deals in more detail with this community, and Bradley (1973) has some ideas concerning the actual organism responsible for the *Zoophycos* structure.

CHAPTER 7

Biogeographic Framework

Provincialism or Environment; Biogeographic Unit or Ecologic Unit?

Differing biotas of the same age may be explained in two ways. First, the environments may be different. Second, the difference may be produced by provincial barriers to faunal migration and movement, so that many taxa remain restricted to their areas of origin. Conditions conducive to production of both environmentally-controlled and provincially-controlled faunas must have existed side by side at least since the beginning of the Phanerozoic. Our task is to try and assess the relative importance and dominance of the two factors, time interval by time interval and region by region.

Basically this question is dealt with by assuming that if differing biotas of the same age are widely scattered over the world (essentially the continents when sampling shallow water Silurian-Devonian rocks) then environmental control rather than provincialism resulting from barriers is the responsible factor. If, on the contrary, differing biotas are not widely scattered, barriers to migration and movement may have entered into the picture. In an earlier paper (Boucot, 1970) I mentioned that inspection of the "clumping" or "non-clumping" behavior of coeval faunas was important in making these distinctions. Another way of phrasing the problem is to note whether particular taxa have a wide, intercontinental distribution (cosmopolitan distribution) or whether taxa with common ancestors have a narrow, restricted distribution (possibly provincial). For example, if a widely distributed taxon gives rise to taxa that are also widely distributed it may be concluded that their local distribution is only environmentally-controlled. If, however, the widely distributed ancestral taxon gives rise to a number of taxa with narrow geographic distributions it may be concluded that barriers have resulted in provincialism. This conclusion is based on the assumption, which should be true as an overall generalization although wrong for a small percentage of cases, that the closely related descendents of an individual taxon will not diverge far in their broad environmental requirements. As the time interval between ancestor and descendent increases this assumption will become increasingly unreliable;

after a sufficient interval has passed the assumption may become essentially useless.

Biogeographic Units

Kauffman (1973) has set up a quantitative scheme for defining biogeographic units as follows: Realms 75% endemic genera or over; Regions
50-75% endemic genera; Provinces 25-50% endemic genera; Subprovinces
10-25% endemic genera; and Endemic Centers 5-10% endemic genera.
Kauffman's efforts at quantification are the first on a worldwide basis. In
dealing with the Silurian-Devonian I am partly following in his footsteps
with, however, a few modifications suggested by the nature of the data as
discussed below. It is important to keep in mind that Kauffman's percentages do not consider the cosmopolitan genera. Table III (p. 156) lists
the total number of taxa, percentages of endemic genera, and percentage
of cosmopolitan genera.

In addition attention is paid to the number of genera missing from a
biogeographic unit (cosmopolitan and endemic) as in many instances
absences are construed to be fully as important (if not more important)
than presences. This question of absence is particularly important **for**
faunas thought to occur in cold regions. For example, today the polar
flora and fauna is really more characterized by absences than by presences
(the absence of reptiles and amphibians plus many major placental groups
is more important than the presence of endemic taxa of groups present in
other regions). Obviously, the significance of absences depends on one's
having a thoroughly representative sample (see Fig.28 and accompanying
discussion).

The attempt to adhere to Kauffman's scheme of classification results in
some anomalies. (1) certain Realms have not been subdivided as they
appear to be very cosmopolitan in their contents; (2) certain Realms may
not have precisely the "right" percentages of endemic taxa specified in
Kauffman's scheme for their subdivisions (including Regions, Provinces,
and Subprovinces) but the terms have been used (in some cases it is
thought that future work will change the percentages to accord with
Kauffman's scheme; see Fig.28 and accompanying discussion) for other
reasons.

For example, the Malvinokaffric Realm of the Devonian does not have
75% endemic taxa, but the unusually low diversity (small total number of
taxa) as contrasted to the allied Eastern Americas Realm suggests that low
diversity is an important consideration as well. Additionally it will be
noted that the Eastern Americas Realm has no intermediate level Regions
or Provinces, only Subprovinces — this anomaly is a result of employing

Kauffman's scheme requiring that a fixed percentage of endemic taxa be present before each of the units may be employed: it does not imply that the missing Regions and Provinces are actually present. This lack of uniformity in units is probably in largest part related to the complexity of the actual situation.

The experience gained to date with Silurian-Devonian biogeographic units suggests that Kauffman's scheme needs some modification to take care of the low taxic units. For these low taxic units it seems clear that a mere percentage of endemic taxa fails to adequately emphasize the uniqueness of the units extremely low taxic diversity. What may be needed is a biogeographic ranking of units that considers both percentage of endemic taxa and also total number of taxa. Possibly what we are observing is merely a contrast between cold- and warm-water regions which will require a somewhat different biogeographic treatment.

Fischer's (1960) treatment of benthic invertebrate diversity along the east coast of North America provides an excellent illustration of provincialism for shallow-water benthos being paralleled by a striking decrease in diversity going from warm to cold water.

An additional problem encountered in trying to apply Kauffman's scheme is the question of what to do with a biogeographic unit in which the percentage of endemic taxa changes rapidly and consistently in time from one of his categories into another. One observes a Subprovince developing into a Realm. These changes make for nomenclatorial difficulties. Should one use the same name or different names for a developing biogeographic unit? More experience in other parts of the Phanerozoic is probably needed before hard and fast rules are put forward. In any event, Kauffman's insistence that more attention be paid to percentage of endemic taxa in time for the defining of biogeographic units is a definite step forward in the better definition of our problems.

In earlier studies of Silurian and Devonian provincialism (Boucot et al., 1969; Boucot, 1970; Boucot and Johnson, 1973; Johnson and Boucot, 1973) the terms Province and Subprovince, respectively, were used in a very qualitative manner. The change in terminology used in this work does not indicate any change in concept; merely an effort recommended by Kauffman (written communication, 1972) to try and use the same nomenclature employed for the Mesozoic and Cenozoic.

Biogeographic unity in the marine environment

There is no intrinsic reason why the shallow-water benthos, bathyal benthos, abyssal benthos, various depth necton and plankton in the oceans should be governed by precisely the same isolating factors. This being the case it is critical in any biogeographic analysis covering a number of groups known from a large region that groups potentially subject to different isolating factors be kept apart when trying to determine biogeographic units. For example, it is not hard to imagine that tropical reef organisms might exhibit a very different distribution pattern than would abyssal benthos occurring over the same region. In other words, we have the potentiality for overlapping, but contemporaneous, biogeographic units that may be largely unrelated to each other.

Causes of Provincialism and Faunal Barriers to Migration

The basic cause of provincialism in most instances is the presence of faunal barriers to migration. However, in trying to uncover the cause or causes of provincialism in specific examples, geologists, biologists, and oceanographers each approach the problem differently. The variety in approach understandably results from the conditioning each specialist receives from his own background and experience.

Land barriers, if sufficiently large, can act as very effective barriers to interbreeding and mixing of faunas in the marine environment, particularly if coupled with temperature differentials as in the present New World. However, temperature barriers unrelated to land barriers in a direct, intersecting manner can be just as effective. Examples of such temperature barriers are the present-day polar faunas of both land and sea and the fauna strongly affected by the Gulf Stream and by the Labrador Current. Fischer's (1960) compilation of both diversity and provincial data for the east coast of North America that correlates well with temperature changes makes this point very clear. Distance may act as an important factor in isolating marine, shallow-water faunas with a short enough larval life if broad deep seas separated regions with favorable shallow-bottom environments. Distance in this sense, if enough deep seas are interposed, may act as a filter that excludes shallow-water benthic taxa progressively across a relatively great distance with the result that faunas inhabiting similar environments at either end are essentially isolated from each other. Distance between platforms acts as a filter in this sense for the widely distributed level-bottom communities, and distance on even the same platform acts as a filter in this sense for the far less widely distributed reef communities. MacArthur and Wilson (1963) have carefully considered the questions involved with distance, immigration, rate of evolution, and rate of extinction that affect land dwelling organisms living on islands of varying size and distance from each other. Many of their conclusions may well apply directly to shallow-water benthic invertebrates. Barriers may result from lack of suitable bottoms for benthic organisms, lack of adequate nutrients or oxygen, large areas of excessively hypersaline or brackish water, or any one of several other variables, both physical and biological, within the environment. A biologic barrier may even result from the existence of a well-adapted taxon in a given area that acts effectively to prevent the spread of an ecologically similar taxon. Regions characterized by certain diseases or infested with certain parasites may also act as effective faunal barriers restricting the migration of organisms independently of any physical barriers.

Conclusions about the causes of provincialism and the type of faunal barriers responsible are unusually difficult when dealing with materials from the geologic record. All that can be done to arrive at such conclusions is to try to understand what mechanisms have been held responsible as faunal barriers at the present time, to search for geologic evidence suggesting the presence of such barriers, and also to search for geologic evidence suggesting the presence of faunal barriers different from those thought to be operative at the present time. Finally, caution should be used before accepting the correlation between two events, in this case a potential faunal barrier to migration occurring between two faunas, as the actual cause unless the evidence becomes very great.

Salinity barriers to faunal dispersal for marine organisms

The distribution of shallow-water marine animal communities and Benthic Assemblages has the potential for providing information about uniformity of the environment.

One may conclude that areas with strata of the same age, and containing the same animal communities, were probably subject to the same range of physical environments. Benthic Assemblage and animal-community (sic., animal-community) maps by Ziegler and Boucot (in Berry and Boucot, 1970) and by Boucot and Johnson (1967a) show that certain animal communities have geographic distributions extending nearly across North America (the *Virgiana* and *Pentamerus* Communities of the Llandovery are excellent examples from the Silurian, the *Hipparionyx* Community is a good example from the Early Devonian). Counterparts of these widespread Silurian and Early Devonian communities are found in the Late Cambrian and Ordovician of the North American Platform and also in the Early Paleozoic Platform faunas from other continents. There is no reason to suspect that these widespread communities had broader tolerances environmentally than other communities, which in turn suggests that widespread relatively homogeneous environmental conditions prevailed during long intervals of time on the platforms of the Early Paleozoic. In an attempt to understand the environments present on the vast platforms of the Cambro-Silurian and Devonian, community distribution provides an elegant tool.

The conclusion about time intervals with widespread homogeneous communities is very difficult to reconcile with the inferences made about the nature of the Early Paleozoic platforms and the epicontinental seas on them. Many of these Early Paleozoic platforms were of great lateral extent, some up to 2,000-3,000 miles in diameter, low-relief, essentially featureless structures lying at or near sea level. During those time intervals when a platform was slightly below sea level (most estimates of water depth do not exceed a few hundred feet as evidenced by the presence of shelly invertebrate faunas in abundance, reef-like structures, and calcareous algae occurring in the zone of photosynthesis) one cannot expect mechanical mixing of oceanic sea water from the edges of the platform (Shaw, 1964) to have maintained the almost constant salinity across the platform indicated by such widespread communities. In the absence of a constant movement of oceanic, normal-salinity sea water from the platform margins into the internal parts, one could infer that due to evaporation one should find an abundance of evaporitic deposits across the interiors of platforms. However, although present on most Early Paleozoic platforms, evaporitic deposits are not invariably widespread either in time or space. When we do encounter such deposits they are commonly characterized by low-diversity faunas easily interpreted as having lived in a specialized environment (the *Prosserella* Community of the Eifel Detroit River Group is a good example). These low-diversity faunas are no more widespread than are the evaporitic deposits themselves in either time or space. The continuity of a single-invertebrate animal community over vast reaches of platform can be taken as evidence for a remarkably uniform environment; the reverse is equally true. Therefore we can infer that for much of Early Paleozoic time many of the platform areas were somehow able to maintain relatively normal salinity. It must be emphasized that the vast continental platforms of the Early Paleozoic are not comparable to the relatively small open-shelf conditions characterizing the present as well as much of the post-Paleozoic.

Thus, one of the puzzling questions about these epicontinental seas is how a normal marine salinity was maintained in the central part of a large sheet of very shallow sea water. To maintain normal salinity far into the internal parts of the platform the effects of evaporation would be continually requiring an admixture of fresh sea water that one could not expect to be delivered, or an offsetting amount of fresh water. This being the case one would predict that the epicontinental seas in the more interior parts (of those lacking large riverine sources of fresh water) of platforms, with restricted circulation, should have become progressively hypersaline

(Shaw, 1964); a well-established salinity gradient should have been present. During some time intervals on some platforms one does in fact find evaporitic deposits in the internal parts, but during other intervals such deposits are absent.

It is tempting to suggest that the chief source of fresh water on the platform, serving to freshen the interior region of the epicontinental seas will be the runoff from low-lying land areas (Shaw, 1964), but this mechanism can only be important if a large low-lying area is present. Boucot et al. (1969) tried to maximize the Early Devonian platform land areas but the new reconstructions (Fig.37-40) are more realistic and indicate that such land areas for the North American Late Silurian-Givet were probably either absent or not very large; certainly not large enough to freshen the platform. Rainfall appears to be the only mechanism adequate for the task of sufficiently freshening the water on an entire platform, in the absence of good circulation, to account for similar community distributions across the platform and in adjacent geosynclines. Rainfall evidently was insufficient during the Late Silurian-Givet interval, as shown by the widespread occurrence of evaporites.

For the areas bearing evaporite deposits there has been a time interval of restricted circulation during which the rate of evaporation exceeded the rate of inflow of normal-salinity sea water or the addition of enough rainwater[57] and mixing to maintain a normal salinity. Shaw (1964) felt that restricted circulation in the interior parts of a platform was due chiefly to the friction present in the very thin sheet of water. A further contributory factor to restricted circulation certainly has been the presence of reef-like masses. In some regions of evaporites the presence of reef-like masses has been viewed as the chief factor that restricted circulation, but on a platform-wide scale it probably should be thought of as contributory in some but not all cases. I conclude following Shaw that inflow and mixing of normal-salinity sea water into the interior of a platform is probably a very minor factor in the maintenance of normal salinities. Evaporites require low rainfall, although the absence of evaporites does not prove high rainfall, particularly if circulation has been efficient. Using this approach one can then conclude that the marine invertebrate animal community distributions (normal-salinity and hypersaline communities) and evaporite deposits left behind by epicontinental seas should provide evidence about the relative rate of evaporation and circulation restriction of the sea water in the internal parts of platforms. Shaw (1964) dealt ably with the epicontinental sea question and the lithofacies belts to be expected under conditions in which evaporation exceeds rainfall input of fresh water. The discussion here differs only in using animal community distributions as an index of salinity (normal or hypersaline)

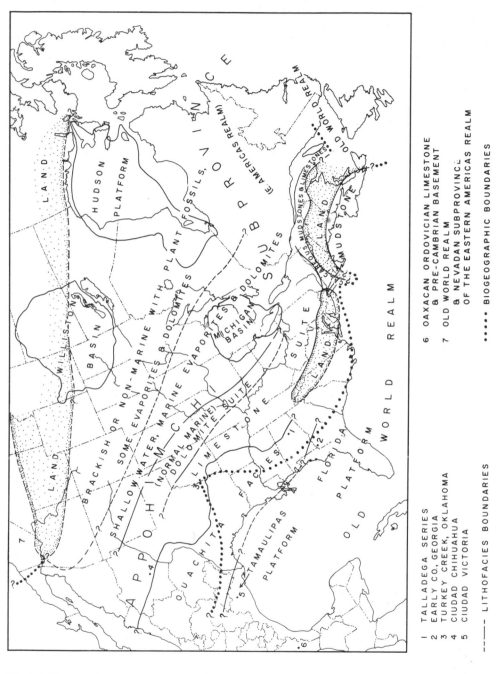

Fig. 37. For caption, see p. 276.

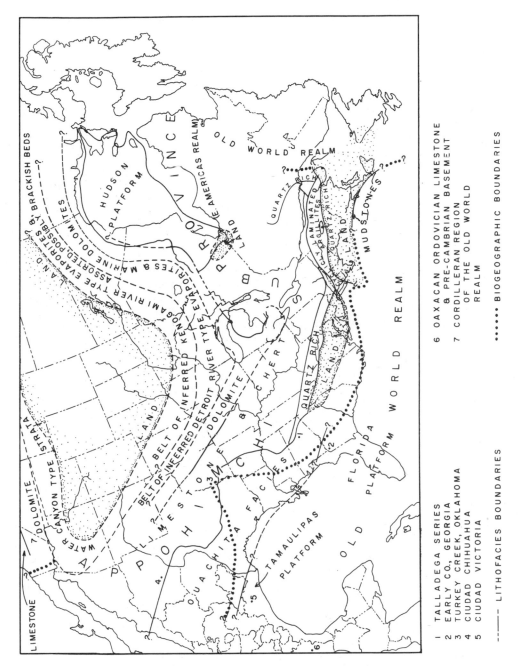

Fig. 38. For caption, see p. 276.

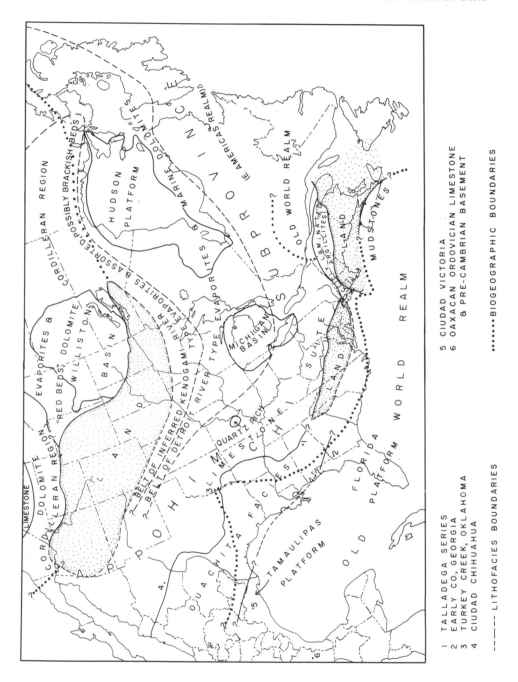

Fig. 39. For caption, see p. 276.

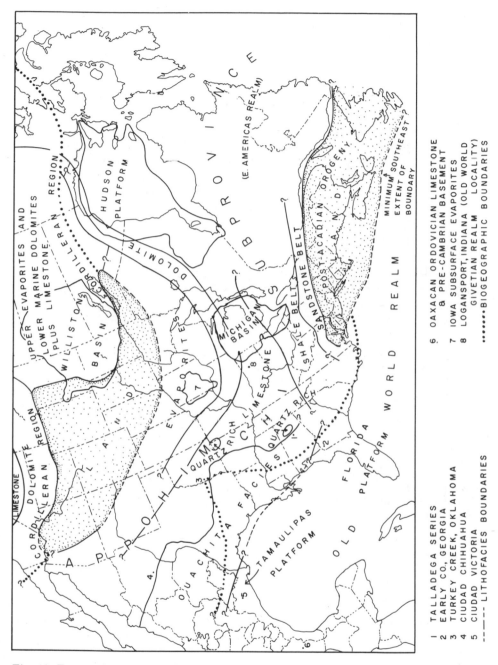

Fig. 40. For caption, see p. 276.

that may indicate those time intervals and locations where rainfall and circulation has been adequate to maintain normal salinities far from a platform margin.

My assumption is that where an animal community containing a "normal" assortment of taxa spans an entire platform, normal salinity was maintained. This would be particularly so if the same animal community occurs within the geosynclinal environment where normal salinity would be expected because of readier access to the oceanic reservoir. For these cases one can conclude that there has been enough rainfall and circulation internally on the platform to maintain a normal salinity. Where vast evaporites occur on a platform associated with animal communities characterized by a greatly reduced number of taxa (far from the "normal"), one may conclude that rainfall and circulation have not been adequate to maintain salinity and that evaporation has greatly exceeded input of fresh water from rainfall.

I postulate that in those times when extensive evaporitic deposits were precipitated on the platforms, large masses of hypersaline water may have served as effective barriers to organisms adapted to normal sea water particularly at the larval stage, i.e., communities occurring peripheral to the hypersaline regions, and may be at least partly conducive to the devel-

---

Fig. 37 Gedinne lithofacies and paleogeography for eastern North America (amplified and modified after Boucot and Johnson, 1967a).

The occurrence (Jones,1944) of 400 ft. of Devonian bedded chert in Pecos County, Texas, suggests the presence of an Arkansas Novaculite-Caballos Formation chert facies to the south of the Limestone Suite and distinctly north of the Ouachita Line, which helps to further tie in the Ouachita Facies to the Platform.

Fig. 38. Ems lithofacies and paleogeography for eastern North America, amplified and modified after Boucot and Johnson, 1967a.

The occurrence (Jones,1944) of 400 ft. of Devonian bedded chert in Pecos County, Texas, suggests the presence of an Arkansas Novaculite-Caballos Formation chert facies to the south of the Limestone Suite and distinctly north of the Ouachita Line, which helps to further tie in the Ouachita Facies to the Platform.

Fig. 39. Eifel Lithofacies paleogeography for eastern North America.

The occurrence (Jones,1944) of 400 ft. of Devonian bedded chert in Pecos County, Texas, suggests the presence of an Arkansas Novaculite-Caballos Formation chert facies of the Ouachita Line, which helps to further tie in the Ouachita Facies to the Platform.

Fig. 40. Givet (Hamilton) lithofacies and paleogeography for eastern North America.

The occurrence (Jones, 1944) of 400 ft. of Devonian bedded chert in Pecos County, Texas, suggests the presence of an Arkansas Novaculite-Caballos Formation chert facies to the south of the Limestone Suite and distinctly north of the Ouachita Line, which helps to further tie in the Ouachita Facies to the Platform.

opment of provincialism. An examination of the Silurian-Devonian record is very illuminating in this regard.

The Early Silurian is a time in which the Northern Hemisphere has very cosmopolitan faunas, and also in which individual animal communities (the *Virgiana* Community of the early Middle Llandovery, the *Pentamerus* Community of the Late Llandovery-Early Wenlock) extend from one side of a platform to another (North America is the prime example). The Early Silurian is also a time for which we lack extensive evaporitic deposits. In contrast, in the Late Silurian through the Middle Devonian, numerous and major evaporitic deposits are present. An increasing degree of provincialism correlates with the increasing area of evaporitic deposits from the beginning of the Late Silurian into the Middle Devonian. Some, but not all, provincial boundaries occur adjacent to evaporite deposits. Meyerhoff (1970a,b) reviewed evidence regarding the distribution of Late Silurian and Devonian evaporites and also pointed out the tendency for evaporitic deposits to be minimal in extent during times of major continental glaciation and vice versa. His suggestion is compatible with the evidence for the Silurian and Devonian. Within the Silurian-Devonian there is some evidence for a coincidence of times of provincialism among marine invertebrates and the occurrence of extensive evaporitic deposits; and the reverse as well, a correlation between times when individual marine animal communities are widespread and evaporitic sediments are nearly absent but the reverse is true during the provincial Late Ordovician.

Meyerhoff's inverse relation between abundant glacial deposits and abundant platform evaporites may be explained in terms of Northern Hemisphere pluvial intervals being counterparts of Southern Hemisphere glacial episodes (e.g., Gondwana); the pluvial intervals are concluded to have provided enough rainfall to the platform interiors to have prevented the formation of large masses of hypersaline water (the Late Ordovician-Early Llandovery laterites of Nova Scotia support this conclusion). Evidence bearing on this hypothesis may be provided by careful study on a worldwide basis of animal community distributions for other time intervals.

The occurrence in the Northern Hemisphere (Arisaig, Nova Scotia; see J.F.Dewey, in Boucot et al., 1973) of lateritic beds interlayered with andesitic volcanics, both of either Late Ordovician or Early Llandovery age, is compelling evidence for the existence of an area subject to a tropical or subtropical possibly monsoonal, very humid climate. The presence of uncommon reef-like masses over some of the Northern Hemisphere Late Ordovician and Early Silurian (as well as common Late Silurian reef-like masses) is consistent with this observation. These data combined with those for a Southern Hemisphere glacial climate for Africa, Arabia and at least a large part of South America south of Venezuela (see Berry and Boucot, 1973a) indicate the presence in Late Ordovician-Early Silurian times of a profound climatic gradient.

The relative abundance in the Northern Hemisphere nearshore to non-marine Late Ordovician-Early Silurian of abundant red beds (units like the Queenston Formation, Thorold Formation, Sequachie Formation, much of the Scottish Midland Valley Inlier Silurian, etc.) is consistent with this conclusion. Red Beds of this age are not present in any abundance within the glaciated regions of Africa and South America (the Zapla Iron Ore of northern Argentina is easily related to the specialized conditions of near-shore marine iron precipitation that may occur in either a warm or a cold climate). The brackish-water nonmarine Ordovician and Silurian sand-stones of North Africa are not characterized by red colors of primary origin (chiefly yellows and oranges much of which is due to relatively recent oxidation and surface weathering under the influence of a very different climatic regime).

For the Silurian one may outline a region with a colder climate — the Mediterranean region, Africa and South America from Lake Titicaca and the Amazon region South — on the basis of both ecological evidence (summarized by Berry and Boucot, 1972a; their Malvinokaffric Realm — sic., Province — of the Silurian) and the presence of glacial and glacially derived deposits between the Late Ordovician and the Early Silurian (summarized by Berry and Boucot, 1973a).[58] This evidence indicates that the first half of the Silurian (Early Llandovery through Early Wenlock) was characterized by a region of colder climate in the area described above, plus a warmer region to the north (North America, northern South America, northern and eastern Europe, most of Asia and Australia) in which carbonate rocks (chiefly of bioclastic origin) are abundant. This northern region is concluded to have had, during the first half of the Silurian, a rainfall regime (a pluvial interval and circulation) permitting the maintenance of salinities on the continental platforms adequate to allow individual animal communities to span the platforms from both north to south and east to west. This situation changes after the end of the lower half of the Silurian to a regime in which the colder region still is present on the south, but in which the rainfall and evaporation pattern, as evi-denced by the distribution of evaporitic sediments and also the lesser degree of community continuity across platforms, becomes more variable (a non-pluvial interval or a more complex distribution of areas subject to high and low rainfall). Needless to say, community boundaries do not cross evaporite facies boundaries.

Additional evidence of a circumstantial nature is provided by the abundance of unweathered white mica and biotite flakes in the Silurian

clastic beds of both North Africa and the central Andean region of Peru
and Bolivia. Many geologists have routinely commented on the obvious
abundance of detrital mica flakes in those two large regions in the course
of describing Silurian strata. Such detrital mica is known, of course, in
clastic Silurian rocks elsewhere in the world but the abundance appears to
be much lower than present in the Malvinokaffric region. The presence of
this abundant detrital mica may be taken as evidence for a climatic regime
in which the weathering of mica proceded more slowly than would be the
case in a warm region (unless mechanical weathering, transportation and
deposition had occurred very rapidly). We lack the kind of quantitative
data about this detrital mica question on a regional scale that would be
completely conclusive, but the qualitative impression is consistent with
what we have deduced about Malvinokaffric region Silurian cold climate.

It has previously been difficult to account for the increasing degree of
Silurian provincialism. Correlative with it in a very general way is the
coming into being of several regions of low-lying land within the Northern
Hemisphere, i.e., regression following Llandovery-Ludlow transgression
and the advent of extensive evaporitic deposition, indicating that hypersaline bodies of water were present, as well as an abundance of reef and
reef-like structures in the Wenlock and Ludlow. Thus, the increasing provincialism may be the resultant of a number of factors.

We have evidence for southerly migration of the Malvinokaffric Realm
northern boundary (Fig. 2, 25, 41) near the Silurian-Devonian boundary.
The presence in the Devonian of a Southern Hemisphere Malvinokaffric
Realm (in the Devonian it includes Antarctica as well as South Africa and
the southern half of South America) is similar to that for the Silurian
except that the northern boundary of the realm is located between the
Amazon Basin and the Parana Basin. North of the Malvinokaffric Realm
region are widespread areas (containing highly provincial faunas) of both
marine and nonmarine sedimentary rocks and also numerous evaporitic
basins (Fig. 2 and 41). The coincidence of high provincialism among marine invertebrates and the presence of widespread evaporitic deposits suggests that a different pattern of evaporation rate, rainfall and circulation
existed from that of the Lower Silurian. Riverine input is still considered
to have been small. Because of the absence of reefs or limestone and the
presence of low taxonomic diversity the Malvinokaffric region can still be
considered as comprising a relatively cool region compared with the rest
of the known Devonian. However, during the Devonian we have no evidence for the existence anywhere of widespread glacial conditions such as
were observed near the Ordovician-Silurian boundary in North and South
Africa and in South America's mid-Andean region from southern Peru to
west-central Argentina.

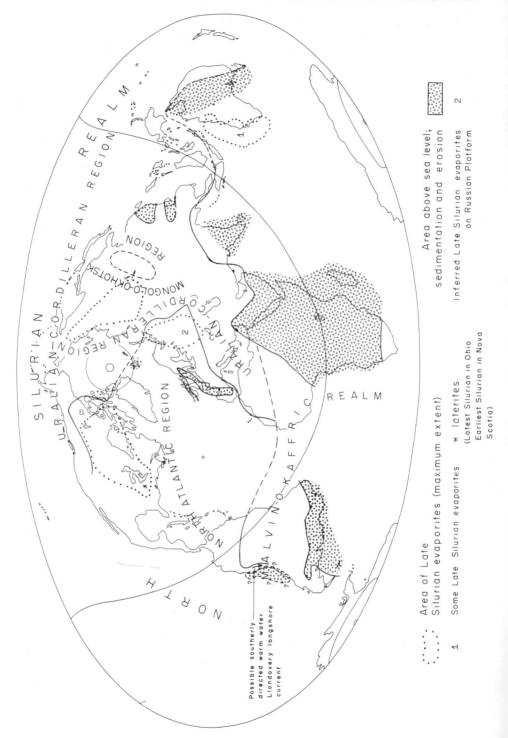

The presence of these Late Silurian-Middle Devonian evaporitic deposits in the relatively central parts of the North American, Siberian, and Russian Platforms favors the concept that these platforms existed in a form generally similar to today.

The existence of a North American "continental backbone" largely separating provincial faunas from northern Mexico to the Boothia Arch region of northern Canada might be interpreted as the environmental "cause" of the provincialism (e.g., the barrier to faunal migration) for the Early Devonian and part of the Middle Devonian. In terms of the geography present on the south (Fig. 2, 37—40) one may still view this continental backbone as correlating well with the presence of the Nevadan Subprovince Lower Devonian, Cordilleran Region, and Appohimchi Subprovince shelly faunas. However, the presence of large masses of Late Silurian through Middle Devonian hypersaline waters, extending far beyond the limits of the inferred land mass, within the interior of North America (Fig. 37—40) may have provided a far larger barrier to the migration of faunas, possibly supplemented by the continental backbone, than any land "barrier" around which marine larvae could (if current and temperature distributions were adequate) be drifted. The causal mechanism for Late Silurian-Middle Devonian endemism affecting the shelly invertebrate faunas might have been a more complex meteorological situation in which hypersaline water masses were produced on the platforms and actually acted in part as the barriers by which faunal migration was interrupted. However, there is still the possibility, as we are dealing here only with circumstantial evidence, that even these postulated salinity barriers were too small to act as effective faunal barriers; that a current-driven temperature barrier was the real faunal barrier with both land and salinity anomalies being only coincidence or possibly supplemental rather than causative agents. The cosmopolitan condition of the platform shallow-water animals returning during the Late Devonian also correlates with a significant decrease in the abundance and distribution of evaporites as well as with continuing transgression, but this may be merely a coincidence rather than a causal relation.

The notable reduction in areas of Late Devonian evaporites is not correlated with glacial phenomena as is the situation for the Early Silurian. However, widespread black shales indicating the presence of greater water

Fig. 41. Silurian biogeography. (North African relations modified after Cocks, 1972). Areas of Late Silurian evaporites indicated. The Uralian-Cordilleran Region boundary with the North Atlantic Region in the New World Arctic area should be moved to a position south of the North Greenland coastal region (Hall Land, Nyboe Land, Peary Land) and into the central portion of Cornwallis Island based on unpublished information (see text) that has recently become available.

depth (more offshore and deeper than, Benthic Assemblage 5, here interpreted as near 500—600 ft. below sea level) over large portions of some
platforms during the Late Devonian raises the possibility of a water circulation sufficiently improved to have prevented the formation of very
extensive regions of hypersaline water and dolomite. The widespread
black shales of the Late Devonian-Early Carboniferous (the Chattanooga
Shale, sensu lato, and its environmental correlatives) are characterized by
Pelagic Community faunas consistent with the presence of vast expanses
of water on the platforms deeper than was present earlier in the Period.
Although concluded to be "deeper" these depths are still concluded to be
in the shelf category-possibly equivalent to the outer shelf. Such an improved circulation is also consistent with the cosmopolitan nature of the
benthic fauna present in the shallower regions of the platforms, and the
limited amount of dolomite.

Silurian-Devonian evaporites and the dolomite problem

Any consideration of the factors involved in establishing barriers to
faunal migration for the lower half of the Paleozoic must consider the
nature and origin of dolomite. Dolomite is far more widespread and abundant on the platforms during the Cambro-Silurian, and to a lesser extent
during the Devonian, than is the case for the rest of the Phanerozoic or for
the Pre-Cambrian. Berry and Boucot (1970, pp. 87—88) have discussed
the remarkable regularity in distribution of Silurian dolomite suite and
limestone suite rocks occurring on the North American Platform. Boucot
(1969) has discussed the same regularity during the Silurian for the Russian and Siberian Platforms. For all three cases a large central region of
dolomite exists that shows evidence of having been made up in large part
of biogenic calcium carbonate before dolomitization occurred. Peripheral
to this large central region of secondary dolomite occurs a region characterized by abundant limestone and calcareous shale. The transition zone
between these two lithofacies is commonly, although not invariably, narrow. Peripheral to the limestone suite rocks occur geosynclinal suite beds
of varying, noncarbonate lithology that represent a variety of environments. Dolomite is very rare in the geosynclinal suite although minor
limestone is present. Evaporite suite rocks, when they are present as in the
Late Silurian, occur interior to the dolomite suite rocks. The taxonomically impoverished nature and relatively low absolute abundance of fossils in the evaporite suite rocks indicates the presence of an environment
inimical to the varied marine life represented by the abundant fossil invertebrates of both the limestone and dolomite suites. Berry and Boucot
(1970) point out that there is no essential difference between the faunas

occurring in the dolomite suite and those occurring in the limestone suite. Therefore, they conclude that the sea water covering the area of the dolomite and limestone suites must have been uniform in composition. The similarity of the invertebrate faunas of both the limestone and dolomite suites to the shelly faunas associated with the geosynclinal suite from place to place suggests that normal-salinity sea water was associated with all three.

It might be contended that the occurrence of similar benthic faunas in areas of limestone subject to dolomitization, areas of limestone not subject to dolomitization on the continent, and also to geosynclinal areas reflects a high degree of euryhaline behaviour on the part of the benthos of the Silurian-Devonian. This argument is weakened by the absence of these benthic faunas in an interbedded condition with evaporitic rocks where it is certain that hypersaline conditions obtained. Thus we are reduced to considering whether or not dolomitization might have occurred under conditions of only moderate hypersalinity, i.e., inadequate concentrations of brine adequate to permit precipitation of evaporitic minerals. The strongest argument against this view of conditions of moderate hypersalinity associated with a moderate degree of euryhalinity on the part of the benthos is that it is unlikely that so many organismal groups would have varied so uniformly in their salinity tolerances (corals, bryozoans, echinoderms, brachiopods, arthropods, molluscs, etc.). There does not appear to be any consistent segregation of animal groups as one passes from the geosynclinal environment, through the limestone suite and eventually through the dolomite suite up to the limits of the evaporite suite (at the last there is, of course, a marked faunal change and precipitous diminution in number of taxa).

Lithofacies relations for portions of the North American Devonian (Fig. 37—40) indicate a similar situation to that of the Silurian, except that less dolomite is present and there is provincialism as well. It is germane, therefore, to ask whether or not the high or low abundance of these dolomite suite rocks is evidence of some factor involved in establishing and maintaining a barrier to faunal migration conducive to provincialism. This question involves the problem of the origin of dolomite.

The origin of sedimentary dolomite has been discussed for many years. Zenger (1972) has summarized many of the present views. These present views involve the following three mechanisms: seepage refluxion, capillary concentration or evaporative pumping, and supratidal dolomitization. All three mechanisms in essence depend on the presence of brines enriched in magnesium due to evaporation of one type or another in different seawater environments. They will be briefly discussed below followed by a fourth proposal that I feel is capable of accounting for the bulk of this

Cambro-Devonian dolomite through what is here termed "biotic dolomitization."

Zenger considers the evidence for both modern and ancient supratidal dolomitization. He points out that although supratidal dolomite is certainly being produced at the present time in certain restricted areas, and that evidence for its production in a few places in the Lower Paleozoic is strong, it is equally certain that most of the Early Paleozoic dolomites were produced under subtidal conditions. The distribution of animal communities, almost entirely subtidal in nature (Benthic Assemblages 3 through 5 at any rate), over the greatest part of the Early Paleozoic platforms that are characterized by such abundant dolomite confirms Zenger's point.

Cloud and Barnes (1948) extensively review the occurrence and nature of sedimentary dolomite. They conclude that the bulk of the evidence strongly favors a penecontemporaneous replacement process of previously existing calcium carbonate sediment. Their conclusions are in agreement with our own studies (Berry and Boucot, 1970) of Silurian dolomite in relation to the animal community problem.

Evidence for the widespread nature of the capillary concentration or evaporative pumping type of mechanism for dolomite formation discussed by Hsu and Siegenthaler (1969) is no better than that favoring supratidal dolomitization due to the bulk of these dolomites being subtidal.

Next I will consider the evidence favoring dolomitization by seepage refluxion. The Silurian evaporite residue minerals found in the Michigan Basin, Hudson Platform, Williston Basin, as well as those indicated to be present in northern Baffin Island (Trettin, 1965), northern British Columbia (Gabrielse, 1963) and the Yukon (Hume, 1954) have abundances entirely inconsistent with their representing volumes of normal salinity sea water evaporated in toto in situ. The absence in these deposits of large amounts of both potassium and magnesium evaporite minerals is particularly noticeable as is the overabundance of gypsum and anhydrite relative to halite. In other words, some mechanism must be invoked for the removal of very large volumes of brine enriched in potassium and magnesium and enrichment in calcium sulfate as opposed to sodium chloride. Several mechanisms may be envisaged to account for this situation. First there is the possibility that heavy brines enriched in these elements, as well as other bittern constituents, might have flowed back to the oceanic reservoir along the sea floor. This possibility is considered unlikely due to the problems of maintaining a density stratification over a vast region between the area where the brines became concentrated and the oceans proper in the face of chances for continual mixing with an overlying layer of normal-salinity sea water through the agency of wind- and tide-driven currents and waves. Another objection to this first possibility is that the presence of such high-density hypersaline brines moving back along the shallow sea floor to the open oceans would imply the existence of channels inimical to the presence of normal-salinity benthos; such channels or areas have not been recognized. A second possibility is the selective removal of more soluble minerals through later solution, particularly during the time of relatively widespread Middle Devonian uplift. This second possibility is considered unlikely as although it would help to account for selective removal of certain evaporite minerals exposed near or at the surface during the early Middle Devonian it could not account for their deficiency at some depth in areas where the Late Silurian evaporites are relatively thick and inaccessible to solution. In addition this second mechanism would not account for the great deficiency in both potassium and magnesium evaporite minerals found in the Middle Devonian and minor Late Devonian evaporite suite rocks which are also inaccessible to solution. A third possibility described in some detail by Adams and Rhodes (1960) for the relatively large Permian Basin and by Deffeyes et al. (1965) for the relatively small Caribbean island of Bonaire, as well as by Fisher and Rodda (1967) for the Texas Cretaceous, is dolomitization by seepage refluxion of evaporitic brines through loosely consolidated, high-porosity, high-permeability, relatively contemporaneous biogenic calcium carbonate skeletal debris at a time prior to lithification into limestone. In essence these authors propose that heavy brines left behind in lagoons, after the precipitation of halite, gypsum, calcite and anhydrite, find channels either on the bottom, in the bottom, or marginal to the lagoon through which to percolate through unaltered calcium carbonate biogenic accumulations. While percolating through the biogenic material the magnesium-enriched brines replace much of the calcium in the biogenic material to form dolomite, following which the brines ultimately find their way back to the normal-salinity marine reservoir. These papers do not discuss the fate of the potassium excess present in these brines on their path back from the lagoon where evaporation took place to the open-sea reservoir. It is conceivable that a significant portion of this potassium might enrich clay-mineral fractions associated with the organogenic accumulations of calcium carbonate or with clay-mineral rich layers beneath such material if the former were pervious. In any event this mechanism of seepage refluxion if employed on a grand scale rather than merely to explain the relations observed in a single sedimentary basin or small lagoonal area is qualitatively capable of explaining the widespread occurrence of dolomite suite rocks rimmed by limestone suite rocks on the Early Paleozoic platforms of the world (the results reported by Berry and Boucot, 1970, for the North American Silurian differ very little from those for the Cambro-Ordovician of North America as outlined by Harris (1973, fig. 1) or by Lochman-Balk (1971), or the Devonian of North America as outlined in Fig. 37—40 or for the Early Paleozoic of the Russian and Siberian platforms). The chief question to be answered is whether or not it is rheologically conceivable to move heavy brines for hundreds of kilometers through loosely consolidated biogenic material resting on an extremely low slope!

Another method of dolomitizing these rocks, which would be consistent with the paleoecologic evidence indicating similar marine environments within both the limestone and dolomite suites, as well as within the geosyncline in places, would be to appeal to the activity of ground-water during intervals of uplift well after the deposition of the sediments (Sonnenfeld, 1964; Garrels and Mackenzie, 1971). However, this concept of dolomitization well after deposition during intervals of uplift, although undoubtedly correct in some special cases, presents problems that do not accord well with the known distribution of dolomite. It is recognized that all of the widespread platform carbonate deposits have undergone intervals of uplift since their formation. It is probable that the Early Paleozoic deposits on the North American, Siberian, Russian and Australian platforms have undergone many such intervals of uplift to positions above sea level where they would be potentially accessible to ground-water activity. The first problem raised by this concept of ground-water alteration during intervals of uplift is the great concentration of sedimentary dolomite within the Cambrian-Silurian interval (little is present in the Precambrian), far less being present in the Devonian (the bulk of that in the Lower Devonian), and very little compared to the Cambro-Silurian in the post-Devonian. It is difficult to think of a process of uplift and groundwater activity that would necessitate such a radical drop-off in dolomite production after the Devonian, essentially the Lower Devonian, as there is good evidence for many widespread intervals of uplift and erosion on a very broad scale during the Permo-Carboniferous and the Mesozoic. The small dolomite limestone ratio of the Precambrian is also puzzling if this groundwater alteration mechanism is important. Additionally, there is the problem of the age-correlated symmetry observed for the successive positions in time of the limestone-dolomite lithofacies boundary. This boundary appears to move in and out relative to the central part of the continent on a time basis, i.e., for some time intervals the boundary is located further out than for other time intervals.

If uplift accompanied by alteration induced by groundwater activity were the chief cause of dolomitization one would expect to find an areally-concentrated center or centers of dolomitization unrelated to time. The rarity of dolomite in geosynclinal rocks, regardless of age, is another piece of evidence suggesting that post-uplift dolomitization of limestone is not the chief cause of the widespread Cambro-Silurian dolomites. (Although limestones are not the most important rocks in the geosynclinal suite, they are present in enough abundance and scatter both geographically and temporally to make this rarity of geosynclinal dolomite significant.)

After having reviewed these four dolomitization mechanisms, each of which certainly may operate under certain sets of conditions, we are faced with the problem that they fail to explain the bulk of the widespread, subtidally deposited, now dolomitized platform carbonate rocks unassociated with evaporites. We must have a mechanism for such dolomitization capable of operating over the bulk of a platform which does not involve changes in the chemical constitution of the sea water because of the essential identity of the fossils recovered from the limestone suite, the geosynclinal suite and the dolomite suite (except where dolomites occur interbedded with evaporitic rocks). This mechanism must be capable of explaining the unusually high concentration and abundance of such dolomite in the Cambro-Silurian, its lesser extent during the Devonian, and its much lower abundance and concentration during the remainder of the Phanerozoic as well as during the Precambrian. Such a mechanism termed biotic dolomitization, is given below.

In the epicontinental seas of the Siluro-Devonian, in areas of poor circulation and communication with the oceanic reservoir, it is reasonable to assume that the removal of calcium by calcium carbonate secreting organisms will have had an appreciable effect on the magnesium:calcium ratio. It is conceivable that in such areas the magnesium/calcium ratio will increase enough to permit at least partial dolomitization of permeable organogenic carbonate debris present on the sea floor. During time intervals, like the Lower Silurian, when we have no evidence for the presence of widespread evaporites and cannot therefore deduce the presence of magnesium-rich brines produced from evaporite formation, this biotic activity alone might be capable of producing sea-water magnesium/calcium ratios adequate to dolomitize at least partially the organogenic debris resting on the sea floor. In a sense such biotically induced dolomitization might tend to act as a buffering effect in maintaining a near-equilibrium magnesium/calcium ratio and the living, lime-secreting shells would be protected from dolomitization by their periostracum-type outer layer until its removal, normally after death. However, dolomitization of organogenic debris on the bottom, in addition to solution of non-dolomitized material, might act to restore the normal magnesium/calcium sea water ratio before it could reach a point having unfortunate physiological effects on living organisms. Thus we might think of such biotic dolomitization as a self-buffering mechanism capable of maintaining a near-normal, equilibrium magnesium/calcium ratio under conditions of restricted circulation. Such partial dolomitization of calcareous debris resting on the sea bottom would probably be inadequate by itself to account for the extensive dolomites of the Cambro-Silurian platforms. An additional important consideration in the process is the solubility relative to that of dolomite of the calcite, magnesian calcite and aragonite, making up the hard parts of the calcareous organisms. Sea-water solubility decreases in the order (magnesian calcite <5% Mg)—(aragonite)—(calcite)—(dolomite). These differing solubilities will all act in the direction of producing a dolomitic bottom residue as calcareous material on the bottom is recycled into solution in the overlying sea water. Over a period of time, after many such cycles it is reasonable to expect the production of a highly dolomitic residue. I have already noted the reasons for thinking that the preserved calcareous organic material preserved on the platforms is but a small fraction of that actually produced, due to re-solution. This last conclusion is consistent with the calcareous organic debris on the sea bottom being driven inevitably towards a dolomite-rich end product. The presence of numerous voids in these secondary dolomites with the form of various invertebrate megafossils that might well have had either aragonite, magnesian-calcite or calcite as their shell material (trilobites, brachiopods, tetracorals, molluscs) is consistent with the differing solubilities of the organic materials laid down on the sea bottom.

Widespread dolomitization by either seepage refluxion and/or biotic agencies is clearly dependent on a serious restriction of normal circulation of marine waters across and into the platforms. The far more abundant and widespread dolomite of the Cambro-Silurian as contrasted with the remainder of the Phanerozoic and Pre-Cambrian (unless the Precambrian limestones are non-biogenic) must also be considered (the Devonian is transitional in this regard; a large amount of dolomite present in the Lower and Middle, far less in the Upper). The platforms of the Cambro-Silurian are far greater in extent than those of the remainder of the Phanerozoic (the Devonian is again transitional). These larger platforms are an inheritance from the widespread peneplanation of the Late Precambrian, and they persist through the Cambro-Silurian, only beginning to be destroyed in the Devonian. This information may be interpreted as indicating that the widespread dolomitization characteristic of the Early Paleozoic platforms, if accomplished by biotic agencies, may have involved a restricted circulation[59] dependent in largest part on the sheer size and extent of the platforms alone.

Land masses of significant size were present intermittently on the Phanerozoic platforms from place to place but they do not correlate well with the widespread, concentrated dolomitization characteristic of the Early Paleozoic.

The nearly complete absence of sedimentary dolomite within the geosynclinal environment is consistent with its presumed better circulation and access to the oceanic reservoir, i.e., an excess of magnesium would not develop. The nearly complete absence of dolomite within the geosynclinal environment during the Cambro-Devonian, a time of maximum dolomite production in the geologic record, serves to underline this conclusions.

Another problem encountered by Berry and Boucot (1970) was to explain the presence in many places of a sharp boundary between the dolomite and limestone suites. It is noticeable that the amount of clay-size and silt-size insoluble material in the limestone suite rocks, 30—75%, is higher by at least an order of magnitude than the 3—5% in the dolomite suite. This statement is based on the solution of many tons of dolomite and limestone from both suites in many regions in the extraction of silicified fossils. It is logical to conclude that interstitial water circulating in the high-permeability, high-porosity biogenic calcium carbonate debris would be impeded by large amounts of silt- and clay-sized material filling most interstices between the biogenic debris. This suggests that areas lacking a sharp boundary between the two suites might also show a gradual rather than a rapid outward, lateral change in clay-size and silt-size residue. An additional consideration is the position of the limestone suite peripheral to the dolomite suite and adjacent to the geosynclinal suite. The geosynclinal suite is thought to represent at least the margin of the oceanic environment, the oceanic reservoir in other words. Thus it is reasonable to conclude that circulation of sea water in the limestone suite area and mixing with the normal oceanic water mass would not have permitted biotic dolomitization to have taken place. In this view the very presence of the limestone suite indicates areas where circulation was better.

Cloud's (1962) monographic study of Bahama Banks carbonate sedimentation has great bearing on the problem of dolomitization. Cloud's work makes very clear that dolomite is not being formed in any significant quantity over the present-day area of the Bahama Banks in a subtidal position. This absence clearly indicates that a carbonate platform with small dimensions (no more than a few hundred miles in diameter) occurring in an area of reasonably good marine circulation (even if circulation across the small-dimension platform is relatively poor for much of the year) will not be a site of subtidal dolomite formation. The degree of hypersalinity attained in the Bahama Bank region is evidently not enough to seriously disrupt marine organisms over much of the region, i.e., this region is not characterized by a hypersaline flora and fauna. The key to this problem locally in the Bahamas may lie in the existence of some threshold for short-term water circulation over a platform, below which dolomitization takes place. It is not clear whether or not a significant amount of time is necessary for dolomitization to occur, although Cloud's detailed account certainly makes clear that if time is significant for a platform similar to the Bahama type then far more time than would be encompassed in a few life cycles of most marine organisms is involved. The important question may be the effect of poor short-term water circulation on the interstitial processes in the bottom sediment.

Ladd et al. (1970) have discussed the dolomitized limestone occurring on a number of Pacific islands. They conclude that seepage refluxion is a reasonable mechanism that accounts for the dolomitization. If seepage refluxion is not appealed to, one is left with processes operating in the sediment interstices or partially consolidated-sediment interstitial water as an explanation of these localized occurrences of dolomite. The rare occurrences of geosynclinal dolomite may have formed in a manner similar to that deduced by Ladd et al. for these Pacific islands.

In summary, one may appeal to secondary formation of areally widespread dolomite (not supratidal, groundwater-replacement or capillary-concentration dolomites) in the interior portions of Early Paleozoic platforms by biotic dolomitization, or by biotic dolomitization assisted in areas of evaporite formation by seepage refluxion dolomite formation. Biotic dolomitization is viewed as a process in which lime-secreting organisms remove calcium from the sea water to raise the normal Mg/Ca ratio at the same time that solution of calcareous debris with the highest solubilities (aragonite, magnesian-calcite and calcite as opposed to dolomite) is striving to restore that normal ratio. These effects are possible under conditions of poor circulation between the shallow, epicontinental seas' interior parts and the oceanic reservoir. Slow replacement of magnesium from the oceanic reservoir should occur for that permanently locked into the bottom sediment. The process of biotic dolomitization is viewed as one in which dolomite very slowly accretes on the bottom as a residue from the re-solution of countless generations of lime-secreting organisms. Biotic dolomitization is consistent with the small thicknesses of carbonate preserved on the platforms compared to what would have been expected if the remains of all the generations of organisms had been preserved. R.M. Garrels (written communication, 1972) writes: "There is a hint that the dolomitization process may be helped by anaerobic conditions. There are some cores in which it looks as if the sequence of events involved de-aeration, increase in CO_2, in the interstitial water and the more or less continuous formation of dolomite as a replacement of calcite, with the magnesium diffusing downward from the bottom waters and the calcium diffusing out as a counter-ion." Garrels' comments would appear to be consistent with the concept of biotic dolomitization.

Friedman and Sanders (1967) extensively review the dolomite problem and conclude that "Dolomite is an evaporite mineral. . ." (p. 338) in the sense of requiring brine-like conditions for its production or formation by replacement. I see no reason why their conclusions are not in accord with those reached here *if* the brine-like, "evaporitic" conditions occur in the area immediately below the sediment — water interface in such a manner that normal-salinity sea water is present at and above the sediment — water interface.[60]

The presence of limestone suite peripheral to the dolomite suite may be viewed as related to markedly reduced permeability and porosity caused by the presence of large amounts of silt- and clay-size materal plus better circulation of the overlying sea water with the oceanic reservoir. An over-all view of the distribution of Phanerozoic dolomite strongly suggests that biotic dolomitization under conditions of shallow water and restricted circulation on the large platforms of the Lower Paleozoic is the most likely explanation for the bulk of the sedimentary dolomite. Such conditions are met with in the Cambro-Silurian, to a slightly smaller extent in the Early and Middle Devonian, but not over large areas in the Late Devonian or subsequent intervals.

Ronov (1973) makes the assumption that the changing bulk compositions, in time, of sedimentary rocks provides a measure of the changing bulk composition, in time, of crustal source rocks. This assumption, at least in part, is not borne out if one carefully considers the calcium/magnesium ratio information presented by Ronov. Ronov plots two calcium/magnesium ratio curves against time: one for the Russian Platform, the second for the North American Platform. The North American data shows a marked change in the calcium/magnesium ratios (relatively low in the Cambro-Devonian; far higher in the post-Devonian) near a Devonian inflection point, whereas the Russian Platform curve shows a similar inflection point in about the Jurassic. These two curves actually are integrating the relative abundances of sedimentary dolomite and limestone. For the North American Platform there is a distinct breakup of the Early Phanerozoic pattern, with its topographic form conducive to dolomitization of pre-existing limestone, in the Devonian, whereas the Russian Platform does not show this change until the Jurassic. What Ronov is "measuring" here is not a change in nature of the source materials in the crust with time, but rather the factors that govern the rate of formation of dolomite as compared to limestone. In other words, Ronov's curves tell us something about the vertical stability, topography and marine circulation patterns on the North American and Russian platforms. This conclusion about the significance of calcium/magnesium ratios not being necessarily related to changing crustal abundances also indicates that the relative abundances of other elements correlated strongly with calcium and magnesium should be viewed with great caution if they vary in the same direction as the calcium/magnesium ratios. My alternate interpretation of the significance of the calcium/magnesium ratio data raises the basic question of whether or not relative elemental abundance changes derived from study of sedimentary rocks do indeed provide a true measure of crustal abundances in time. The assumption that they do provide a true measure implies that a multitude of geologic processes and distribution patterns operated in a very similar manner and at about the same rates during the past; an assumption that I feel is unjustified at the needed level.

Transgression-regression

Kauffman (1972), Keen (1972) and Wiedmann (1973) present evidence suggesting a strong correlation between transgression-regression and rate of production of taxa. Kauffman, dealing with North American Cretaceous molluscs, finds that major increases in rate of speciation and radiation occur during regressive intervals. Keen, dealing with Tertiary ostracods from northwestern Europe, finds major evolutionary radiation associated with transgression.[61] Mamet (1971) finds that information for Lower Carboniferous foraminifers conforms to Keen and Wiedmann's findings of major evolutionary radiation associated with transgression. The different position of increase in rate of production of taxa with respect to transgression or regression might, in principle, be due to either differing characteristics of Mollusca as contrasted to ostracodes and foraminifers, or to ultimate control by a factor or factors unrelated directly to transgression-regression. In considering this question attention should also be paid to the outmoded concept suggesting that times of orogeny are times of transgression of epicontinental seas on the continental interiors. If this concept accorded well with the facts then it would be possible to combine the data for orogeny[62] with that for transgressions and regressions on the platform to arrive at even more information correlating with times of increased production of fossil taxa.

First let us consider some of the data bearing on the orogeny-transgression correlation concept. Major regression at the end of the Ordovician (Berry and Boucot, 1973a) is correlated with the continental glaciation affecting much of Africa and at least the central Andean portion of South America (southern Peru through northern Argentina). Late Ordovician orogeny is trivial in both nature and distribution. A continuing, worldwide transgression occurs from the beginning of the Llandovery through the Ludlow, *but* there is no evidence for any Silurian orogeny of significance in the world. Regression begins in Pridoli time in many parts of the

Northern Hemisphere followed by a beginning of renewed transgression during the Ems (Ems marine deposits are more widespread than those of any other interval of Lower Devonian time). A short transgressive-regressive Ems-Eifel episode affected the Southern Hemisphere Malvinokaffric Realm area (Boucot, 1971) unaccompanied by evidence of orogeny. Transgression continues during the remainder of the Devonian with most widespread transgression present from the Givet through the Famenne (Famenne deposits are probably more widespread than those of the Frasne on a worldwide basis, and both are certainly more widespread than those of the Givet). The time of major orogeny (as detailed in the following section) for the Devonian is the Middle Devonian. The Middle Devonian is a time of maximum Devonian orogeny, but of only moderate, middling transgression. The maximum transgression for the Cretaceous occurs in the Cenomanian and Albian followed by another in the Coniacean-Santonian (E.G. Kauffman, written communication, 1972) whereas the maximum for orogeny occurs in the latest Cretaceous-Early Tertiary interval especially the earliest Tertiary — a time of distinct regression and less widespread transgression than is encountered during parts of the earlier Cretaceous. In summary, the concept of times of orogeny corresponding neatly with times of maximum transgression on the continental interiors (platforms) does not accord with the facts of latest Ordovician through Devonian geology or for those of the Late Cretaceous-Early Tertiary. Finally, the widespread orogeny of the middle and later Miocene plus Pliocene, responsible for so many of the major mountains of the earth (Alps, Himalayas, Andes, main chains of Central Asia) is conspicuously unaccompanied by any widespread transgressive flooding of the continental interiors of any affected continent (the reverse happens to be the case; strong continentality sets in toward the end of the Miocene for all the present continents). Lack of correspondence for these significant time intervals prevents acceptance of the concept.

Next follows a consideration of the correlation between transgression-regression and rate of evolution. Kauffman (1970, 1972) has undoubtedly made the most detailed, exacting and reliable scheme of fossil-based zones in existence. Kauffman's fossil-based zones are assisted by a detailed web of reliable stratigraphic information and a mass of radiometric age determinations. It would be foolish to question the reliability of Kauffman's conclusions about either production and appearance rate of taxa or their correlation with times of transgression or regression. The correlations he has made must be accepted at this time, with there being little possibility that they will be seriously modified in the future. However, as Kauffman (1972) himself points out the correlation he has observed between rate of production of taxa and transgression-regression may possibly be a coincidence rather than indicating a causal relationship. It would be critical to the conclusions in both his paper (1972) and this treatment to have tests of area applied to the various taxa and communities present in the world's Cretaceous to learn whether or not there is also an inverse correspondence between rate of evolution and area occupied by an evolving group of taxa. The same comment is true of Keen's (1972), Wiedmann's (1973) and Mamet's (1971) conclusions.

Next a consideration of transgression-regression against Silurian-Devonian data to see what correlations exist. There is a marked increase and replacement in the number of taxa (at the generic and family levels) present in the Upper Llandovery of the world as compared to the Lower and Middle Llandovery. This marked increase occurs during a time of gradual, worldwide transgression. I interpret this marked increase as largely being the result of the spread of a number of taxa, previously restricted to southeast Kazakhstan due to environmental factors not known to us at present. We have a marked evolutionary increase in number of taxa in the Wenlock-Ludlow, a time of continuing transgression. During the regression beginning in Pridoli time and continuing through to the Ems we have an even more marked increase in number of taxa beginning in the Gedinne (transgression begins again in the Ems and continues through the remainder of the Devonian). Therefore we have a good correlation in the Silurian-Devonian between times of both regression and transgression co-occurring with maximum production of taxa. The Late Ordovician can also contribute some information to the problem. There is a very marked decrease in number of shelly marine megafossils during the Late Ordovician into Early Llandovery interval. Regression characterizes the Late Ordovician (correlated with continental glaciation) followed by transgression during the Llandovery. This information does not correlate well with the concept of regression invariably being the most important factor controlling high production rate of taxa.

Summation of the above information suggests that there is correlation in increasing production rate of fossil taxa between intervals of both regression and transgression. In view of the excellent inverse correlation found to exist between size of a population and rate of production of taxa I suggest that in those cases of transgression or regression for which increased production rate of taxa is recorded it would be well to investigate the size of the populations in question as well as the correlations with regression or transgression before arriving at a final conclusion. Regression and transgression, in each specific case, may affect the environment in unique ways either conducive to smaller or to larger interbreeding populations with appropriate consequences. I can think of no a priori principle that would force all examples of either transgression or regression within the Phanerozoic record to have identical effects on population size; chiefly because of the widely differing and ever changing climatic regimes and geographies on which regression and transgression present during the Phanerozoic have been superimposed. It is conceivable that during one time interval a regressive interval, because of differing geography and/or climatic regime on which it is superimposed, may have the opposite effect on rate of production of taxa to that experienced during a completely separate, far removed interval of time.

Other things being equal one would expect that rate of evolution during transgressive intervals would slow down in the shallow-water, marine environment due to the increased area (over low-slope continental regions) of similar environment made available (with consequent increase in population size), and that the reverse should be true during regression. However, this simple consequence of changing population size related to regression and transgression does not take into account the multitude of potential changes that may accompany transgression or regression over different topographies, or the consequences inherent in differing climatic regimes, that may in turn give rise to differing size populations of different types as well as change in number of communities. Therefore, one should not expect a simple relation between population size and transgression or regression during the Phanerozoic, although there may be an overall statistical trend favoring higher rate of evolution during pronounced regressive episodes. The apparently conflicting conclusions reached by Kauffman, Keen, Wiedmann, and Mamet (discussed above), plus the data for the Silurian and Devonian make this point clear.

Combining these deductions for transgressive-regressive event effects on interbreeding population size with what we have learned of the effects of cosmopolitan versus provincial time intervals it is evident that a coincidence of a provincial time interval with a regressive time interval will statistically raise the possibilities for overall increased rates of evolution whereas the coincidence of a transgressive interval with a cosmopolitan time interval should have the reverse effect, i.e., tendency towards a far lower rate of evolution. The data for the Silurian and Devonian accord reasonably well with these deductions. It will be of great interest to learn how the distribution of provincial and cosmopolitan time intervals integrate with that for regressive-transgressive events and rates of evolution within the Cretaceous.

Orogeny and provincialism?

In exploring the possible mechanisms responsible for provincialism it is worth considering the timing of orogeny. As used here the term orogeny

refers to those brief, intense intervals in earth history when certain localized regions have been subjected to regional metamorphism, mineralization
and extensive plutonic activity accompanied by structural activity resulting in the formation of very complex fold and fault complexes. Orogeny
is not synonymous with epeirogenetic uplift resulting in the formation of
large clastic deposits peripheral to the area of uplift, although prior to,
concurrent with, subsequent to and independent of orogeny in many areas
we have good examples of uplift that resulted in extensive clastic deposition.

There is no evidence for significant orogeny during the Silurian. During
the Early Devonian there is some evidence (Kwangsi Orogeny) for orogeny
in southeastern China. However, on a worldwide basis there is evidence for
an orogenic peak in the Middle Devonian, chiefly in the Late Eifel-Early
Givet interval.

North American expressions of this peak are to be found in the Acadian
Orogeny of the Northern and Central Appalachians (metamorphism and
plutonic activity affecting Eifel-age beds, but not unconformably overlying Givet-age beds; plutonic and metamorphic rocks yield radiometric
ages in the span 360—380 although the geologic evidence suggests that the
orogeny is of no more than a few million years duration; Boucot, 1968b).
The platform margin and adjacent platform sedimentary expression of the
Acadian Orogeny is the sharp change from the carbonate deposition of the
Onondaga Limestone and its correlatives below to the terrigenous clastics
of the Marcellus Formation (the Acadian source area continued to provide
terrigenous debris to the west throughout the remainder of the Devonian
and Carboniferous, but this fact should not be confused with the time of
Acadian Orogeny. Minor intervals of localized faulting occurred within the
Acadian source area during the later Devonian and Carboniferous).

There is no evidence for Devonian orogeny within the southern Appalachians or in the extensive Ouachita Geosynclinal Belt.[63]

The only evidence for Middle Devonian orogeny in western North
America to date is afforded by radiometric ages from the Central Metamorphic Belt of the Klamath Mountains in Northern California (Lanphere,
1967, p. 118) of 380 million years for the Abrams Mica Schist. Lanphere
suggested that this 380 m.y. metamorphic event might correlate with the
Antler Orogeny, but the great disparity in time between this point near
the Eifel-Givet boundary as opposed to the Famenne-Visé boundary precludes this possibility. There is no widespread evidence on the far removed
platform in Nevada and Utah of terrigenous, clastic sedimentation related
to this metamorphic event, but the distance between the Klamath Mountains and eastern Nevada is such that even an extensive apron of material
might not have extended as far east as Nevada and Utah. However, there is

no evidence in the geosynclinal central Nevada Western Assemblage Devonian rocks for orogeny of Middle Devonian age either, although their structural complexity might still conceal such evidence if it exists. The stratigraphic relations of known Early and Middle Devonian rocks in the Klamath Mountains of northern California, as well as between the Silurian and Mississippian rocks of the northern Sierra Nevada are consistent with the above conclusions.

Thorsteinsson (1968) discusses Devonian orogeny in Arctic Canada. He finds radiometric ages of 360 plus or minus 25 m.y. on Axel Heiberg and 335 ± 25 m.y. from Ellesmere Island. The marked change from Eifel-age carbonate sedimentation (Blue Fiord Formation and equivalent) to Late Eifel and Givet terrigenous clastics (Bird Fiord Formation and equivalents) is concluded to correlate with an orogenic event including at least the 360 m.y. Axel Heiberg plutonic rock in a manner similar to the 360—380 m.y. ages in the Appalachians with the nearby Onondaga-Marcellus change in lithofacies. The Ellesmerian Orogeny bracketed between Famenne-Visé time does not appear to involve any evidence for regional metamorphism (unless the 335 m.y. date pertains). Like the Antler Orogeny (also lacking any undoubted evidence for contemporaneous regional metamorphism or plutonism) of similar age in Nevada the Ellesmerian Orogeny is concluded to be a localized structural event quite distinct from any earlier metamorphic and plutonic, orogenic events. There is no evidence for continual metamorphism or plutonic activity in the Appalachians, western North America or the Canadian Arctic between the Eifel-Givet and Famenne-Visé boundaries.

In any event it is clear, on a worldwide basis, that the mid-Devonian orogenic maximum does not coincide with the upper Lower Devonian maximum for benthic marine provincialism. The relatively limited regions (probably well under ten percent of the present continental areas) affected by Devonian orogeny are consistent with this conclusion. Orogenies in principle have, however, the capability of being associated with uplifted areas that may independently alter current circulation in such a manner as to contribute to the formation of faunal barriers if other factors are also conducive (temperature distributions, etc.). Newell (1956) has summarized earlier views in which orogeny is discarded as an important factor influencing rate of evolution.

Silurian-Devonian climatic gradients

A knowledge of climatic gradients is essential for making an estimate of the number of communities potentially present during any time interval. The presence of a strong climatic gradient similar to that of the present or of the type thought to be present during the Ashgill provides insight into the possibilities for the presence of barriers to faunal migration related to regional temperature gradients with the resultant provincialism induced by the strong temperature gradients. Conversely, absence of a strong temperature gradient should correlate with an absence of provincialism induced by surface temperature barriers (although not excluding the possibility of temperature gradients brought about by upwellings of cold water from depth). During intervals of worldwide equable temperature it can be predicted that the deeper oceanic waters should be warmer than during times of strong temperature gradients with a consequent weakening of the gradients attainable at the surface due to upwelling of deep oceanic water.

The temperature criteria employed here are the following: (1) presence of areas subject to continental glaciation; (2) presence of areas in which laterites and bauxites are present; (3) presence of areas in which limestone is absent; (4) consideration of worldwide level bottom community taxonomic diversity gradients; (5) consideration of areas in which reefs and reef-like bodies are present; (6) occurrence of widespread red beds deposited in the non-marine or brackish environment.

Near the Ordovician-Silurian boundary there is ample documentation for widespread continental glaciation[64], bordered by regions in which glaciomarine sedimentation is abundant, in the Southern Hemisphere and the present Mediterranean regions (all of Africa, the Mid-Andean region of South America, much of central and western Europe; see Berry and Boucot, 1973a). In Nova Scotia (Boucot et al., 1973a) laterites are present in the Arisaig Volcanics of latest Ordovician or earliest Silurian age. Bauxites are known in the Ludlow of the Urals (Arkhangelsky, 1937). Limestones are abundant and widespread outside of the areas where glaciomarine and glacio-fluviatile rocks are present (including portions of the west-central Andes for the Ordovician, northern and northeastern Europe including the Russian Platform, North America, Australia and much of Asia). Within the areas subject to glacial influence limestone is either rare or absent. Reefs and reef-like bodies are known from the areas in which limestone is abundant, but absent elsewhere. Taxonomic gradients have not been carefully studied, but it is my impression that those regions subject to glacial influence have lower taxonomic diversity during the Late Ordovician than do areas lacking glacial influence. During the Llandovery through Wenlock there is a pronounced taxonomic gradient from rich faunas associated with the non-glacial regions to faunas poor in taxa in the regions (Table V) subject to glacial influence. Red beds are widespread during the Late Ordovician-Early Silurian, as well as Late Silurian interval outside of the regions for which we have knowledge of glacial influence.[65]

During the Upper Silurian the distribution of limestones and reef or

TABLE V
Upper Llandovery articulate brachiopod superfamily occurrences

Realm	Superfamily																								Total number of superfamilies	Tetracorals	Stromatoporoids	Tabulates
	Orthacea	Enteletacea	Triplesiacea	Eichwaldiacea	Plectambonitacea	Strophomenacea	Davidsoniacea	Chonetacea	Porambonitacea	Pentameracea	Rhynchonellacea	Atrypacea	Dayiacea	Retziacea	Athyridacea	Cyrtiacea	Spiriferacea	Reticulariacea										
North Silurian	X	X	X	X	X	X	X	X	X	X	X	X	X	X	X	X	X	X							18	X	X	X
Malvinokaffric	X					X		X		X	X	X													6			X

reef-like bodies is similar to that known for the Ordovician-Silurian boundary region.

This information for the Late Ordovician-Silurian adds up to the presence of an intense temperature gradient (from continental glaciation to climate suitable for laterite and red bed formation) near the Ordovician-Silurian boundary region to a less marked, but still marked climatic gradient (no evidence for later Silurian continental glaciation) during the bulk of the Silurian.

Information for the Devonian is similar to that for the Silurian but suggests that the area of cooler (or cold) climate was smaller and possibly the climatic gradient was not as strong. The region of lowered taxonomic diversity (Table VI; the Malvinokaffric Realm) is smaller. The Malvinokaffric Realm boundary with other realms retreats from a position extending from the Istanbul-North African region (fig. 41) during the Silurian to a Devonian position south of the Sahara (Fig. 2) for the Old World. In the New World this same boundary retreats south from a Silurian position between the Amazon region and Venezuela to a new position well south of the Amazon: between the Parnaiba Basin and the Parana Basin in Brazil. No widespread glacial deposits of Devonian age are known. Limestone deposition, reefs and reef-like bodies are widespread outside of the Malvinokaffric Realm during the Devonian, but are absent within the Malvinokaffric Realm. Non-marine red beds (the famous Old Red Sandstones) are widespread outside of the Malvinokaffric Realm during the Devonian but absent within this realm.[66] A lateritic-type soil is present in Ohio between beds of latest Silurian or possibly Gedinne-age and beds of Eifel age (Summerson, 1959). Bauxites of Lower Devonian and Eifel age occur in the northern Urals (Harder, 1949). Bauxite of Eifel age occurs in Spain (Font-Altaba and Closas, 1960). Bauxites of possible Devonian age occur in China (Harder, 1949). All of these data add up to the presence of a pronounced climatic gradient in the Lower and Middle Devonian similar in position to that known for the Silurian. Our information for the Late Devonian is less well known for the Malvinokaffric Realm region; a possibility exists that the climate of the Late Devonian was more equable than that of the Lower and Middle Devonian. The provincial Malvinokaffric Realm Devonian terrestrial flora (Edwards, 1973) supports this conclusion about Devonian climatic diversity.

The absence of evidence favoring continental glaciation in the Malvinokaffric Realm area except in the Ashgill does not necessarily indicate that this region was significantly warmer during the subsequent Silurian and Devonian. Meyerhoff (1970a, b) has made a good case for times of extensive evaporite formation not coinciding with times of extensive continental glaciation, but this correlation does not necessarily indicate that times

TABLE VI
Ems-Eifel articulate brachiopod superfamily occurrences

Realm	Orthacea	Enteletacea	Plectambonitacea	Strophomenacea	Davidsoniacea	Chonetacea	Strophalosiacea	Productacea	Porambonitacea	Pentameracea	Rhynchonellacea	Stenoscismatacea	Atrypacea	Dayacea	Retziacea	Athyrisinacea	Athyridacea	Cyrtiacea	Spiriferacea	Reticulariacea	Stringocephalacea	Dielasmatacea	Cryptonellacea	Total number of superfamilies	Tetracorals	Stromatoporoids	Tabulates
Old World	X	X	X	X	X	X	X	X	X	X	X	X	X	X	X	X	X	X	X	X	X	X	X	23	X	X	X
Eastern Americas	X	X	X	X	X	X	X	X	X	X	X	X	X	X	X	X	X	X	X	X	X	X	X	23	X	X	X
Malvinokaffric	X			X	X	X					X		X	X			X		X		X	X		11	R		X

of extensive evaporite formation were characterized by less pronounced climatic gradients than those characterized by the presence of extensive continental glaciation. Therefore, we must face the possibility that the absence of evidence for extensive Southern Hemisphere continental glaciation during the Silurian and Devonian may merely indicate a climatic regime unfavorable to the accumulation of enough ice for such glaciation rather than the presence of a warmer Southern Hemisphere climate.

Finally, as suggested earlier, there may well be a regional gradient in the abundance of unweathered detrital mica from clastic beds relatively rich in this component within the Malvinokaffric Realm to beds relatively poor or lacking in this component elsewhere.

Significance of disjunct distributions of taxa

An ever-puzzling phenomenon encountered when plotting the distribution of organisms is a laterally disjunct pattern (time-disjunct patterns may be related to migration). By observing the distribution patterns through time of such organisms fossil zone by zone, and by determining their evolutionary antecedents and descendents, one may gain some understanding of the significance of such data. For some groups one may infer a former widespread distribution which in time has become restricted to a few disjunct remnants (reductional relicts; Fig. 42). For some groups one may observe a former, very restricted area of occurrence followed by an interval of time during which descendents have become very widespread and cosmopolitan (Fig. 25, 34, 42). For some groups one may observe a central region of widespread distribution in both time and space from which colonies have emanated with a very short life (commonly increasingly filtered out with distance from the source). For still other groups one may deduce a pattern of continuing migration in time that results in disjunct distributions, each area being of slightly different age; which, if any, of these patterns pertains in a given case must be determined by trial and error. For example, the Eastern Americas Realm elements for the Early Devonian that have been encountered in Kazakhstan, eastern Australia, New Zealand and Manchuria, appear to be explained best as having emanated from and having been filtered from the Eastern Americas Realm of North America into the Old World Realm during a few times of short duration within the total time range of the taxa involved. There is no reason to invoke parallel evolution of these taxa within the Old World Realm. Ecologically workable dispersal pathways existed along which certain Eastern Americas Realm taxa were filtered during parts of Early Devonian time from the Eastern Americas into the Old World Realm.

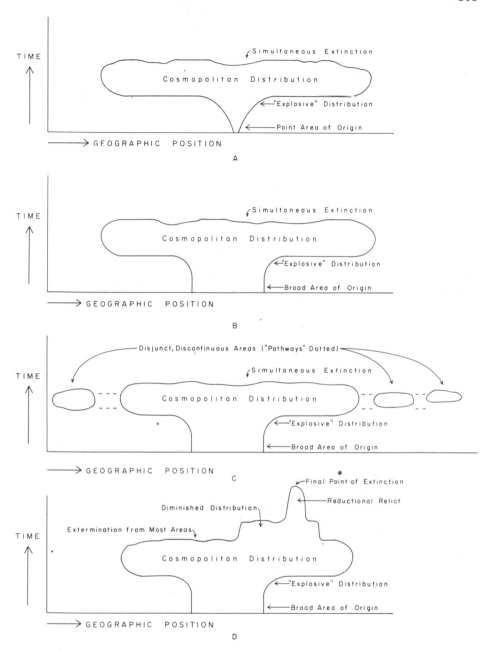

Fig. 42. Diagrammatic illustration of various restrictive and widespread faunal distributions in time; note both wide and small original source types.

Despite our limited knowledge of Asiatic Lower Devonian brachiopods, it is still possible to consider the significance of disjunct Eastern Americas Realm brachiopod taxa occurring outside of that Realm.

In the Rhenish-Bohemian Region immediately adjacent to the Appohimchi Subprovince in Nova Scotia we find *Meristella*, and in northwest Africa *Amphigenia* and *Plicoplasia*, the last two occurring in Eifel-age beds. Other Appohimchi Subprovince brachiopod taxa have not been recognized with certainty in the Rhenish-Bohemian Region (Oliver, 1968, 1973b, has recognized a few Eastern Americas Realm coral taxa in the Western part of the Rhenish-Bohemian Region).

In the Tasman Region *Megakozlowskiella*, *Pacificocoelia*, and *Meristella* are recognized as Eastern Americas Realm elements.

In the Mongolo-Okhotsk Region *Pacificocoelia*, a true chonostrophid, and *Discomyorthis* are recognized. The small number of Eastern Americas Realm taxa may possibly be partly explained by the small amount of attention devoted to the study of Lower Devonian fossils in this Region.

In the Dzhungaro-Balkhash Complex of Communities of the Uralian Region *Pacificocoelia*, *Meristella*, true *Leptostrophia*, *Megakozlowskiella*, *Coelospira*, and *Orthostrophia* are recognized.

The relatively low number of Appohimchi elements in the Rhenish-Bohemian Region certainly suggests the presence of a very effective barrier (the presence of the two known elements in the western part of the Region emphasizes the problem, as does the presence of a few Old World elements in the northeastern part of the Appohimchi Subprovince in Gaspe and eastern Quebec). This situation is contrasted with central and east-central Asia where a relatively large number ·(far larger than those known in the Tasman Region) of Eastern Americas Realm taxa are known. It is evident that the marine conditions of Lower Devonian time made transport from ·the Eastern Americas Realm to central and east-central Asia relatively easy, to the Tasman Region more difficult and into the Rhenish-Bohemian Region very difficult except in the boundary region.

This type of "filtering" contrasts strongly with the gradual or rapid restriction into disjunct, relict distributions found for some taxa that possessed widespread antecedents (Fig. 42). Examples, also from the Silurian-Devonian, of this last phenomenon are to be found in the so-called "Silurian Holdovers" (relicts) of Boucot et al. (1969). They interpreted these holdovers as representing relict populations of a formerly almost worldwide group of antecedents in the North Silurian Realm. An example of an early restricted range is found in some of the pre-Late Llandovery taxa restricted to southeastern Kazakhstan (Boucot and Johnson, 1973) that in Late Llandovery time become almost worldwide in their distribution. Neuman (1972) presents similar data for the Early and Middle Ordovician (Early-Ordovician restricted taxa that give rise to Middle Ordovician widespread taxa).

Therefore, disjunct, discontinuous distribution patterns of taxa in time may give us information about the restricted, or nonrestricted place of origin for taxa, the availability of viable migration routes for taxa from other biogeographic units and for patterns of diminution and eventual extinction of taxa. Deductions employing these explanations will only be as good as the data on which they are based and should be used with caution. The genus *Tropidoleptus* is a good example.

Tropidoleptus occurs in Gedinne through Ems beds in the western part of the Old World Realm in Europe, but is absent in Eifel and Late Devonian beds in Europe. It is present in the Lower Devonian through Frasne of North Africa (Sougy, 1964; Freulon, 1964). The absence in the post-Ems of Europe is due to the post-Devonian erosion of the nearer-shore facies (the carbonate "Bohemian" facies for the Eifel, Givet and Frasne do not contain environments suitable for *Tropidoleptus* whereas the North African equivalents, which belong to the same Rhenish-Bohemian Region biogeographic unit do contain these environments). *Tropidoleptus* also occurs in probable Ems age beds in northern South America (Amazon-Colombian Subprovince of the Eastern Americas Realm). It is also present in the Late Eifel through Frasne of the Appohimchi Subprovince and even into the Frasne of New Mexico. It is reasonable to assume that it migrated from the western part of the Old World Realm or from northern South America (Amazon-Colombian Subprovince) near the end of the Early Devonian into the Appohimchi Subprovince of eastern North America. *Tropidoleptus* also is a prominent element in the Malvinokaffric Devonian fauna, but owing to uncertainty regarding the precise age of the Malvinokaffric faunas (I conclude that these faunas are Ems or Eifel), it is difficult to be certain as to when or whether *Tropidoleptus* migrated into this region during the late Early Devonian or later, and for how long it persisted.

Silurian Biogeography

The Llandovery part of the Silurian is an interval of relative cosmopolitanism (Fig. 25) at least for brachiopod genera (too few data are available for species). The most widespread biogeographic entity is the North Silurian Realm[67], followed by the far more restricted Malvinokaffric Realm (present in both the Early and Late Silurian; see Fig. 41, modified following Cocks, 1972, for North Africa). Wenlock-Pridoli times are characterized by a moderate degree of provincialism, particularly in Benthic Assemblages 4—5 equivalents (see Fig. 41). The Late Silurian Uralian-Cordilleran Region extends from the east slope of the Urals, through northern and central Asia (including the Tien Shan, Altai-Sayan), eastern Australia and New Zealand to Arctic North America, the Yukon, and Nevada plus northern California; the Late Silurian North Atlantic Region (Fig. 41) includes the region from the west slope of the Urals, to the circum-North Atlantic regions of Europe, North America, and northern South America, to central Nevada, and much of Arctic Canada (Fig. 25, 41).

Malvinokaffric Realm

The Malvinokaffric Realm of the Silurian includes (Fig. 41) much of South America and Africa. The Malvinokaffric Realm Silurian is characterized by a limited number of endemic brachiopod genera including *Clarkeia*, *Heterorthella*, *Harringtonina*, *Castellaroina*, *Amosina* and *Australina* together with a limited number of poorly studied brachiopod taxa (chiefly rhynchonellids). The small number of taxa in the fauna is partly a result of only the *Clarkeia*, *Amosina*, *Australina*, *Heterorthella* and *Harringtonina* Communities having been studied (it is not certain that many other communities are present). Despite the limited number of taxa the noncorrespondence of this fauna with anything known elsewhere is impressive. The splitting off of a Malvinokaffric Realm for the Silurian (Boucot and Johnson, 1973) is consistent with the subsequent Devonian history despite the fact that most of the provincial Malvinokaffric Silurian taxa appear to have left no descendents. One exception is the genus *Harringtonina* (A.J. Boucot in Berry and Boucot, 1972a; Fig. 34A). The Silurian leptocoelid *Harringtonina* known from Late Llandovery through Ludlow-age beds from the Precordillera de San Juan, Argentina, to southern Peru, is a far more likely ancestor for *Australocoelia* of the Devonian Malvinokaffric Realm and Tasman Region and other Devonian leptocoelids (*Leptocoelia*, *Leptocoelina*, and the two groups of *Pacificocoelia*) than the other widespread Silurian leptocoelid genus *Eocoelia*. This is an important exception to the earlier generalization (Boucot et al., 1969) that the Malvinokaffric brachiopod taxa of the Early Devonian were derived chiefly from the Eastern Americas Realm (there is, however, a real possibility that the Pridoli-Siegen intermediates between *Harringtonina* and *Australocoelia* developed in an unknown Eastern Americas Realm enclave, possibly after migrating from the Tasman Region, and then reinvaded the Malvinokaffric region after a lengthy absence necessitated by the absence of marine, megafossil-bearing beds from Pridoli through Siegen time, unless the megafossil-barren, intervening strata ultimately yield megafossils).

A Llandovery fauna belonging to the North Silurian Realm (Fig. 41) is recognized in the western part of the Malvinokaffric Realm in the Pojo region and elsewhere in central Bolivia. This fauna is interpreted as the result of a brief incursion of warmer water during Llandovery time (actually only during the deposition of a part of the Llallagua Formation; see Berry and Boucot, 1972a, for details of the fauna and localities) to the south from the northern regions of South America. These warm-water conditions ceased after the Late Llandovery as evidenced by the presence in all regions of the Malvinokaffric Realm of endemic Malvinokaffric Silurian faunas above those of the Late Llandovery. The presence of well-dated Early Llandovery Malvinokaffric Realm faunas in Brazil and Paraguay plus South Africa is evidence of the restriction of this North Silurian Realm fauna to the westernmost part of the Malvinokaffric Realm. The presence of Zapla Tillite beneath the North Silurian Realm fauna indicates a temporary oscillation, possibly due to the presence of a marine warm-water current from the north (see Fig. 41) paralleling the same Devonian western source land (Isaacson, 1973b), during at least part of Llandovery time. The fauna recovered (see Berry and Boucot, 1972a, p. 27) includes *Leptaena "rhomboidalis"*, *Atrypa "reticularis"*, *Orthostrophella*, *Dalejina?*, *Strophochonetes?* which are consistent with a *Striispirifer* Community assignment, or a slightly more offshore position.

Mongolo-Okhotsk Region

The Mongolo-Okhotsk Region of the Silurian (Fig. 25) is restricted to a small region in central Asia, and represents only a single, shallow-water community (*Tuvaella* Community) as presently understood. The peculiar nature of this community suggests that it should be considered as a separate provincial entity because of its geographic restriction. The atrypacean genus *Tuvaella*, its most abundant and characteristic taxon, is concluded to have developed from a local Ordovician antecedent. What is here termed the Yangzte Valley Region? may also be a part of the Mongolo-Okhotsk Region (different communities) during the Silurian, but our data are inadequate for decision-making.

The biogeographic term Mongolo-Okhotsk Region has not previously been applied in this region to beds of Silurian age, although employed for the Lower Devonian. I feel that as well developed *Tuvaella (T. gigantea)* Community is present in the Devonian of this region it is logical to conclude that the Silurian faunas of similar type belong to the same Region.

North Silurian Realm

Early—Middle Llandovery
The initial interval of the Silurian, the Early through Middle Llan-

dovery, is a time of fairly widespread shallow-water marine sedimentation on large segments of the major continents. Much of the interior of North America during this interval is characterized by dolomitic beds containing abundant virgianid brachiopods assigned to the *Virgiana* Community. This community extends from Arctic Canada (northern Baffin Island and Somerset Island) to West Texas and adjacent New Mexico on the south; from Anticosti Island on the east to central Nevada on the west. The great expanse occupied by this community and its taxonomic monotony are mute testimony to equable environmental (there is no evidence that *Virgiana* is unusually adaptible) and tectonic conditions for much of the continent during this time interval. The same is true of much of the Siberian Platform, from Kolyma on the east through the Sette Daban, to Taymyr and the Yenisey on the north and west. The existence of great areas occupied by Early-Middle Llandovery virgianids and their apparent lack of recognizable morphological change throughout the time interval correlates well with the concept of slow change in large populations. Contrasting with this situation is the division of the stricklandid brachiopods of Eurasia and North America during the same time interval into three very distinctive subspecies. The stricklandids (*Stricklandia* Community) occupy only a small fraction of the area characterized by virginianids. Other communities during this time interval are, as yet, too poorly known to provide relevant data about population size and evolutionary change, and are also areally restricted.

Late Llandovery

The Late Llandovery animal communities represent the best understood section within the Silurian Period. Similarly, the evolution (Fig. 18) of several brachiopod groups is well understood for this time interval. Areally speaking, *Pentamerus* Community faunas are far more widespead in North America and Eurasia than are all the other shelly communities combined. Simple inspection of the Late Llandovery Community maps (Berry and Boucot, 1970) makes this relation clear. Taxa derived from Benthic Assemblages 1 and 5 are as yet too poorly known from an evolutionary point of view to contribute to the problem of population-size effect on evolutionary change. Taxa from both Benthic Assemblages 2 and 4, however, are very informative. Within the *Stricklandia* Community three taxa (two at the subspecific level, one at the generic level) are inferred to succeed each other in the *Stricklandia* lineage (Fig. 27), as do four species within the closely related *Microcardinalia* lineage stricklandids. Within the *Eocoelia* Community, four species form an evolutionary sequence assigned to the genus *Eocoelia*. Within the *Pentamerus* Community the genus *Pentamerus* is relatively unchanged throughout the time interval, although the very similar related genus *Pentameroides* does appear in about the middle of the time interval.

Provincialism affecting the fauna of Benthic Assemblages 3—5 appears within at least the Early Wenlock (zone of *Monograptus riccartonensis*) as judged by Uralian-Cordilleran Region faunas from both Nevada (lower part of the Roberts Mountains Limestone) and Baillie-Hamilton Island (Cape Phillips Formation) within the Canadian Arctic. It is probable that this endemic development preceded the Early Wenlock, but non-recognition of Llandovery shelly faunas from the Uralian-Cordilleran Region that belong to Benthic Assemblages 3—5 prevents answering the question.[68]

Uralian-Cordilleran and North Atlantic Regions

Wenlock-Pridoli

By Wenlock time regional developments are present throughout the North Silurian Realm (see Fig. 25, 41) in the form of the Uralian-Cordilleran and North Atlantic Regions. These two regions are not assigned realm rank because their level of faunal distinctiveness is comparable to what are termed regions within the Devonian rather than to what are termed realms. Within the Wenlock-Pridoli part of the Silurian a degree of provincialism manifests itself in Benthic Assemblages 3—5. With this initiation of provincialism (see Fig. 25, 41), is developed a greater number of taxa than were present earlier in the period during an equivalent length of time. Among the chief brachiopod groups involved are the smooth and plicate pentamerinids evolved from the *Pentamerus* and related Pentamerinae communities of the Llandovery. Many new genera belonging to this group appear in both of the provincial units.

The Subrianid pentamerids, occurring in Benthic Assemblages 3 and 4, are characteristic of the Late Silurian Uralian-Cordilleran Region. A relative profusion of forms indicates rapid development of the various taxa (Fig. 27). Endemic pentamerinid genera, mostly undescribed, also are present in the Late Silurian Uralian-Cordilleran Region, whereas in central North America a somewhat parallel set of endemic undescribed pentamerinid genera is present within the North Atlantic Region. In both areas these forms are associated with Benthic Assemblage 3, rough-water, commonly reef-related beds. The worldwide distribution during this time interval of the pentamerinid *Kirkidium* contrasts strongly with the aforegoing, and suggests that its ecologic tolerances were either broader or that its specific environment was far more widespread than those of other pentamerinid taxa in the same Benthic Assemblage. Evolution within the genus *Kirkidium* itself during this time interval appears to have been almost nil in contrast to other similar members of the subfamily.

A local development in a small portion of the North Atlantic Region is the abundance in the *Striispirifer* Community St. Clair Limestone of Arkansas, Clarita Limestone of Oklahoma and equivalent units in the Permian Basin subsurface of Texas (see Berry and Boucot, 1970) of an extraordinary abundance of both endemic triplesioids and more cosmopolitan triplesioids. The reasons for this local endemism and abundance are unknown, but the presence of abundant *Onychotreta, Eilotreta, Lissotreta,* and *Placotriplesia* in association with abundant *Brachymimulus, Streptis,* and *Oxoplecia* is indisputable and notable. No other area of Silurian rocks elsewhere in the world contains such an unusual abundance of triplesioid specimens and taxa.

The Wenlock-Pridoli genus *Gracianella* is represented from the Urals to Nevada and the Canadian Arctic plus the Tasman Region (Berry and Boucot, 1973d) in Benthic Assemblages 4 and 5 by at least six species (Fig. 24); the morphologically similar North Atlantic Region genus *Coelospira* is represented by no more than two taxa; as is the central and northwestern European genus *Dayia* of somewhat similar form. *Coelospira* and *Dayia* both have a far wider areal distribution than the restricted linear areas yielding *Gracianella* (Fig. 24). The relatively long-ranging, areally widespread Silurian species of *Coelospira* gave rise to four or more species in the more highly provincial and areally restricted Early Devonian.

A latest Wenlock or earliest Ludlow collection from Cornwallis Island, provided by Dr. Raymond Thorsteinsson, contains a typical Uralian-Cordilleran Late Silurian Region brachiopod fauna (including *Gracianella* and *Conchidium*).[69] The boundary between this Region and the North Atlantic Region can be drawn northward from central Nevada through the central part of British Columbia, and then northeastward to central Cornwallis Island. A similar Uralian-Cordilleran Early Wenlock fauna is known on Baillie-Hamilton island to the north of Cornwallis and in Nevada. It is important to note that the Uralian-Cordilleran Region of the Late Silurian (developed by at least Early Wenlock time) persisted through the Pridoli, and at least initially affected only Benthic Assemblage 5 or possibly part 3 plus 4 and 5. This last statement is based on the occurrence in eastern Nevada, Utah, eastern British Columbia, and the southern part of the Canadian Arctic (south of central Cornwallis Island) of typical Benthic Assemblage 3 faunas belonging to the North Atlantic Region.

North Africa, Czechoslovakia and the Istanbul-region Silurian brachiopod occurrences, together with limited information from Iran, indicate that all of the Mediterranean Region during the Late Silurian belonged to the Uralian-Cordilleran Region.

During the Pridoli, the Appalachian *Eccentricosta* Community (see

Berry and Boucot, 1970) can be recognized; it is known from Gaspe, southwest through northern Maine and the Eastern Townships of Quebec, to the central Appalachians (recognized in New York, New Jersey, Pennsylvania, Maryland, Virginia and West Virginia). The *Eccentricosta* Community probably belongs in Benthic Assemblage 3 and is evidence of the gradually increasing degree of provincialism from the North Silurian Realm during the Early Llandovery to the highly provincial late Early Devonian.

The earliest intimation of an Appohomchi Subprovince (ancestral to the Eastern Americas Realm) differentiating from the Late Silurian North Atlantic Region is provided by this Pridoli *Eccentricosta* Community (Fig. 25) in both the central and northern Appalachians. Although the number of provincial taxa in the *Eccentricosta* Community is small it is distinctive (see Bowen, 1967, for the Keyser fauna). Ecologically the fauna is similar to that in the *Cyrtina* Community of the Appohimchi Subprovince Early Devonian. This similarity is underlined by the former difficulties in correlating the upper, *Cyrtina* Community and lower, *Eccentricosta* Community Keyser faunas with the New York Lower Helderberg.

During the Pridoli the presence of the Benthic Assemblage 2 *Quadrifarius* Community from Podolia (Skala Horizon) through the Baltic region to Nova Scotia (Stonehouse Formation) to coastal Maine and southern New Brunswick (Pembroke Formation) heralds the separation of a distinct (Fig. 25) Rhenish-Bohemian Region fauna.

Yangzte Valley Subprovince?

The shelly facies Wenlock and Ludlow of the Yangzte Valley in China and some of the Japanese Silurian is characterized by the eospiriferid *Nikiforovaena tingi* as well as other characteristic megafossils. It is possible that the fauna in this area is evidence for another Late Silurian Region, but our ignorance of the Chinese Late Silurian raises the possibility that it is merely another community of the Uralian-Cordilleran Region. In any event it is clear that the Late Silurian of the Yangzte Valley and Japan has differences from that recognized elsewhere in the world. These faunas may alternatively merely represent separate communities of the Mongolo-Okhotsk Region during the Late Silurian but our data is inadequate for decision making.

Devonian Biogeography

The Lower and Middle Devonian are times of high provincialism as contrasted to the preceding Silurian, particularly the Early Silurian, and the succeeding Late Devonian. The interval of highest provincialism coinciding with the highest production of new taxa per unit time and greatest development of recognizable biogeographic units is the Siegen—Ems—Eifel (Fig. 31). Preceding and following these times the Late Silurian and Givet (Fig.17,20,21,22) are transitional intervals of either lower rate of production of new taxa or smaller number of recognizable biogeographic units. Within the shallow-water animal communities of the highly provincial Early Devonian, the number of brachiopod taxa shows a marked increase (Fig. 22) over that in an equivalent part of the Silurian, particularly the Llandovery and Pridoli. Within equivalent communities of the Early Devonian the evolution of equivalent groups and numbers of taxa in the biogeographic units may be observed. Although the taxonomy and evolutionary relations of many groups of Early and Middle Devonian brachiopods have still not been worked out it is clear that each Devonian biogeographic unit is largely characterized by its own set of taxa; for example, the various sets of provincial spiriferids derived from common Silurian ancestors. This last statement also appears to be true of most other groups of megafossils present at this time.

Since the last published summaries of Silurian-Devonian biogeography (Boucot et al., 1969; Boucot and Johnson, 1973; Johnson and Boucot, 1973) many new data have accumulated. These are summarized below, biogeographic unit by biogeographic unit.

Malvinokaffric Realm

Since our most recent summary of Malvinokaffric Realm faunas and their origins (Boucot et al., 1969), R. Wolfart (in Wolfart and Voges, 1968) has considered the highly endemic trilobites of this Realm in some depth. In the present book the origins of the provincial Malvinokaffric leptocoelid genus *Australocoelia* from the Silurian Malvinokaffric genus *Harringtonina* via hypothetical intermediate taxa developed outside of the Malvinokaffric region are considered. The Eastern Americas Realm source of the Malvinokaffric genus *Meristelloides*, considered later in this book, serves to further substantiate the largely Eastern Americas Realm derivation of most of the Malvinokaffric Devonian brachiopod genera. The relatively short time duration (Ems—Early Eifel; Boucot, 1971) and widespread, large communities of the marine Malvinokaffric Devonian did not permit a significant amount of evolution to have taken place; thus accounting for the relatively monotonous nature of the Malvinokaffric fauna from section to section and from top to bottom.[70] Benthic Assemblages 3—5 in this Realm are notable for their abundance of conularids and hyolithids, an abundance unknown elsewhere in the World. The absence of stromatoporoids and bryozoans plus the virtual absence or extreme rarity of corals (both tetracorals and tabulates) is notable.

Edwards (1973) has summarized the botanical evidence suggesting that the land flora of the Malvinokaffric Realm differs taxonomically from that known elsewhere in the Devonian. This floral difference is consistent with the Malvinokaffric Realm having been characterized by a cooler climate during the Devonian than was present elsewhere.

Old World Realm

Since 1969 (Boucot et al., 1969) there has been a steady accumulation of information regarding the Old World Realm of the Devonian and its Regions, but the basic outlines appear unchanged.

Rhenish-Bohemian Region

The most significant developments affecting the Rhenish-Bohemian Region have to do with better definition of its boundaries with the Appohimchi Subprovince. Oliver (1968, 1973) has pointed out several areas in northwest Africa and western Europe where some Appohimchi Subprovince (Oliver earlier employed the term "Eastern North America Province" for what we both now term Eastern Americas Realm with an included Appohimchi Subprovince) tetracoral genera occur together with typical Rhenish-Bohemian Region taxa. Oliver's work indicates that some mixing consistent with the presence either of means of easy migration for some taxa or of a nearby Realm boundary existed. Ormiston (1968) has described Rhenish-Bohemian Region trilobites from Turkey Creek, Oklahoma (see Fig. 38 for location) that indicate the presence of the boundary between the Eastern Americas and Old World Realms in this region (essentially on the North American Platform; well northwest of the Ouachita Facies Geosynclinal Belt). T.W. Amsden (oral communication, 1971) finds that the brachiopods from the same Turkey Creek occurrence are also of Rhenish-Bohemian aspect (Bohemian Complex of Communities, largely smooth, articulated shells).

The new proschizophorid brachiopod *Baturria mexicana* and associated brachiopods from the Cuidad Victoria region of northeastern Mexico (see Fig. 38 for location) is indicative of the presence of Old World Realm, probably Rhenish-Bohemian Region (Rhenish Complex of Communities) shells to the southeast of the Ouachita Facies Geosynclinal Belt. Berry and Boucot (1970) included these beds within the Silurian, although the evidence was not discussed by them. Subsequent study shows the presence of a new species of *Baturria (B. mexicana* of Appendix I) with a flat brachial valve, *Nucleospira* sp., a large *Delthyris (Ivanothyris)*, *Strophonella* sp., a non-parvicostellate leptostrophid, an orthotetacid, *Leptaena* "*rhom-boidalis*", an eatonioid, and *Dalejina* sp. This faunule is difficult to date. However, the fossils present, and those unknown in the Eastern Americas Realm (*Baturria* and *Ivanothyris*), except in eastern Gaspe, suggest the Old World Realm and Gedinne age. But, the Pridoli cannot be ruled out. The fauna may be Late Gedinne, as shelly marine Late Gedinne faunas are almost unknown within the Rhenish-Bohemian Region (suggested by its geographic proximity as the most likely for this locality). The general aspect of the shells (i.e., their stage of evolution) is consistent with this conclusion.

All of this information, taken together with the presence of Rhenish-Bohemian Region faunas in southwestern Nova Scotia, indicates the presence of a boundary between the Old World and Eastern Americas Realms that extends (Fig. 37, 38) from southern New Brunswick southwest to northeastern Mexico. The presence of Pridoli faunas of Rhenish-Bohemian Region type (characterized by abundant *Quadrifarius*) in Nova Scotia (Stonehouse Formation), southeastern Maine and New Brunswick (Pembroke Formation) helps to substantiate this interpretation (Fig. 25).

Fig. 38 and 39 indicate the presence of an Eastern Americas Realm/Old World Realm boundary near the St. Lawrence River. For the Ems (Fig. 38) this conclusion is based on the occurrence of Rhenish-Bohemian taxa (*Rhenorensselaeria*, *Schizophoria*, some Rhenish-type spiriferids) in Gaspe, where they are mixed with a preponderance of Eastern Americas Realm taxa. For the Eifel this conclusion is based on the occurrence of Rhenis-Bohemian brachiopods (see Plate I, pp. 338, 339) and trilobites in several carbonate units, together with some Appohimchi Subprovince type tetracoral genera (Touladi Limestone, Famine Limestone, Mountain House Wharf Limestone). This information suggests, when taken together with that from the marginal North American Rhenish-Bohemian fauna, the presence of easy communication, and a realm boundary, northeast of Newfoundland, or a realm boundary situated during some parts of the Devonian on the North American Platform.

Uralian Region

In the preceding section on Devonian communities both the Uralian Complex of Communities and the Dzhungaro-Balkhash Complex of Communities have been discussed from an ecologic viewpoint. Boucot et al. (1969) recognized the endemic distinctiveness of the Uralian carbonate faunas, but little attention was devoted to the fauna occurring in the Kazakhstan terrigenous facies. The Uralian Region carbonate rocks occur around the Kazakhstan terrigenous facies in the Urals (to the northwest), Russian and Chinese Turkestan (to the southwest) and the Altai (to the northeast). The Dzhungaro-Balkhash Complex of Communities is also characterized by a mixture of some Appalachian elements (including *Pacificocoelia*, *Coelospira*, *Meristella*, *Leptostrophia*, and *Orthostrophia*) as well as the Tasman Region genus *Maoristrophia* together with some undescribed endemic genera (observed by Boucot in various Soviet collections). Ormiston (1972) has emphasized the Eastern Americas Realm trilobites present in the Dzhungaro-Balkhash Complex of Communities.

Mongolo-Okhotsk Region

The recent contributions by Modzalevskaya (1968, 1969) and Hamada (1971) allow a Mongolo-Okhotsk Region to be defined within the Old World Realm (Fig. 2). This region is based on fossils derived from terrigenous facies, suggesting that subsequently it may be found to contain different faunas in the carbonate rocks, as do both the Uralian and Rhenish-Bohemian Regions. It must be pointed out again, however, that this prediction is based on the coincidence in much of the Rhenish-Bohemian and Uralian Regions of carbonate and terrigenous lithology with differing faunas, *not* with the control over faunal content of carbonate as contrasted with terrigenous lithology. Soviet workers (for example, Maksimova et al., 1972) have employed the term Mongolo-Okhotsk Province but they have confused (in my opinion) what I conclude to be part of a biogeographic entity with an ecologically distinct unit elsewhere (my Dzhungaro-Balkhash Complex of Communities that equals their Dzhungaro-Balkhash Province).

The Mongolo-Okhotsk Region fauna, as described by Hamada (1971) and Modzalevskaya (1968, 1969), does not fit easily into any of the previously defined Old World Realm Early Devonian Regions (I agree with Hamada that this fauna is approximately Ems). The fauna is of unusual importance biogeographically, because it helps to explain the anomalous occurrence in Kazakhstan of taxa that are common in the Eastern Americas Realm. Hamada reports the first Asian chonostrophid (previous validated reports are solely from the Eastern Americas Realm) and a *Discomyorthis* (his *Dalejina kinsuiensis* may have a slightly more convex pedicle valve than is average for the genus, but the general form of the pedicle valve muscle field strongly argues for including it in *Discomyorthis*) a genus otherwise restricted to the Eastern Americas Realm. Hamada (1971) also mentioned that *Pacificocoelia* has been reported from the Lesser Khingan Mountain region (*Leptocoelia* aff. *biconvexa* Bublichenko). Hamada's record of *Fimbrispirifer* cf. *divaricatus* may be an additional Eastern Americas Realm element, but the material is too limited for certain identification and should not be a basis for decision as to biogeographic affinities of the fauna. The occurrence of *Reeftonia* (Hamada, 1971) in both Asiatic

Russia and Manchuria provides evidence of another genus (in addition to *Maoristrophia*) common to both the Australian-New Zealand and central Asian Devonian. The bulk of Hamada's fauna consists of Old World Realm elements (*Bifida*, for example), a few provincial elements (*Sinostrophia*) and relatively cosmopolitan taxa.

Both the Dzhungaro-Balkhash Complex of Communities of the Uralian Region and the Mongolo-Okhotsk Region contain Eastern Americas Realm genera (in contrast to their scarcity in the Rhenish-Bohemian Region). One can only conclude that marine communications and/or environments were more favorable for migrations of Eastern Americas taxa between southwestern North America and eastern Asia than back and forth across the Atlantic. These observations between eastern Asia and southwestern North America as opposed to the Atlantic are unexpected, and possibly are related to vagaries of oceanic circulation, and other physical environmental factors of the time.

The *Tuvaella* Community is concluded to be a quiet-water low-diversity Benthic Assemblage 2 equivalent belonging to the Mongolo-Okhotsk Region that contrasts markedly with the "Normal" Community of this region.

The very low percentage of endemic taxa currently recorded from the Mongolo-Okhotsk Region is concluded to reflect the very limited sample studied to date rather than the true state of affairs (see section on "Biogeographic statistics" for the strong correlation between sample size and number of endemic versus cosmopolitan taxa).

Eastern Americas Realm[71]

Amazon-Colombian Subprovince of the Eastern Americas Realm

Since their first description and discovery the Devonian faunas of the Amazon region (summarized in Clarke, 1900), Colombia (summarized in Morales, 1965) and Venezuela (Weisbord, 1926; Amos and Boucot, 1963) have been recognized as belonging to the Eastern Americas Realm and to be very distinct from those of the Malvinokaffric Realm Devonian faunas characterizing the remainder of South America. Several problems of a puzzling nature are presented, however, by these Eastern Americas Realm Devonian faunas of northern South America. First is the question of their precise age. Opinions have differed over the past 100 years within an age span from about Ems (Schoharie) to Givet (Hamilton), but with Eifel (Onondaga) being the most commonly accepted. The faunas from the Amazon region of Brazil, Colombia, and westernmost Venezuela cannot be correlated precisely with New York equivalents although their overall similarity is striking. Almost all of the contained genera are present within the Schoharie to Hamilton span in New York, but not all occur together in the Devonian of New York (or other correlative Eastern Americas Realm units within eastern North America). Most vexing is *Tropidoleptus* which has been found in the Amazon region, Colombia and Venezuela. *Tropidoleptus* does not occur with certainty in pre-Hamilton beds in the Appohimchi Subprovince of eastern North America (it does, however, occur in the Lower Devonian Rhenish Complex of Communities Old World Realm faunas of Nova Scotia). Its occurrence in Brazil, Colombia, and Venezuela with genera that in the Appohimchi Subprovince in North America are restricted to either Schoharie or Onondaga equivalents is taken to indicate that *Tropidoleptus* occurs earlier in northern South America than in the Appohimchi Subprovince of eastern North America. The occurrence of *Anoplia* in the Devonian of the Amazon as well as *Plicoplasia* and *Discomyorthis* of the *D. musculosa* type is more consistent with an Ems age than with an Eifel age unless the ranges of these genera eventually are extended upward. The occurrences of *Eodevonaria* of Schoharie type in Colombia and Venezuela, of *Plicanoplia* in Colombia, of *Pacificocoelia* in Venezuela, of a prionothyrid in Colombia, of mutation-ellinids in the Amazon region (the *Derbyina* of that region) also are more consistent with an Ems than with an Eifel correlation. However, W.A. Oliver Jr. (oral communication, 1971) considers that the tetracorals from Venezuela possess great similarities with those of the Eifel age Edgecliff Formation of the Onondaga Group. The presence in both the Amazon Devonian and Colombian Devonian of small productids suggests a more Middle Devonian aspect, although Early Devonian productids are known within the Northern Appalachian Early Devonian as well as elsewhere in the world. The *Pustulatia* (a characteristic Givet taxon) of both the Amazon region and Colombia are misidentified *Plicoplasia* (an Eifel and older Devonian taxon).[72]

In summary, it appears best, based largely on the evidence afforded by the brachiopods, to consider these Amazon-Colombian faunas as of Ems age and more likely correlatives of the Schoharie faunas of eastern North America than of anything else (the tetracorals in this view would be similar precursors to the Eifel taxa of the Edgecliff; ecologically distinct from the Appohimchi Subprovince "Zone B", Ems coral fauna).

The second question relates to the zoogeographic affinities of these Brazilian, Colombian, and Venezuelan Devonian faunas. They clearly possess many taxa in common[73] with the Appohimchi Subprovince of eastern North America, with just about all genera being common, as well as many of the species (in strong contrast to the specifically distinctive faunas of the Nevadan Subprovince). However, the presence of *Pacificoelia* in Venezuela (characteristic of the Nevadan Subprovince), absence of *Lepto-*

coelia, *Costellirostra* and *Eatonia*, presence of mutationellinids in the Amazon region, of a very different species of *Plicoplasia* (*P. curupira*) in the Amazon region, and of abundant *Tropidoleptus* in both Brazil and Colombia in beds considered to be older than the *Tropidoleptus*-bearing beds of the Appohimchi Subprovince localities justifies separation of these faunas into a separate Subprovince. The difficulty of correlating these faunas in detail with the Appohimchi Subprovince faunas of eastern North America substantiates their distinctiveness and is additional justification for setting them up as a separate zoogeographic unit, whatever the final age turns out to be.

The occurrences of mutationellinids and the unique species of *Plicoplasia* in the Amazonian Devonian may be considered as Malvinokaffric elements occurring close to the Realm boundaries. The occurrence in Colombia of spiriferids possessing ribbed fold and sulcus, but possibly not being Appohimchi genera of the *Costispirifer* or *Fimbrispirifer* type, also is distinctive of this Subprovince (the affinities of these spiriferids may be with the Rhenish Complex of Communities, but this point is presently very uncertain; they also could be endemic taxa).

The absence of earlier Devonian faunas within this Amazon-Colombian Subprovince (that is, faunas of Helderberg, Oriskany or Esopus type) also is very puzzling. It probably can be interpreted best as indicating the presence of a major unconformity in northern South America between the Ems or Eifel and Silurian and older beds. Morales (1965) showed an unconformity to be present in Colombia between the Devonian and a basement complex in some areas and between the Devonian and fossiliferous Ordovician in other areas. The presence of fossiliferous Late Silurian in Venezuela (A.J. Boucot, in Berry and Boucot, 1972a) indicates that the uncomformity may represent a break between the Ludlow and the Ems or Eifel. In the Amazon region (see Lange, in Berry and Boucot, 1972a) the presence of a major disconformity between the Lower Llandovery and the Devonian supports the presence of a similar break to the north. The general absence in South America of Gedinne and Siegen beds containing megafossils may even indicate that the entire continent was above sea level during this time interval. This situation contrasts strongly with the megafossil-rich, transitional nature of the Silurian-Devonian boundary in North Africa, except for Ghana where the Devonian (Ems or Eifel) rests unconformably on the Precambrian. The Ghana occurrence is closer on a continental drift model to the South American Devonian than is any of the other Devonian of North Africa.

Finally it is possible that the Parnaiba Basin Devonian (Kegel, 1953) of eastern Brazil with its reported *Amphigenia* and *Eodevonaria* may well belong to this Amazon-Colombian Subprovince, and also be of Ems age.

The Parnaiba Basin Devonian is also reported to contain *Tropidoleptus* and the mutationellinid *Derbyina*.

The unconformable relations between the Ems and the pre-Ems rocks in the Amazon-Colombian Subprovince also is consistent with the rapid Ems (and possibly also Early Eifel) onlap—offlap regime postulated by Boucot (1971) for the Malvinokaffric Realm Devonian (Africa from Ghana to South Africa; Antarctica; South America from the Lake Titicaca region and Matto Grosso south). Ignorance of the precise sequence of faunas within the Amazon-Colombian Subprovince, the lack of measured sections, etc., make it impossible at present to determine whether the Devonian of this region has a basal non-marine sequence overlain by: (*1*) marine strata representing a progressively deepening set of marine animal communities; (*2*) a shallowing sequence capped by another nonmarine sequence (as is concluded to exist within the Malvinokaffric Realm; Boucot, 1971; Isaacson, 1973b). The very limited data from the Amazon-Colombian Subprovince are, however, not inconsistent with this view. The faunal barrier separating the two Realms was unrelated to this marine—nonmarine history.

Nevadan Subprovince

The Nevadan Subprovince is defined to include a sequence of Gedinne and possibly Siegen faunas found in central Nevada and southern California that include a preponderance of Eastern Americas Realm genera represented by species restricted to this region.[74] The other two Eastern Americas Realm Subprovinces (the Appohimchi Subprovince from Gaspe to Chihuahua and the Amazon-Colombian Subprovince) share many common species by contrast with the Nevadan Subprovince. Johnson's (1970a) summary makes clear that during the Early Devonian the Eastern Americas Realm-Old World Realm boundary oscillates east-west through time in central Nevada. The Nevadan Subprovince includes Johnson's (1970a) *Spinoplasia* Zone fauna, *Trematospira* Zone fauna, and to a lesser extent the *Acrospirifer kobehana* Zone fauna. Johnson's *Quadrithyris* Zone has a few Eastern Americas Realm elements, as does the fauna of the *Eurekaspirifer pinyonensis* Zone (Ems), but the overall Old World Realm aspect of the former and the more strictly unique aspect of the latter (assigned to the Cordilleran Region of the Old World Realm; see Fig. 25) remove them from the Nevadan Subprovince category.[75] The *Spinoplasia* and underlying *Quadrithyris* Zones now are assigned to the Middle and Upper Gedinne, following Carls' (1969) work on co-occurring Rhenish Complex of Communities brachiopods and conodonts common to Bohemian Complex of Communities and other Old World Realm regions, including Nevada. Diagnostic conodonts are not available for correlation of the *Trematospira*

Zone; but the material described as *Rensselaeria* sp. (Johnson, 1970a) with relatively non-obsolescent dental lamellae differs from Oriskany types and suggests a Late Helderberg age. As this *Rensselaeria* of Late Helderberg aspect occurs in the *Trematospira* subzone, possibly this subzone spans both Helderberg and Oriskany times. Carls assigns beds with *Icriodus pesavis* (*Quadrithyris* Zone) and *I. latericrescens* (*Spinoplasia* Zone) to the Late Gedinne. Compatible with Carls' correlation of the zone of *I. pesavis* with the Late Gedinne, R. Thorsteinsson (written communication, 1971) found Lower Dittonian vertebrates (i.e. Upper Gedinne), Uyeno (written communication, 1971) found *I. pesavis*, and J.G. Johnson (oral communication, 1971) found *Quadrithyris* Zone brachiopods, associated on Prince of Wales Island. The *Trematospira* Subzone may prove to be latest Gedinne, placing all of the Helderberg within the Gedinne and indicating a corresponding revision of howellellid evolution during the Gedinne-Siegen of the Eastern Americas and Old World Realms (Boucot and Johnson, 1967a). These changes emphasize the difficulty of making detailed correlations between strongly differing biogeographic units; more such changes are to be expected in the future.

Evidence provided by the trilobites (A.R. Ormiston, written communication, 1973) is in strict agreement with that of the endemic brachiopods "Ems-Eifel Cordilleran Subprovince" (sic. Region) embracing *kobehana* through *circula* Zones for my purposes): "*Scutellum; Prionopeltis?; Ganinella;* and *Kasachstania (K. ulrichi)* plus "new genus" *meeki* (Walcott). Coeval Appalachian (sic. Eastern Americas) endemics include *Odontocephalus; Anchiopsis; Synphoria; Synphoroides; Coronura; Trypaulites; Odontochile; Dalmanites; Corycephalus, Dechenellurus; Teretaspis; Mystrocephala; Ceratolichas; Phacopina; Echinolichas; Crassiproetus; Greenops?.* Genera common to the two areas include: *Calymene, Phacops, Isoprusia, Leonaspis* and *Otarion.*" "As with the brachiopods the trilobites of the *Trematospira* and *Spinoplasia* Zones are more similar to the Appalachian Province (sic. Eastern Americas Realm) even down to species group similarities." These comments from Ormiston emphasize the Old World Realm nature of the *kobehana* through *circula* Zones and the Eastern Americas Realm nature of the *Trematospira* and *Spinoplasia* Zones.

Nevadan Subprovince faunas of Siegen and Ems as well as Early Gedinne age would be predicted to occur to the east of central Nevada, but both dolomitization of the Early Devonian in this region and removal of Early Devonian beds by Middle Devonian erosion have probably destroyed such material.

The Hamilton fauna
The Eastern Americas Realm, Late Eifel-Givet Hamilton fauna has been

a problem ever since it was first described. It consists of a mixture of both Eastern Americas Realm earlier Devonian derivatives and Old World Realm, Rhenish-Bohemian Region earlier Devonian derivatives. The presence of abundant Rhenish-Bohemian taxa like *Athyris* and *Tropidoleptus* as well as the trilobite *Greenops* (an asteropygid) together with abundant *Pustulatia* (a *Plicoplasia* descendent), *Trematospira*, meristellids, and *Protoleptostrophia* illustrates my point. If we assume that the Rhenish-Bohemian taxa lived under somewhat different physical conditions than the Eastern Americas Realm taxa during the Early Devonian, it is puzzling to see their derivatives living together during the Givet. One must conclude that either the faunal barriers of the Early Devonian did not separate regions having a different physical environment, or that one or both groups of taxa evolved enough to tolerate a common Givet environment. This mixing might be viewed as a type of homogenization tending towards a lower level of endemism, but the lowering of provincialism within the Late Devonian is actually accomplished by the wide spreading of a more limited number of taxa rather than the wide spreading and mixing of taxa combined from a number of previously existing biogeographic units. Thus the actual number of taxa decreases markedly in the later Devonian whereas the Hamilton fauna is extraordinarily rich in taxa and is anomalous in this regard.

Limits in North America of the Eastern Americas Realm

South. Fig. 2 indicates very diagrammatically a boundary between the Amazon-Colombian Subprovince and the Appohimchi Subprovince. It is important to keep in mind that the Amazon-Colombian Subprovince does not include any pre-Ems strata. Therefore, it is reasonable to conclude that the entire Eastern Americas Realm of the pre-Ems probably covered only half of the area occupied during the Ems-Givet interval. It is also important to keep in mind that the large region between northern Mexico and Colombia has not yet yielded any Silurian or Devonian faunas.

East and southeast. The occurrence in both central Texas and the Nevadan Subprovince of *Pacificocoelia* and *Llanoella* suggests closer faunal communication between these regions than between other parts of the Eastern Americas Realm. The *Pacificocoelia* also indicates closer faunal communication with the Amazon-Colombian Subprovince in Venezuela and Colombia. The sampling north and east of Texas is good enough to give this limited information significance.

In an earlier paper Boucot and Johnson (1967a, Fig. 1, 2, 3a, 4 and 5) outlined Early Devonian paleogeography and lithofacies for the region occupied by the Appohimchi Subprovince faunas. Since then considerable new information has accumulated. The outlining of a Ouachita Facies Belt for the Early and Middle Devonian (Fig. 37—40) gives one a southeastern limit for most Appohimchi Subprovince faunas (except for the Turkey Creek, Oklahoma occurrence) that coincides well with the lithologic information.[76] Southeast of the Ouachita Facies Belt, which extends to Zacatecas (Cordoba, 1964) on the southwest, the shelly faunas (at least Silurian through Mississippian) of Ciudad Victoria occur in an area that has a stratigraphy different from anything to the northwest (Fig. 2, 37, 38, 39). The presence of Silurian shelly faunas in the Cuidad Victoria area in rocks (Caballeros Canyon Formation) unlike anything known in the northwest shows the presence of an area of Early Paleozoic shallow-water, terrigenous, shelly deposition (the Tamaulipas Platform of Fig. 37—40). Devonian novaculite above the Silurian indicates a tie with the Ouachita facies. The Old World Realm aspect of the Pridoli or Gedinne faunule from Cuidad Victoria also fits this picture. The Tamaulipas Platform, admittedly based on limited data (crystalline basement of possible Precambrian age is present in this area), may be a part of the "Llanorian" source region postulated by many geologists to account for the terrigenous material present in the Ouachita Geosynclinal Belt. The Florida Platform may be roughly equivalent to the Tamaulipas Platform as indicated by its geographic position relative to the Ouachita Facies (Fig. 2, 37, 38, 39). The sedimentary rocks of Early Ordovician to Devonian age present on the Florida Platform are of non-geosynclinal facies and only 3,000—6,000 ft. thick and belong to the Platform Mudstone facies. Should rocks of Early Paleozoic age be present in the metamorphic basement of Cuba (as concluded by Khudoley and Meyerhoff, 1972), it is possible that they could have been part of a similar Llanoria complex. Possibly the Platform carbonate-type Early Ordovician (Tremadoc) (Pantoja-Alor and Robison, 1967) resting on typical Precambrian basement complex of central Oaxaca is a fragment of a formerly extensive shield area bordered by the terrigenous, shelly mudstone of the Tamaulipas Platform region on the northwest. The Oaxaca Tremadoc is lithologically very different from either the terrigenous Early Ordovician of the Florida subsurface or that of North Africa.

The lithofacies information from northern Mexico and the southwestern part of the United States implies that the Ouachita Facies Belt (Fig. 37, 38) has a far greater breadth in northwestern Mexico (essentially between the limestone of the Hermosilla, southern Sonora area, and central Zacatecas), or that a large area of carbonate deposition was present in this region during the Early Devonian but has not yet been recognized, or that subsequent structural complications have moved various blocks about so as to produce the present relation. The relatively narrow width of the

Ouachita Facies Belt between the Tamaulipas Platform (Fig. 37) and the Platform Carbonates of Chihuahua and the Llano Uplift may well be considerably less than the original predeformational width, but the great increase in width made possible by the data from northwestern Mexico does appear to require an explanation based on data that are not yet known.

McLaughlin's (1970) data on land plants and marine acritarchs of possible Ems age (A.R. Ormiston, written communication, 1972, suggests that these beds may be Eifel rather than Ems) from southwest Georgia (part of the Florida Platform) are consistent with the termination well east of the Mississippi Embayment of the large island extending south from eastern New York through eastern Tennessee to the northern Alabama region (i.e., having terminated between northern Alabama and the marine rocks of the Ouachita Facies Belt).

Westward from the southern Appalachians evidence is lacking for nonmarine Devonian beds. Neither the Missouri Mountain Slate nor the Arkansas Novaculite (at least in part of Devonian age) of the Ouachita Belt contain nonmarine Devonian beds, nor do their correlatives along strike toward the east or west in the surface or subsurface. Thus Devonian strata west of the southern Appalachians are exclusively marine. T.J. Carrington's (1972a; unpublished report) discovery of Early Devonian marine fossils in the Butting Ram Sandstone of the Talladega Series in northern Alabama below the Early Devonian, fossiliferous Jemison Chert[77] provides an additional reason for terminating the Appalachian land area north of this region (Fig. 37—40).

The presence in the Nova Scotia Early Devonian Rhenish Complex of Communities characterized by abundant *Meristella*, a typical Eastern Americas Realm taxon, favors the concept that this part of the Old World Realm was close to the Eastern Americas Realm or at least had an easily accessible pathway available for migration of an Eastern Americas taxon (*Meristella* is unknown elsewhere in the Rhenish Complex of Communities although present in the Tasman and Uralian Regions).

In summary (Fig. 2, 37—40) it appears that the eastern and southeastern limits of the Eastern Americas Realm are formed by a series of islands in the Nova Scotia-Newfoundland to northern Alabama-eastern Tennesee region and coincide largely (except for the Turkey Creek, Oklahoma locality) with the Ouachita Geosynclinal Facies Belt from there southwest.

North. Knowledge of the northern limits of the Eastern Americas Realm is very unsatisfactory. Fig. 37—40 indicate that Appohimchi Subprovince faunas are known as far north as the Hudson Platform and Gaspe during portions of the Early and Middle Devonian. North and east of these positions in Canada and Greenland we have no information. Old World Realm faunas are known in Arctic Canada indicating that the boundary must be to the south of the Boothia Arch region of Arctic Canada and also south of Ellesmere Island. Fig. 37 for Gedinne time indicates an Old World Realm boundary located to the northeast of Gaspe. This boundary placement is based on the occurrence of an undescribed notanopliid, *Ivanothyris*, *Lissatrypa*, *Amphistrophia* (the youngest known occurrence of this genus of Silurian-type stropheodontid), *Merista* and *Camarium* in northeastern Gaspe (P.A. Bourque, 1972, and oral communication, 1972) in the Roncelles and Lefrancois Members of Gedinne age. Fig. 38, 39 indicate evidence for Ems and Eifel time of an Old World Realm boundary in the St. Lawrence River-Gulf of St. Lawrence region. On Fig. 38 a boundary between the Old World and Eastern Americas Realms is shown northeast of Gaspe. This boundary is placed here because of the presence in Gaspe of Schoharie equivalent beds (the 4-Mile Brook Beds, for example) that contain Rhenish elements in addition to typical Eastern Americas taxa, indicating relatively nearby access to the Old World Realm. This interpretation is supported further by the presence in Eifel beds of the Eastern Townships of Quebec (Fig. 39) of typical Old World Realm brachiopod taxa (Plate I), together with some Eastern Americas Realm tetracoral genera and *Brevispirifer* the only brachiopod shared (Plate I), with little relation to the Eastern Americas Realm brachiopods of the coeval Onondaga Limestone and its correlatives. The presence of Old World Realm elements like *Paraspirifer* and *Brevispirifer* in the Eastern Americas Realm during Eifel time is entirely consistent with this interpretation.

The boundary of the Eastern Americas Realm north and east from Hudson Bay is unknown. The presence of Old World Realm faunas and faunal elements in parts of the Eastern Americas Realm Early Devonian, Eifel and Givet (see the positioning of the boundary shown in Fig. 37—40) shows that there was some communication adjacent to the platform between the two Realms. If the sea covered the Canadian Shield during the Devonian the Eastern Americas Realm-Old World Realm boundary north and west of Hudson Bay would had to have been determined by some faunal barrier to migration in the marine realm, as is true from the southern Appalachian region southwest into Mexico.

It is alternatively possible that an extensive land area occupied the region north of Hudson Bay and extended to the Boothia Arch region as well as east across most of Greenland, but we lack any evidence. The oldest non-marine Devonian in east-central Greenland is possibly of Eifel age, but is restricted to a relatively limited region indicating nothing of its former extent.

West and northwest. There is little to add to our knowledge of the western limits of the Eastern Americas Realm since publication of Boucot and Johnson's (1967a) summary.

A typical Givet Old World Realm faunule containing *Stringocephalus* has been described by Cooper and Phelan (1966) from Indiana. This information indicates communication with the widespread Old World Givet faunas over most of the Northern Hemisphere.

Lithofacies boundaries on the west and northwest side of the Eastern Americas Realm

Fig. 37–40 show several lithofacies boundaries not dealt with by Boucot and Johnson (1967a). The following paragraphs provide the information on which these additional boundaries are based.

Gedinne facies for eastern North America are shown in Fig. 37. On the northwestern side of the area occupied by the Eastern Americas Realm faunas there is good reason (Kjellesvig-Waering, 1958) for placing the platy dolomites of the Put-In-Bay Dolomite of the Michigan Basin with their Eurypteridae Community fauna into an Early Helderberg position. McGregor et al.'s (1970) find of Gedinne plant spores in the upper part of the James Bay Kenogami River Formation (a unit that may be interpreted as either very shallowest marine or possibly estuarine) enables us to terminate the Early Helderberg shoreline in the James Bay region and also to infer the presence of a belt of coastal-type environment between the Great Lakes region and the continental backbone land-area farther west.

The Risser Beds (Berry and Boucot, 1970) of the Williston Basin (Fig. 37) may be in part of Early Devonian age as suggested by Wilson and Majewske (in Berry and Boucot, 1970). They include dolomite and anhydritic limestone similar to those of the Kenogami River Formation of the Hudson Platform (plant spores have not yet been reported from the Risser Beds). The absence from the Risser of any marine fossils or of an evaporitic sequence similar to that of the Michigan Basin Salina Group would seem to strengthen the correlation with the Kenogami River rather than with any part of the Bass Islands Group. However, Norris and Uyeno (1971, p. 212) noted that the thin red bed Ashern Formation of the Williston Basin region, although unfossiliferous, is lithologically similar to the middle member of the Kenogami River Formation of the Hudson Platform (this middle member may include the Silurian-Devonian boundary within itself or at its top).

One should note that in Fig. 27–39 two belts of evaporitic suite rocks have been outlined adjacent to the continental backbone (with one to the west of it for Eifel time). During the Gedinne interval the wholly marine evaporites of the Bass Islands Group provide evidence for one type whereas the Kenogami River Formation evaporites that lack marine megafossils provide evidence for the other type. During the deposition of both the Oriskany and the Schoharie there is no evidence for the presence of evaporitic rocks in either the Michigan Basin or Hudson Platform (the uppermost Kenogami River Formation might be partly equivalent to the Oriskany, but this is not completely certain); their presence to the west of these basins is inferred for the sake of symmetry on Fig. 38. The symmetry is that afforded by the evaporites present in the underlying Gedinne rocks and those present in the overlying Eifel and Givet rocks. Fig. 38 is an attempt to organize the available data regarding Eifel age beds. The Detroit River Group rocks of Fig. 39 contain typical marine Evaporite Suite rocks bearing a taxonomically very reduced fauna (*Prosserella* Community). Fig. 40 indicates the presence of the Givet-interval evaporite rocks of the Wapsipinicon Group in Iowa and adjacent states. It is logical to conclude that during the deposition of the Oriskany and Schoharie, Evaporite Suite rocks were originally present to the west of the Michigan Basin and Hudson Platform but were removed by pre-Eifel Devonian erosion. During Eifel and Givet time it is also reasonable to infer that the Evaporite Suite rocks east of the relatively small, emergent continental backbone were continuous on the north with those present on the west side of that land area. It is possible that these evaporites continued as far north as northern Baffin Island and its environs, but there is no evidence at present on this point.

Fig. 40 is a synthesis of Hamilton, Givet-age paleogeography and lithofacies. The most significant post-Eifel change is the large post-Acadian Orogeny land mass in the Appalachian region. Typical Hamilton Group sandstone containing typical Eastern Americas Realm Hamilton brachiopods is present in the Saint Helen's Island breccias at Montreal. Evaporites are absent within both the Michigan Basin and Hudson Platform areas, but are present in the subsurface further southwest. Evaporites are widespread in the Williston Basin and the Prairie Provinces of Canada during much of this time interval. The post-Givet North American Devonian shows evidence for the presence of fewer large bodies of evaporites (chiefly in the Prairie Provinces) or of any significant land masses in the interior portions of the continent. However, the post-Acadian Orogeny land mass persists through the Carboniferous.

Based on lithofacies data Poole and Hayes (1971) suggested that Early and Middle Paleozoic shelf clastic carbonate rocks of northwestern Sonora (Early Cambrian through Devonian, and Late Precambrian) represent a southern continuation of the Cordilleran miogeosynclinal facies pattern of western North America (Fig. 2, 37—40). If true, as seems likely, faunal communication between the Nevadan Subprovince and the Appohimchi Subprovince may have been near the southwestern extremity of the continental backbone. This may well have terminated within northern Mexico as implied by Poole and Hayes.

In all of these paleogeographic and lithofacies reconstructions (Fig. 37—40) the placement of shorelines has not been influenced by the position of the zero isopach, unless lithofacies data and animal community data so indicate. For example, the structurally isolated Devonian marine strata in the Hudson Basin and Williston Basin obviously are remnants of formerly far larger bodies of Devonian marine sediment. Fig. 40, the diagram for Givet time, indicates continuity of marine beds between the Michigan Basin, the Hudson Platform and Montreal based on isolated occurrences in all three places. There would be little point to placing a large land area, or land areas, between these structurally isolated bits of marine Givet unless lithofacies data indicated such to have probably been the case (which it does not). The presence as well of Eastern Americas Realm faunas in the Hudson Platform and James Bay regions would make the placement of a major east—west land mass extending from New York to the continental backbone an additional, unrequired problem in trying to explain the distribution and isolation of Eastern Americas Realm faunas occurring elsewhere in North America. Overstepping relations do not necessarily indicate, unless lithofacies and animal community data concur, that the overstepped unit was originally absent in the region where it is now absent. The multitude of small disconformities now recognized on platforms makes it clear that many stratigraphic sections can incorporate previously unrecognized disconformities that in an overstep situation may prove to be of significance in trying to understand whether or not the overstepped unit was originally deposited well away from its present last area of occurrence.

In Utah and southeast Idaho the vertebrate-bearing dolomites of the Water Canyon Formation are interpreted as being either non-marine or brackish in origin. The lack of marine fossils and the many vertebrates represent an environment similar to that of the Kenogami River Formation (Fig. 38). Some of the Sevy Formation of western Utah and possibly even portions of eastern Nevada may fit into this same category.

Eastern Americas Realm barriers to migration with the Old World Realm

The foregoing information about the location of Eastern Americas and Old World Realm faunas in the Late Silurian, Lower and Middle Devonian plus the discussion of lithofacies boundaries involving evaporites to the west and northwest of the Eastern Americas Realm permits a discussion of faunal barriers that controlled the boundaries of the Eastern Americas Realm.

The potential barriers to faunal migration considered are land barriers, temperature barriers and hypersaline water barriers, or combinations of the three. In addition the role of currents in controlling the distribution of either hypersaline water masses or water masses with differing temperature must be considered, as well as the location of deep ocean basins separating shallow-water regions.

During the earlier part of the Late Silurian we have no evidence for the presence of any large body of land in the continental backbone region despite the presence of endemic taxa to the east and west (Fig. 25, Uralian-Cordilleran and North Atlantic Regions). During the latest part of the Silurian, and possibly during the Gedinne-Siegen interval, a large landmass may have been present in much of the continental backbone region (Fig. 37). By Ems time the size of this landmass (Fig. 2, 38) may have been considerably reduced, and by Eifel and Givet time (Fig. 39, 40) it was probably fairly small. By Frasne and Famenne time it had essentially disappeared. If closely spaced isotherms occurred normal to this landmass they would have provided a potential thermal barrier. However, the presence of the Nevadan Subprovince to the west of the landmass during part of the Gedinne and Siegen and of Eastern Americas elements in small number during other parts of the Lower Devonian on the west (even to Australia and Central Asia!), plus the occurrence of Appohimchi Subprovince faunas on the east of the land mass from Chihuahua to Hudson Bay, and the occurrence during the Ems (and possibly Eifel time) of widespread Eastern Americas Realm faunas in the Amazon-Colombian Subprovince of South America (Fig. 2) make this possibility appear unlikely.

In addition, the absence of any significant land barriers on the eastern side of the Eastern Americas Realm in North America (the islands indicated in Fig. 37—40 are relatively small features; inadequate to serve as barriers), underline this negative conclusion. Finally, the worldwide synchroneity of decreasing Devonian provincialism with that known from North America is strong evidence for the disappearance of the "continental backbone" during the same time interval being largely coincidence rather than cause.

The inadequacy of the continental backbone as a major barrier is further illustrated by the occurrence of similar Early Cambrian-Early Ordovician (Canadian) faunas peripheral to it. Early Cambrian-Early Ordovician provincialism, where present, is between entire platforms, rather than between eastern and western halves of one (A.R. Palmer, written communication, 1972; Palmer, 1972, 1973). During Middle Ordovician (post-Canadian) through Late Silurian time much of the continental backbone does not appear to have been emergent, and there is no evidence for strong endemism, although endemism peripheral to the platform itself is present during Ordovician time and Late Silurian time.

During the Late Silurian there is evidence for the presence of a large body of hypersaline water not only in the continental backbone region, but also well east (Michigan Basin and Hudson Platform) coinciding in time with the appearance of Late Silurian provincialism (Fig. 25, Uralian-Cordilleran and North Atlantic Regions). During the Gedinne, and possibly into the Siegen, there is evidence for extensive hypersaline water immediately east of the continental backbone that may have existed further to both the north and south than the emergent portion of the continental backbone itself (Fig. 37). By Ems time the size of the body of hypersaline water located east of the emergent portion of the continental backbone may have been more than twice as long north—south as the exposed land area itself (Fig. 38). During Eifel and Givet time (Fig. 39, 40) the emergent portions of the continental backbone were fairly small, but the possible region of hypersaline water to the east may have remained fully as extensive as during the Late Silurian—Ems. By Frasne and Famenne time both emergent land area of any consequence in the continental backbone region and large masses of hypersaline water (medium-sized masses present in the Prairie Provinces) had probably disappeared, coincident with the disappearance of large-scale provincialism as well.

It is clear from the above deductions regarding accordance of times of provincialism and extensive bodies of hypersaline water, and from similar situations for the Late Silurian—Middle Devonian of the Siberian Platform, for the Middle Devonian of the Russian Platform, and possibly for the Late Silurian—Early Devonian of the Russian Platform, that hypersaline water masses may serve as faunal barriers, or as contributory factors, within platform interiors. However, the clear evidence for the absence of hypersaline water bodies on the eastern margins of the Appohimchi Subprovince suggests that this mechanism by itself was not responsible for the existence of the Eastern Americas Realm.

The most plausible explanation would appear to be the existence of a current-driven water mass differing enough in temperature to provide the faunal barrier. During the Ems-Eifel interval such a water mass, extending from the Amazon region to the Hudson Bay-Gaspe region and possibly beyond might have served as the faunal barrier. There is no evidence for a land mass between the Malvinokaffric Realm Devonian and that of the Amazon-Colombian Subprovince, but a thermal barrier (colder to the south in the Malvinokaffric Realm) might have existed. Using this interpretation the combined hypersaline water body adjacent to the continental backbone, as well as the emergent portions of the continental backbone might have helped in North America (north of the Gulf of California region at least) to have channeled the current. The northeasterly termination of such a current might have been in the region of the present North

Atlantic or in a "cul de sac" well north of both Hudson Bay and Gaspe, but to the west of eastern Greenland.

These conclusions depending on better circulation across the eastern half of the North American Platform might, at first, appear to be in conflict with the concept of biotic dolomitization presented in this book. Biotic dolomitization is conceived of as requiring a very limited amount of circulation on the platform areas covered by secondary, subtidal dolomite. However, inspection of Fig. 37—40 indicates that far more limestone (indicating good circulation and exchange of water with the oceanic reservoir) is present in the Devonian of the Eastern Americas Realm area of North America than was present during the Silurian in the same regions (see Berry and Boucot, 1970, plate I).

It is clear that the Amazon-Colombian Subprovince and the Appohimchi Subprovince were in closer faunal communication because of their greater taxonomic similarity, than were either with the more provincially distinct Nevadan Subprovince. This suggests that the Nevadan Subprovince faunas may have been in more discontinuous communication with the Appohimchi Subprovince either about the southern end of the continental backbone or to the south into the Amazon-Colombian Subprovince.

In summary then, I propose a faunal barrier model that calls for a large mass of hypersaline water within the platform interior combined with a current-driven water mass having sufficient temperature differential to adequately isolate the Eastern Americas Realm from the Old World Realm occurring to the west, north and east. The Malvinokaffric Realm to the south, interpreted as a colder region, may be inferred to have been isolated from both the Eastern Americas and Old World Realms on the basis of temperature alone (the Malvinokaffric elements known in Tasmania, Victoria and New Zealand in the Tasman and New Zealand Regions respectively, are consistent with this interpretation). Testing of this model for Ordovician and Permo-Carboniferous intervals should give evidence as to its validity.

If one concludes that the Malvinokaffric Realm Devonian fauna existed in a relatively cool or even cold environment then some conclusions about the Eastern Americas Realm may be made. It is clear from consideration of the Malvinokaffric taxa that the small number present are in largest part either endemic forms derived from Eastern Americas Realm precursors or taxa shared with typical Eastern Americas Realm endemic taxa. In other words, one can make a case for the Eastern Americas Realm and the Malvinokaffric Realm having much in common taxonomically. If this taxonomic closeness reflects environmental closeness as well it is not difficult to go one step further and to conclude that the Malvinokaffric

Realm environment was more like that of the Eastern Americas than like that of the Old World Realm. Possibly the shared Malvinokaffric Realm-Eastern Americas Realm taxa were eurythermal on the cool end, whereas the Eastern Americas Realm taxa shared with the Old World Realm (a group different from those shared with the Malvinokaffric Realm) were eurythermal on the warm end. If such were the case one may deduce that the real "boundary" between the Eastern Americas Realm and both the Malvinokaffric and Old World Realms of the Devonian was a thermal gradient kept in place by a marine current system. Such a conclusion fits the known facts more easily than any other currently available possibility. Such a marine current would, of course, have been influenced in its position by the bottom topography of the platform over which it flowed and of any land masses adjacent to its track. The inferring of such a current raises grave questions about the nature, or even existence during the Devonian, of the Gulf of Mexico-Central America region as we must have had some means of bridging the gap between South America and the Eastern Americas portion of North America.

Malvinokaffric Realm barriers

The existence of a Malvinokaffric Realm benthic invertebrate fauna during both the Silurian and Devonian is indisputable. The reasons for its separateness are hard to evaluate. Summary of the available climatic evidence is consistent with the Malvinokaffric Realm area having been significantly colder or cooler than areas elsewhere during the Silurian and Devonian. If temperature had been the determining factor in the shallow-water environment one might expect that the cold-water, highly-endemic, Benthic Assemblage 2 through 5 Malvinokaffric Silurian and Devonian faunas would appear at some depth in the northern, warmer-surface water regions. Such is not the case. Therefore, it must be concluded that factors other than temperature, or in addition to temperature, are serving as the barrier to faunal migration. There are many such possibilities but we lack geologic evidence at present with which to make a decision.

A northern limit for the Malvinokaffric Realm is shown in Fig. 2 and 41. A western limit is more difficult. However, during the Llandovery the presence of North Silurian Realm Benthic Assemblage 3 communities in Bolivia indicates that a current of warm water may have come from the north in a southerly direction parallel to the source land region inferred to have existed to the west of the known marine Silurian exposures of the central Andes. During the Devonian a similar current may have come from the north if the little known Devonian of the Arequipa, Peru region is actually Devonian and belongs to the Amazon-Colombian Subprovince. There is, however, no evidence for the presence of such a current during the Late Silurian (Malvinokaffric Realm Late Silurian faunas occur above North Silurian Realm faunas in Bolivia).

Absence of Intermediate Forms

Careful consideration of Fig. 25 indicates several regions where data are lacking. For example, the characteristic Givet Hamilton genus *Pustulatia* is most logically derived from the strongly plicate Early Devonian ambocoelid genus *Plicoplasia*. *Plicoplasia* occurs in the Eastern Americas Realm and Malvinokaffric Realm Early Devonian, but is present in Eifel faunas only in a North African occurrence. An additional example is provided by the Early Devonian species of *Pacificocoelia* (Fig. 34). *Pacificocoelia* in the New World is known from Ems beds in Venezuela and Colombia, from Gedinne beds in Nevada, and from Ems beds in central Texas, all parts of the Eastern Americas Realm, plus rare occurrences in the Ems of west-central Argentina and Bolivia in the Malvinokaffric Realm. The Silurian ancestor of *Pacificocoelia* is *Harringtonina* whose youngest occurrence is in Argentina where it can be no younger than the Ludlow. Therefore, we have a gap for the Pridoli-Siegen in the Malvinokaffric region of South America and South Africa (no marine faunas of this age have yet been proved). One must infer the existence of a region in the New World where these forms developed during Pridoli and possibly parts of both Gedinne and Ludlow times before appearing in Nevada. One then must account for the absence of Siegen intermediates before noting the genus in a widespread position (even as far as eastern Asia and Australia) in the Ems.

The entire Malvinokaffric Devonian is another example of the lack of intermediates of earlier Early Devonian age. The bulk of the Malvinokaffric brachiopods appear to be derived from Eastern Americas Realm ancestors, but no one has found the area in which the pre-Ems Devonian antecedents developed and are preserved. The latest example from this Malvinokaffric Realm is the genus *Meristelloides* proposed by Branisa (1965, p. 130) in a plate description and based on Clarke's (1913) species *Meristella septata* from the Devonian of Parana. The same form is well illustrated by Branisa from Bolivia. *Meristelloides* differs from *Meristella* in having a very stout median septum in the brachial valve that has a triangular rather than a blade-like cross-section. *Meristelloides* and *Meristella* are similar in most other respects. However, even though intermediate forms are unknown, we must infer that somewhere in the New World there was an area where, in pre-Ems Devonian time, transitional forms developed from the blade-like median-septum type to the triangular cross-section type.

Biogeographic Distribution of Other Groups

In a consideration of the problems of Silurian-Devonian biogeography, I have emphasized brachiopods because of their abundance among the megafossil populations. Conclusions based on brachiopods certainly must be checked and compared with those based on other groups.

Forney and Boucot (1973) examined the distribution patterns of several groups of Silurian-Devonian gastropods. During the Silurian, oriostomatid-poleumitids, tremanotids, and euomphalids had a cosmopolitan North Silurian Realm distribution. I have observed a similar pattern for Silurian euomphalopterids. During the Early Devonian the oriostomatid-poleumitids, tremanotids, and euomphalids are present only in the Old World Realm as "Silurian holdovers" (relicts); the euomphalopterids disappear at the end of the Silurian.

Oliver (1968) has emphasized the similarity of Eastern Americas Realm tetracoral distribution during the Early Devonian with that of the brachiopods of that Realm. Some of these Eastern Americas tetracoral taxa do reach the eastern side of the Atlantic (Oliver, 1973b) but do not penetrate very far into either Europe or North Africa. In any event, the Eastern Americas Realm is distinctive for tetracorals, although the precise boundaries with the Old World Realm through various time intervals are not identical for corals and brachiopods. Although based on data from the Eastern Americas Realm only Oliver's (1973a) latest synthesis of Devonian tetracoral data is found to accord well with that based on brachiopods. Oliver finds that the percentage of Eastern Americas Realm endemic tetracorals goes from a Pridoli low to an Ems peak followed by a marked decrease through the Givet to a Frasne low. In other words, provincialism among tetracorals in the Eastern Americas Realm is very similar in intensity to that found among the brachiopods, despite the far smaller number of tetracoral taxa involved. [78]

Lesperance and Bourque (1971) showed that the Eastern Americas Realm dalmanitid trilobites have the same distribution pattern as the coeval brachiopods. Ormiston (1968, 1972) outlined the provincial trilobite distributions in the Early and Middle Devonian of North America as closely parallel with the brachiopod distribution. Wolfart and Voges (1968) analysed the highly provincial Malvinokaffric Realm trilobites. Ormiston (1972) emphasized the Eastern Americas Realm character of some of the Kazakhstan trilobites. K.S.W. Campbell and P.J. Davoren (in: Talent et al., 1972) have indicated the overall similarity of Tasman Region trilobites to those in Europe and North Africa, the uniqueness of the Malvinokaffric Realm trilobites (in accord with Wolfart and Voges), and the separateness of the Eastern Americas Realm trilobites.

Nicoll and Rexroad (1968) report provincialism among earlier Llan-dovery conodonts from central North America (Cincinnati Arch area as opposed to the Niagara Gorge and Austria), but this reported "provin-cialism" may possibly be correlated with depth zonation, i.e., an ecologic rather than a biogeographic explanation for the differing faunas, although such is not yet known to be the case.[79]

P.G. Telford (in: Talent et al., 1972) has outlined four biogeographic entities for the Lower Devonian based on conodonts. These units corre-spond very well with those defined using other fossils, and include a European, a Cordilleran, an Appalachian (sic. Eastern Americas), and an eastern Australian entity. The absence to date of conodonts from the Malvinokaffric Realm Devonian may also be characteristic of that realm.

In Silurian beds W.B.N. Berry (oral communication, 1971) found that, in general, the graptolites at the specific level are predominantly cosmo-politan, but that during part of the Early Devonian (about Siegen—Ems) they tend to be somewhat provincial at the subspecific level with the provincialism in accord with that shown by the brachiopods. Berry (1972, oral communication) finds Early Wenlock Uralian and Yukon plus Nevada graptolites similar to each other (i.e., Uralian-Cordilleran Region types) but different from European Early Wenlock graptolites.

Holland (1971, pp. 70—71) summarized some of the data for the euca-lyptocrinitids and pisocrinids. He reported that eucalyptocrinitids are known from the Silurian of Sweden, England, and eastern North America and from the Early Devonian of Germany, the Urals and Australia. P.M. Sheehan (written communication, 1972) reports eucalyptocrinitids from the Silurian of Utah. This is a distribution pattern consistent with a cos-mopolitan North Silurian Realm fauna followed by "Silurian holdovers" (relicts) during the Early Devonian within the Old World Realm. Holland reported that pisocrinids are known from the Silurian of eastern North America, England, Sweden, Bohemia, and Australia (they are also abun-dant within the Late Silurian of western North America); these crinoid distributions are consistent with a cosmopolitan North Silurian Realm fauna.

The distribution of Silurian ostracodes poses some problems. The Late Silurian beyrichiacean ostracodes appear to follow a provincial pattern (see J.M. Berdan and A. Martinsson in: Berry and Boucot, 1970) consis-tent with the brachiopod patterns that show a Pridoli unit extending from Podolia through the Baltic region, Britain, and thence across the Atlantic to Nova Scotia and coastal Maine, Massachusetts and New Brunswick. Within the Early Silurian of the Appalachians and parts of adjacent east-ern North America, a zygobolbid fauna, associated with Benthic Assem-blage 2 brachiopods, is preserved, but no counterpart has been reported elsewhere, an observation that raises the possibility of provincialism for this group at the generic level during the Early Silurian. Unfortunately, however, knowledge of Silurian ostracod distributions at this time is too limited to help in understanding problems of Silurian zoogeography, and the Devonian data although similarly limited, are consistent with a picture of high provincialism.

I have noted earlier (Boucot, 1970) the major change in pelecypod occurrence that took place within the Devonian. The bulk of the pre-Devonian bivalve taxa occur in abundance only in Benthic Assemblage 1. After the end of the Silurian large numbers of individuals are still most common within Benthic Assemblage 1, but a significantly large number of taxa if not individuals begin to appear in the Early Devonian in far more offshore, deeper positions than Benthic Assemblage 1 (this is well shown in the Rhenish-Bohemian Region, the Appohimchi Subprovince and also the Malvinokaffric Realm). By the Late Devonian, however, a large number of individuals is also encountered well away from Benthic Assemblage 1. The reasons for this diversification and spread are not understood, although they may possibly reflect the beginning of the replacement of the filter-feeding brachiopods by the bivalves that has continued steadily to the present. (This change may coincide with the marked Frasne-Famenne faunal crisis).

House (1971) reviewed the distribution of Devonian ammonoids and concluded that they display no evidence of provincialism. He commented that the environmental controls governing the distribution of Devonian ammonoids, as well as their distribution, may have been different than those governing the highly provincial groups such as the brachiopods, tetracorals, trilobites, etc. There is certainly no a priori reason why all groups should have the same distribution patterns when one can admit that their environmental requirements may have been very different. Also important in determining different biogeographic distributions of different animal groups is the different historical times of origin and diversification.

Ammonoid environments

Hallam (1972) has presented data about Jurassic ammonoid abundance, taxonomic diversity and association with deep-water or shallow-water facies that may be easily interpreted as showing that ammonoids are depth-stratified nekton, at least in the Jurassic (greater taxonomic diversity of deep-water ammonoids as contrasted with shallow-water ammonoids). Scott (1940) has suggested that Cretaceous ammonites from Texas were depth-stratified. I have earlier (Boucot, 1970) suggested that the Wissenbacher Schiefer facies (more offshore, deeper-water) of the German

Devonian with its fauna of dacryoconarid tentaculitids and ammonoids is a deeper-water facies comparable to the agnostid facies of the Cambrian and the graptolitic facies of the Siluro-Devonian. The occurrences of ammonoids in the Bundenbach Schiefer are also highly consistent with this interpretation for the Devonian, as is the Cherry Valley Limestone occurrence in New York (ammonoids occurring in a silty limestone layer bounded on both sides by styliolinid-bearing black shales). It is here proposed that the ammonoids of the late Lower Devonian and younger Devonian be considered, therefore, as a group of depth-stratified forms restricted for the most part to the Pelagic Community or at least to Benthic Assemblages 5 and 6 or deeper positions. Cowen et al.'s (1973) suggestion about a possible correlation of ammonoid ornament with depth might be tested against both Hallam's (1972) and Scott's (1940) data and then extended if found to strongly correlate.[80] It is likely, in view of Hallam's, Scott's and Wiedmann's (1973) observations, that following the Devonian some groups of ammonoids have evolved to occupy positions shoreward of the Benthic Assemblage 5 position. It is entirely reasonable, in view of information about modern nekton, that Paleozoic nekton may in some instances be depth-stratified as are some plankton and benthos. In any event, the distribution pattern of Devonian ammonoids correlates very well with such a conclusion.

Synchroneity in the Initiation and Termination of Provincialism

The reader may be forgiven for concluding that provincialism both begins and terminates synchronously, that it affects all groups simultaneously, after digesting the material presented here. However, buried here and there is information showing conclusively that although the initiation and termination of provincialism is statistically speaking almost simultaneous, in detail, taxon by taxon such is not the case. For example, at the Realm-Region level the Rhenish-Bohemian Region and Eastern Americas Realm (Fig. 25) are shown as differentiating in the Pridoli, whereas the other Old World and Eastern Americas Realms subdivisions differentiate later (most by the beginning of the Gedinne). At the Benthic Assemblage level the Uralian-Cordilleran Region is well differentiated by at least Early Wenlock time in Benthic Assemblages 4 and 5, but probably not in Benthic Assemblages 1 through 3. At the Community level and at both the genus and species level there are similar examples of diachroneity. These observations are consistent with what we infer concerning the factors responsible for provincialism. These observations suggest that the many interacting factors responsible for provincialism affect different taxa in different ways; these observations suggest that some taxa may be more sensitive to certain factors than others.

The statistical initiation and termination of provincialism provide a measure of the reaction of different groups of organisms to different faunal barrier and isolating-type mechanisms operating to one degree or another in the environment with time. Absolutely synchonous initiation or termination of provincialism would indicate the presence of almost cataclysmic events that profoundly affected all organic groups. There are records of a few time intervals that begin to approach this pole such as the Frasne-Famenne faunal extinctions and replacements. The impression gained from a study of the Silurian-Devonian record up to the end of the Frasne indicates the more gradual waxing and waning of a number of factors that may alternatively reinforce and oppose each other in time to produce the resultant of provincialism and cosmopolitanism, differing rates of evolution and differing rates of extinction. We are not even certain whether the initiation of the capability for forming preservable skeletal material at the beginning of the Cambrian was synchronous, although if we had better means of measuring time we might be able to resolve this problem.

The Cambrian-Lower Ordovician Question

Palmer (1972, 1973; written communication, 1972) pointed out to me that for Cambrian trilobite faunas the major Platforms with their provincial shelf faunas show rapid evolution, whereas the cosmopolitan shelf margin and off-shelf geosynclinal-region faunas show far slower evolution. If one concludes that the total area covered by the cosmopolitan off-shelf faunas is much larger than that occupied by the provincial shelf faunas then one would have a Cambrian—Early Ordovician situation in which size of interbreeding population correlates inversely with rate of evolution. Present information, however, does not necessarily indicate that such is the case. However, the paradox raised by Palmer's conclusions, that conditions of provincialism during the Cambrian—Early Ordovician are spatially rather than time-correlated, as is the case with Middle Ordovician and younger faunas, is puzzling. One possible explanation may be related to trilobite environmental requirements. Possibly there were environmental factors present on the isolated platforms, but absent from the intervening geosynclinal regions, that effectively set up conditions conducive to the development of platform endemism. However, the trilobites of the Silurian and Devonian do not appear to display such a control pattern. It would be painful to conclude that Cambrian organisms obeyed ground rules different than those of later ones. Palmer also concluded that during the Cambrian the provincial shelf faunas repeatedly became extinct, to be replenished later (his biomeres, 1965) from the off-shelf,

PLATE I

cosmopolitan reservoir. He pointed out that this behavior is very different from that of the post-Lower Ordovician shelf faunas that display a high degree of continuity in time (some platform faunas do become completely extinguished, but most appear to be part of a continuum). This differing behavior in the Cambrian—Lower Ordovician is hard to understand but eventually must be comprehended into some logical system.

Thus Palmer's biomeres may be regarded as the results of widespread platform extinction events (possibly restricted to individual, isolated platforms) that are not evident in the off-platform record of more continuous evolution lacking evidence for such marked extinction events. It is not apparent just what cause lies behind these biomeric extinction events. Eustatic or epeirogenetic sea-level changes causing major regression with its attendant consequences might be appealed to, but the lack of karst phenomena between beds belonging to adjoining biomeres suggests that the biomeric changes took place beneath sea level.

PLATE I

Eifel-age brachiopods from the Famine Limestone.

1, 2, 4, 5	*Athyris* cf. *A. undata* (GSC Locality No. 6043, south side Famine River near St. George de Beauce).
1.	Impression of interior of pedicle valve (× 2). GSC # 31231.
4.	Posterior view of impression of interior (× 2). GSC # 31231.
5.	Impression of interior of brachial valve (× 2). GSC # 31231.
2.	Anterior view of impression of interior (× 2). GSC # 31231.
3, 6, 7, 10 - 16	*Brevispirifer* cf. *B. gregarius* (GSC Locality No. 6042, south side Famine River near St. George de Beauce).
3.	Side view (× 2). GSC # 21233.
6.	Side view (× 2). GSC # 21232.
7.	Brachial view (× 2). GSC # 21233.
10.	Anterior view (× 2). GSC # 21233.
10.	Anterior view (× 2). GSC # 21233.
11.	Impression of interior of pedicle valve (× 2). GSC # 21232.
12.	Posterior view of impression of interior (× 2). GSC # 21232.
13.	Pedicle view (× 2). GSC # 21233.
14.	Brachial view of impression of interior (× 2). GSC # 31232.
15.	Anterior view of impression of interior (× 2). GSC # 21232.
16.	Posterior view (x2). GSC # 21233.
8, 9.	*Teichostrophia?* sp. (USGS Locality No. 13679, St. George de Beauce).
8.	Impression of interior of pedicle valve (× 2½). Note the peripheral rim characteristic of *Teichostrophia* (also present in *Ancylostrophia*, but that genus possesses somewhat differing internal appearance). GSC # 21234.
9.	Impression of interior of pedicle valve, side view (× 2½). GSC # 21234.

An additional explanation might be that the distribution of the plat-forms during the Cambrian—Lower Ordovician was such that the larvae of shallow-water, platform organisms had little possibility of being swept from one platform to another in order to produce conditions of platform isolation. However, our knowledge of Cambrian paleogeography is cur-rently too poor to permit us to either accept or reject such a possibility at this time.

Palmer's term biomere might be employed in North America when explaining the complete replacement of the North American Realm, Late Ordovician, Richmond faunas after their extinction, by the North Silurian Realm faunas (and possibly the slightly earlier parts of the Edgewood Limestone thought by Amsden (1973) to be of latest Ordovician age), although the biomere "source" is not "off-shelf" — merely another plat-form plus geosynclines.

Robison's (1972, pp. 35—36) observation that the agnostid trilobites of the oceanic environments "appear to have longer stratigraphic ranges than non-agnostid trilobites" is consistent with the concept that the off-shelf fauna covered a larger area than that covered by the more areally restricted, more rapidly evolving (taxonomic evolution producing more genera per unit of time) platform faunas, although Robison (1972, p. 36) appeals to the possibly more uniform nature of the oceanic, off-shelf environment as lowering selection pressure in trying to explain his obser-vations about rates of evolution, rather than appealing to the size of the interbreeding populations.

Jago's (1973) analysis of Cambrian agnostid faunas, and their implica-tions, fits very well with Robison's concepts and helps to extend the idea of benthic assemblages into the Cambrian (as pointed out by Jago in terms of Elles' (1939) work on Ordovician-Silurian units that are essentially equivalent to the benthic assemblages of this book). Jago's work further supports the concept of agnostids having been pelagic organisms.

PART III.
SUMMARY

CHAPTER 8

Conclusions

I have tried to summarize the relation between population size (number of individuals) and rates of evolution and extinction for Silurian–Devonian shallow-water, marine benthos. Population size has been estimated by comparing the relative areas occupied by various taxa and assuming that population size will be a relatively direct function of area occupied. Areas have been estimated by studying the area and the distribution of communities in which the taxa are known to occur, and the area of the biogeographic entities in which the communities themselves are known to occur.

The information for Silurian and Early Devonian marine benthic communities is as complete as present data permits. Limited information has been presented for the Middle Devonian. The available zoogeographic data for both the Silurian and Devonian has been summarized. Because inadequate attention has been given to the distribution patterns of most marine invertebrate groups during this time interval, whereas data are available for the overall abundant brachiopods, the conclusions are based presently almost entirely on the latter. This situation obviously should be rectified as soon as practicable.

I have also summarized the available data on numbers of brachiopod genera currently recognized during each stage of the Silurian–Devonian (almost 100 million years) and have tried to outline their occurrence, either endemic or cosmopolitan, in the biogeographic units recognized for each stage. A summary of the community distribution of these same brachiopods in terms of Benthic Assemblage occurrence of the communities has also been made for the entire time interval.

Preliminary and unsatisfactory as much of the data may be it still permits some conclusions to be made as follows.

Evolution Rate Conclusions

(*1*) The entire time interval Silurian through Devonian shows a good inverse correlation between size of potentially interbreeding population (as measured by area occupied) and rate of change of form (i.e., organic evolution). The size of the area may be judged either as the area of the

Benthic Assemblage during a time of relative cosmopolitanism (for only a few entities can we estimate the area occupied by each community within a Benthic Assemblage), or by the size of the zoogeographic entity plus the area of contained Benthic Assemblages during times of great provincialism. Therefore, on a worldwide basis there is a higher rate of taxonomic evolution (more new taxa per unit of time) during times of provincialism than during times of cosmopolitanism. There is a general tendency for the biogeographic units covering smaller areas (Fig. 2, 41) to have a more diverse fauna than do the larger units (Tasman Region of the Old World Realm and Cordilleran Region of the Old World Realm as opposed to the Rhenish-Bohemian Region, and Eastern Americas as opposed to the Old World Realm), although this tendency is complicated by the problem of trying to ascertain whether or not the various units occur in identical but isolated environments (if they represent greatly differing environments it is very likely that differences in diversity may be partly due to these same differences rather than to pure area because of the difficulty or impossibility of occurrence of certain biotopes that might have supported diverse communities. The absence of limestone and reef environments in the Malvinokaffric Realm is an excellent example paralleled today by the absence of reef environments in polar regions; both examples have far lower taxonomic diversities than those which include the spectrum of reef environments). This appears to be a first-order effect overshadowing environmental factors.

(2)Any study of the factors influencing rates of evolution among marine invertebrates should consider the worldwide size of the potentially interbreeding populations. In other words, natural selection is concluded to play the usual important role in determining what organisms develop, but *not* a first-order role in determining the rate at which they evolve, unless it can be shown that natural selection has determined the size of the interbreeding populations.

(3) There is a tendency for lower rates of evolution to be produced during time intervals when fewer communities (Fig. 22H) are present because of the population-size effect. Fewer communities may be present during a particular time interval because of either physical factors such as climatic uniformity or of biologic factors such as evolution in the direction of greater niche breadth.

(4) Cosmopolitan taxa have a greater time duration (Fig. 29) than provincial taxa, and also are more abundant at any one locality in terms of individual specimens. Thus, cosmopolitan taxa tend to be cosmopolitan in both time and space. Cosmopolitan taxa evolve far more slowly than do provincial taxa. Cosmopolitan taxa occur in a variety of communities (from Benthic Assemblage 2 through 5). There is good evidence that

cosmopolitan taxa are not restricted, or even more abundant, in so-called "unstable" environments.

(5) There is good evidence for rate of evolution not being well correlated with either the so-called "stable" or "unstable" environments.

(6) Intervals characterized by abundant and widespread reef environments (Fig. 22M) correlate well with increased rates of evolution, presumably owing to the multiplicity of isolated, small, reef-related, rapidly evolving populations.

(7) It is concluded that during times of either provincialism or cosmopolitanism the evolutionary effects of the environment have far less impact in terms of rate of evolution than has the size of the interbreeding population.

(8) There is no 1:1 correlation between rate of evolution and either intervals of transgression or regression, although both phenomena during certain specific time intervals in certain specific areas may be interpreted to act in concert with other factors (geography, climate, marine current location) to help produce barriers to faunal migration that in turn will affect the potential size of interbreeding populations. However, if other factors are equal times of regression should increase rate of phyletic evolution.

(9) Competition for food and place does not appear to be a first-order control over rate of evolution.

(10) The high inverse correlation between population size and rate of evolution applies to both marine and non-marine organisms, and to "low" groups and "high" groups. The small residue of organisms that show a marked departure from this correlation (small populations characterized by slow evolution of the bradytelic type; large populations characterized by rapid evolution of the tachytelic type) indicate the importance of additional factors involved as rate controls below the first-order level.

(11) The seeming correlation between higher rate of evolution in terrestrial and "higher" groups of organisms as contrasted with marine and "lower" groups of organisms is easily interpreted as a population-size effect with terrestrial organisms tending to occur as smaller populations than marine organisms, and many "lower" groups as larger populations than many "higher" groups. Thus it is evident that many of the "lower" groups are marine whereas many of the "higher" groups are terrestrial. There are so many exceptions to the earlier generalizations about the correlation between "high" and "low", marine and terrestrial as related to rate of evolution (the ammonites and fusuline foraminifers as prime examples) that the importance of the population-size effect is easy to grasp [81].

Silurian—Devonian Biogeographic Conclusions

(*12*) The trend is from relatively cosmopolitan distribution to more provincial patterns (established after the termination of the more provincial Late Ordovician) for Llandovery age marine benthos (Table III).

(*13*) This benthos shows increasing provincialism during the Wenlock-Pridoli time interval (Table III).

(*14*) The marine benthos show high provincialism during the Early and Middle Devonian (highest during the latter half of the Early Devonian). This provincialism terminates in the Late Devonian, essentially absent by Frasne time, and is replaced by cosmopolitanism (Table III, Fig. 31).

(*15*) There is no correlation between times of high provincialism and times of maximum orogeny during the Silurian—Devonian. Evidence from other time intervals strongly suggests that orogeny does not correlate with provincialism or with rate of evolution (Fig. 22I).

Rate of Extinction

(*16*) Rates of terminal extinction increase at the end of some times of provincialism for the marine benthos, suggesting that competition among similarly adapted forms plays an important role in some cases.

(*17*) Major extinction events may be preceded by the occurrence of either highly cosmopolitan or highly provincial faunas, and may be followed by the occurrence of either highly cosmopolitan or provincial faunas. There is no 1:1 correlation between the factors causing major extinction events and those controlling cosmopolitanism or provincialism.

Hypersaline Water as a Biogeographic Barrier

(*18*) The occurrence of large areas of hypersaline water in the more central parts of the Northern Hemisphere and Australian Platforms in the Late Silurian—Middle Devonian (as shown by the presence of evaporites and hypersaline communities) and of faunal provincialism (Fig. 37—41) possibly suggests that large bodies of hypersaline water on the platforms during this time interval had the potential to act as one of the factors in erecting faunal barriers that produced provincialism. It is possible that the production of hypersaline water was the result of a climatic regime in concert with restricted circulation in which the platform central regions were affected by a greater evaporation rate than input of rainfall necessary to maintain normal, sea-water salinity (in the absence of large riverine input).

Unimportance of Dolomitization as a Biogeographic Barrier

(19) Sedimentary secondary dolomite replacing subtidally deposited limestones on the Cambro—Silurian platforms in such great abundance (as contrasted with the Precambrian and post-Silurian) is thought to have been formed by biotic dolomitization. Biotic dolomitization calls on poor circulation with the oceanic reservoir across vast platforms. The poor circulation makes possible a cycle in which biogenic activity removes calcium which in turn permits the excess magnesium to slightly replace calcareous bottom detritus; the calcareous bottom detritus is being continuously subject to solution; solution affects the differing solubility phases so as to preferentially leave some dolomite undissolved and free to accumulate over an interval of time as a long-term residue. This process of dolomitization does not appear to have any biogeographic or environmental effects on the co-occurring benthos.

Species Diversity

(20) Brachiopod species diversity in a community decreases as dominance of any species increases.

(21) Maximum species diversity is evidently reached during a geologically almost instantaneous interval of time. This effect appears to hold for both reef and level-bottom communities. The time-stability concept, if it actually operates, occurs over a geologically very small time interval before leveling off to a steady state that may be maintained for many tens of millions of years.

Environmental Stability of Taxa during Evolution

(22) Consideration of the Benthic Assemblage position of the Silurian—Devonian brachiopods, and of their community associates, leads one to conclude that the majority have not changed their position during the interval (Fig. 35). A few have evolved in the direction of a narrower Benthic Assemblage spread, a few have evolved in the direction of a broader Benthic Assemblage spread, a few have evolved to form low-diversity communities as opposed to earlier high-diversity associates, and a few have evolved to form part of high-diversity associations as opposed to earlier low-diversity associations, *but* the majority appear to have stayed put. From this information I have concluded that there will be an overall tendency for most taxa to undergo little change in environmental requirements during the course of organic evolution during moderate portions of geologic time. This conclusion is critical in any attempt at evaluating the

similarities and differences between the taxa thought to occupy functionally similar positions in different biogeographic units.

(23) Finally, the overall conclusion is that it is the resultant interplay
in time of all these factors observable in the geologic record (provincialism
versus cosmopolitanism, regression versus transgression, high abundance of
reefs as opposed to absence of reefs, highly differentiated climatic regime
as opposed to a uniform climate, high number of animal communities as
opposed to a low number of animal communities) plus others that we
cannot easily measure or estimate (including such factors as food supply)
from the geologic record at this time, which determines the size of interbreeding populations in time and thus the first-order control over rates of
evolution (Fig. 33).

The review of the Silurian—Devonian benthic record presented here
makes it unlikely that extraterrestrial radiation influences on rate of evolution have had widely differing magnitudes during this time interval although cometary collisions might be investigated. After we have been able
to normalize the Silurian—Devonian data for different population sizes we
may be in a position to analyze the importance and effects on rate of
evolution of various second- and lower-order factors [82].

P.S. All of the conclusions arrived at in this treatment are based on two
assumptions. If these assumptions are seriously in error the conclusions
are probably of little value.

(1) The fossil record of the brachiopods is a reasonable approximation
of their real record.

(2) The ease with which brachiopods have become fossilized and preserved in the geologic record has not varied widely during the Phanerozoic. That is, we assume that brachiopods during one time interval are as
likely to have been preserved as fossils as those during other time intervals.

APPENDIX I

Descriptions of New Taxa Necessary for a Better Understanding of the
Silurian and Devonian Problems Discussed in the Body of the Work

Superfamily Enteletacea
    Family Rhipidomellidae
        Subfamily Rhipidomellinae
            Genus *Discomyorthis* Johnson, 1970

Discussion. Johnson (1970a) introduced the rhipidomellinid genus *Discomyorthis* to receive shells possessing a large diductor field in the pedicle valve that contrast markedly with the closely related genus *Dalejina.* In addition to the internal character, *Discomyorthis* species tend to be much larger than representatives of *Dalejina* (except for *Discomyorthis oblata* which is transitional in this regard). Most, but not all, species of *Discomyorthis* have flatter pedicle valves than *Dalejina.* Following is a list of species assigned to *Discomyorthis,* a genus of great importance in Devonian biogeography (characteristic of the Eastern Americas Realm).

Orthis alsus Hall, 1863, *16th Rep. N.Y. St. Cab. Nat. Hist.*, p. 33.
Rhipidomella musculosa var. *arctisinuata* Schuchert, 1913, *Md. Geol. Surv.*, *L. Dev.*, p. 306, plate 55, fig. 21, 22.
Orthis oblata emarginata Hall, 1859, *Paleontol., N.Y.*, 3, p. 164, plate 10A, fig. 4–6.
Orthis eryna Hall, 1863, *11th Rep. N.Y. St. Cab. Nat. Hist.*, p. 35.
Orthis musculosa Hall, 1857, *N.Y. St. Cab. Nat. Hist. Ann. Rep.*, 10, p. 46.
Rhipidomella musculosa var. *solaris* Clarke, 1907, *N.Y. St. Mus. Bull.*, 107, p. 284, fig. 1, 3 (not fig. 2).
Orthis oblata Hall, 1857, *10th Rep. N.Y. St. Cab. Nat. Hist.*, p. 41, fig. 1–5.
Orthis hartti Rathbun, 1879, *Proc. Boston Soc. Nat. Hist.*, 20, p. 23.
Rhipidomella hybridoides Clarke, 1907, *Bull. N.Y. St. Mus.*, 107, p. 282.
Orthis livia Billings, 1860, *Can. J., N. Ser.*, 5, p. 267, fig. 14–16.
Rhipidomella preoblata Weller, 1903, *Geol. Surv. N.J. Paleontol.*, 3, p. 232, plate 20, fig. 25–26.
Dalejina kinsuiensis Hamada, 1971, *Paleontol. Soc. Japan, Spec. Pap.*, 15, 41, 42, plate 3, fig. 1–12; plate 4, fig. 1–9; plate 29, fig. 3–8.

Subfamily Proschizophoriinae Boucot, Gauri, and Johnson, 1966
    Genus *Baturria* Carls, 1973
        Baturria mexicana, new species.
(Plate II, 1—9).

Diagnosis. Shell plano-convex to concavo-convex in lateral profile with the brachial valve planar or gently concave as contrasted with the gently convex pedicle valve. Pedicle valve muscle field undivided and cordate in outline. The cardinal process is linear, non-lobate, and continuous anteriorly with a broad myophragm. Muscle scars weakly impressed; consist of paired adductor impressions, the anterior pair being the larger. Muscle field bounded posterolaterally by bounding ridges extending out from the brachiophores.

Comparison. B. *mexicana* lacks the highly convex brachial valve and relatively flat pedicle valve of *Proschizophoria* externally; internally it lacks the massive, divided ventral muscle field of that genus. B. *mexicana* differs from *Cordatomyonia* in the possession of a flat or slightly concave brachial valve as opposed to a convex brachial valve. B. *mexicana* lacks the divided muscle field present in the pedicle valve of *Cordatomyonia* as well as the latters cleft cardinal process. B. *mexicana* may be distinguished from *Idiorthis* by the lack of an almost septum-like myophragm in the brachial valve and by the lack of massive brachiophores in the brachial valve, plus a much smaller pedicle muscle field. *Idiorthis* is relatively circular in outline whereas B. *mexicana* is laterally elongate. B. *mexicana* is most like *Baturria simonae*, particularly the early growth stages of that Ludlow and Pridoli species, but B. *simonae* possesses a well differentiated, tripartite pedicle muscle field, a very broad myophragm in the brachial valve and a relatively convex brachial valve in later growth stages as well as a medially cleft cardinal process.

Exterior. Medium to large size shell. Lateral profile plano-convex to gently concavo-convex. Pedicle valve gently convex. Brachial valve bears a broad, weakly developed sulcus and pedicle valve a corresponding fold. Laterally elongate, ellipsoidal in outline. Maximum width slightly anterior of the straight hinge line. Brachial and pedicle valve interareas narrow and apsacline. Delthyrium open, includes an angle of about ninety degrees. Shell surface costellate. Costellae increase by bifurcation, and originate in beak region. Prominent concentric growth lines.

Brachial interior. Brachiophores plate-like and extended anteriorly in ridge of secondary material that antero-laterally bounds muscle field. Cardinal process blade-like, continued anteriorly by a broad, low myophragm to about midlength. Fulcral plates absent. Laterally directed structures present (these are also present in proschizophorids like *Baturria simonae;* their position differs from that of the true schizophorid fulcral plate present in taxa like *Schizophoria, Salopina* and *Megasalopina* which are plate-like structures directed at about a forty-five degree angle to the hinge line and descending towards the midline rather than being roughly parallel to the hinge line and not descending towards the floor of the valve medially). Muscle field feebly impressed; divided into two pairs of adductors. Anterior adductors larger than posterior pair. Interior smooth except for peripheral region crenulated by impressions of costellae.

Pedicle interior. Pedicle muscle field equant, undivided, bounded laterally by short, blade-like dental lamellae. Hinge teeth small, pointed. Muscle field situated on this pad of secondary material raised up slightly from floor of valve. Interior smooth except for impressions of costellae in peripheral region.

Locality. Dr. Grover E. Murray's locality Ca-1775, Peregrina Canyon, vicinity of Ciudad Victoria, State of Tamaulipas, Mexico, collected by Ing. Jose Carrillo B. Petroleos Mexicanos in 1959.

Holotype. USNM No. 181155

Figured specimens (see Plate II, p. 356) USNM Nos. 181156-181159

Superfamily Porambonitacea Davidson, 1853
    Family Camerellidae Hall and Clarke, 1894
        Subfamily Camerellinae Hall and Clarke, 1894
            Genus *Llanoella* new genus
(Plate III, 1—16)

Type species: Llanoella stephensi new species

Diagnosis. Llanoella is created for a camerellid with a strong brachial fold and corresponding sulcus combined with a smooth exterior plus a well developed spondylium and cruralium internally. A few weak costae, rounded in cross-section, may be present peripherally. The cruralium is unique in having the medio-distal margins swollen into the form of two long, subcylindrical structures that barely fail to meet in the midline.

Comparison. Internally *Llanoella* is unique in its brachial cardinalia among the Camerellidae. Externally *Llanoella* is very similar, virtually homeomorphous with *Stenocamara* (Subfamily Stenocamarinae of the Camerellidae), but *Stenocamara* lacks a spondylium in the pedicle valve and has very different brachial cardinalia (the species *S. perplexa* Cooper, 1956 is externally very similar to *L. stephensi*). The absence of alae lateral of the brachial lamellae removes *Llanoella* from the Parastrophinidae. Externally *Llanoella* is also similar to *Camerella* and *Perimecocoelia.*

Discussion. Barnes et al. (1947, p. 131) reported *L. stephensi* from Texas as "*Parastrophinella*, 1 to 3 new species", but consideration of both the internal form (alae present) and the external form (radial ornament present) removes the latter genus from consideration. Johnson (1970a, pp. 93, 94) described a Nevada form as "*Camerella*" sp., based on poorly preserved silicified specimens, that is certainly congeneric with *L. stephensi*, and occurs in beds of about the same age (Gedinne) in Nevada as *L. stephensi* in Texas. *L. stephensi* is far and away the most abundant brachiopod in the Pillar Bluff Limestone of the Llano region, whereas *L. sp.* is rare in both the *Spinoplasia* Zone of the Rabbit Hill Limestone and the *Quadrithyris* Zone of the Windmill Limestone (16 specimens known from the former; two from the latter) in central Nevada. *Llanoella* is the sole Devonian representative of the Camerellidae.

Exterior. The shell outline is elliptical and laterally elongate, width about one and one-third times length. The maximum width is located near the midlength. A palintrope is present that curves gently in an antero-lateral direction posteriorly. The lateral margins are strongly rounded, with a gently rounded anterior margin. The anterior commissure is strongly plicate. The brachial valve bears a strong, high fold with a rectangular cross-

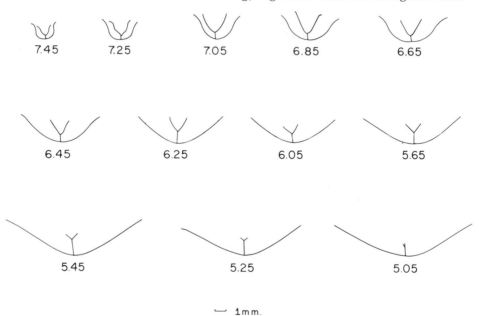

7.45   7.25   7.05   6.85   6.65

6.45   6.25   6.05   5.65

5.45   5.25   5.05

⌣ 1mm.

Fig. 43. Serial sections of *Llanoella stephensi* pedicle valve. Section intervals are measured to the nearest hundredth mm. Peels have been deposited in the U.S.N.M. together with the type specimens.

Fig. 44. Serial sections of *Llanoella stephensi* brachial valve. Section intervals are measured to the nearest hundredth mm. Peels have been deposited in the U.S.N.M. together with the type specimens.

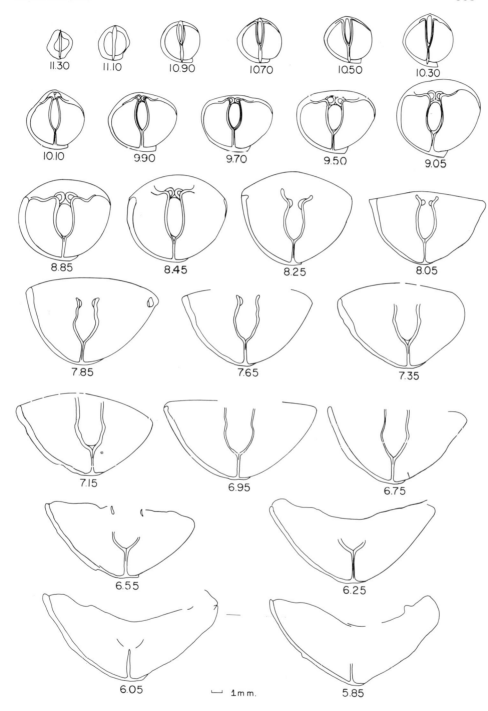

11.30   11.10   10.90   10.70   10.50   10.30

10.10   9.90   9.70   9.50   9.05

8.85   8.45   8.25   8.05

7.85   7.65   7.35

7.15   6.95   6.75

6.55   6.25

6.05   1 m m.   5.85

section. The sides of the fold slope steeply down to the smooth flanks of the valve. The fold expands anteriorly, and originates just anterior of the beak region. The beak of the brachial valve is incurved over that of the pedicle valve. The external ornamentation consists of poorly impressed concentric growth lines and weak, broad peripheral costae (two to three on each flank, two on the fold) separated by interspaces with similar form. The costae are not present on small or medium size shells. The brachial valve has about twice the convexity of the pedicle valve.

Interior of pedicle valve. A median septum supporting a spondylium is present in the pedicle valve (Fig. 43). The relative size and proportions of the septum plus spondylium are similar to those of other camerellids. There is no conspicuous shell thickening in the umbonal regions.

Interior of brachial valve. Fig. 44 illustrates the brachial cardinalia in serial section (made by courtesy of Mr. Michael Stephens). Note the unique nature of the cruralium. It becomes virtually tubular posteriorly due to the rod-like distal structures (note the form of these structures in Johnson's, 1970a, illustrations).

Holotype. USNM No. 188546

Figured specimens (see Plate III, p. 358) USNM Nos. 188543-45, 188547-48

Family Parastrophinidae Ulrich and Cooper, 1938
      Genus *Grayina* Boucot, new genus

Type species: Anastrophia magnifica Kozlowski, 1929, p. 140, text-fig. 42; plate 4, fig. 14—16.

Diagnosis. Coarsely plicate anastrophids in which costal bifurcation near the margins of the fold and sulcus is very pronounced. Shell size small to medium. Long hinge line.

Comparison. Grayina is characteristically smaller and more coarsely plicate than *Anastrophia (Anastrophia).* The difference in shell size has been observed at many localities with a worldwide scatter. *Grayina magnifica* has a longer hinge line than *A. (Anastrophia)* or *A. (Savageina).* (*A. (Savageina) deflexa* may be ancestral to both taxa).

Description. Kozlowski (1929, pp. 140—142) has an excellent description of the type species.

Discussion. A. *(Savageina) deflexa* as illustrated by Schuchert and Cooper (1932, plate 25, fig. 40) has alar projections on the lateral sides of the spondylium, a feature not noted in either *Grayina magnifica* or in *A. (Anastrophia).*

Distribution. Grayina is known from strata of Wenlock through Siegen age in the North Silurian Realm, Uralian-Cordilleran Region and in the Old World Realm.

Ecology. Grayina occurs in high-diversity communities occurring in the Benthic Assemblages 3—5 span.

Species assigned:
Anastrophia magnifica Kozlowski, 1929
Anastrophia magnifica australis Savage, 1971
Anastrophia praemagnifica Kulkov, 1967

Genus *Anastrophia* Hall, 1867
   Subgenus *Savageina* new subgenus

Comment. Savage (1971) has very capably described the morphologic, geographic and stratigraphic relations of the various species previously assigned to *Anastrophia.* He divided the anastrophids into two major groups, the *A. magnifica* Group here formalized as *Grayina,* and the *A. deflexa* Group here subdivided into the subgenera *Anastrophia (Anastrophia)* with *A. verneuli* as the type species and *A. grossa* Amsden, 1958, and *Anastrophia (Savageina)* with *A. deflexa* as the type species.

Type species: Terebratula deflexa Sowerby, 1839,

Silurian System II, p. 625, plate 12, fig. 14.

Diagnosis. Small species of *Anastrophia.*

Comparison. Lacks the long hinge line of *Grayina* and is only about one-half the size of *Anastrophia (Anastrophia).*

Description. See Amsden (1949, for description of a typical member of the subgenus).

Discussion. Savage (1971) illustrates the ontogeny of *Grayina magnifica australis* and comments on its having gone through an *A. deflexa* stage with a short hinge line. Both stratigraphic relations and geographic distribution strongly suggest, following Savage's lead (1971), that both *Grayina* and *A. (Anastrophia)* were derived from Silurian species of *A. (Savageina).*

PLATE II

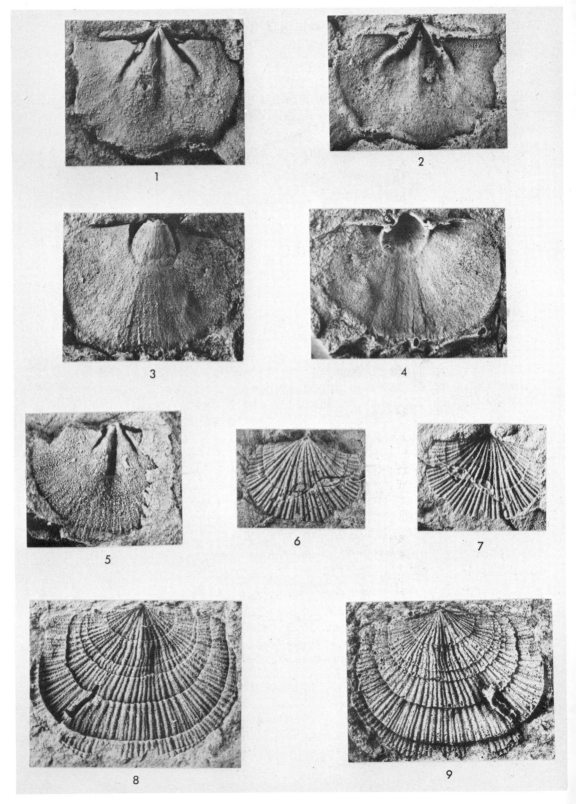

Distribution. See Savage (1971); Wenlock-Ludlow of the North Silurian Realm, North Atlantic Region.

Ecology. Both *A.* (*Anastrophia*) and *A.* (*Savageina*) as well as *Grayina* occur in high diversity communities in the Benthic Assemblage 3—5 span.

Species assigned:
Anastrophia acutiplicata Amsden, 1949
Terebratula deflexa Sowerby, 1868
Anastrophia delicata Amsden, 1951
Anastrophic internascens Hall, 1879
Pentamerus podolicus Wenjukow, 1899 (Savage, 1971, assigned this species a Llandovery age; more recent work has shown that the enclosing Kitaygorod Formation is of Wenlock age).

Superfamily Pentameracea M'Coy, 1844
    Family Gypidulidae Schuchert and LeVene, 1929
        Genus *Amsdenina* Boucot, new genus

Sieberella roemeri Hall and Clarke, 1892 (p. 247, plate 2, fig. 6)
Sieberella roemeri Hall and Clarke, 1892, Amsden, 1949, pp. 49, 50, plate 2, fig. 1—4.
Sieberella roemeri Hall and Clarke, 1892, Amsden, 1951, p. 79, plate 16, fig. 36—40.

Type species: Sieberella roemeri Hall and Clarke, 1892

Diagnosis. Gypidulid brachiopods having a median septum in the brachial valve formed from baso-medially conjunct brachial lamellae (as in *Sieberella)* associated with an ornament on both valves that consists of rounded costae peripherally that are separated by rounded interspaces.

PLATE II
Baturria mexicana, new species. Beds of probable Gedinne age. Vicinity of Ciudad Victoria, Tamaulipas, Mexico (Peregrina Canyon).

1—9   *Baturria mexicana,* new species.
1, 2.   Impression of interior of brachial valve and rubber replica of same (4). Holotype USNM No. 181155.
3, 4.   Impression of interior of pedicle valve and rubber replica of same (3). USNM No. 181156.
5.   Impression of interior of brachial valve (× 3). USNM No. 181157.
6, 7.   Rubber replica of exterior of pedicle valve and impression of exterior pedicle valve (× 3). USNM No. 181158.
8, 9.   Impression of exterior of brachial valve and rubber replica of same (8). USNM No. 181159.

PLATE III

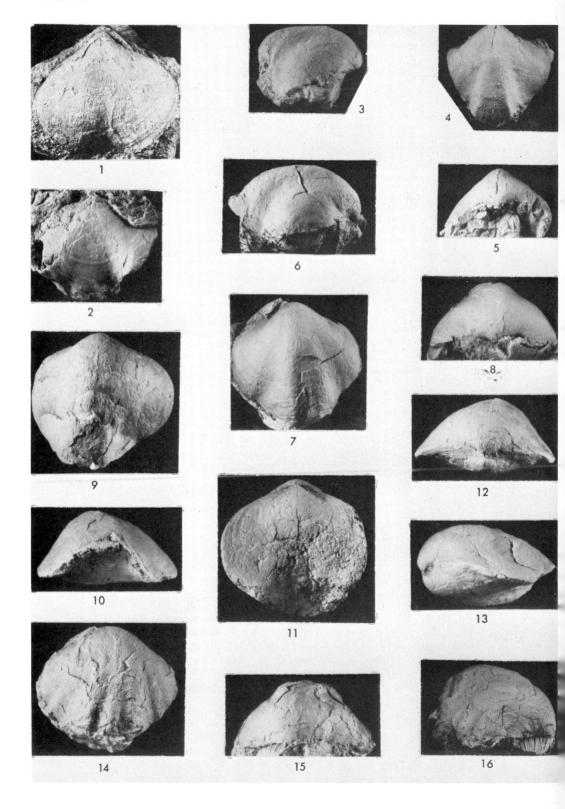

Comparison. Amsdenina is easily distinguished from *Sieberella* as the latter has radial costae with angular cross-sections that are separated from each other by v-shaped interspaces. *Biseptum* has a far more inflated form and a very incurved pedicle valve beak. *Gypidulella* has very weakly developed, small costae. *Gypidulina* is smooth. *Pentamerella* has fold and sulcus on the brachial and pedicle valves respectively as contrasted with *Amsdenina* and most other gypidulids.

Description. Amsden (1949, 1951) gives an adequate description of the type species.

Discussion. Amsdenina roemeri has traditionally been assigned to *Sieberella*, but the ornamentation differences indicate their distinctiveness, and their stratigraphic separation (Late Silurian, Ludlow-Pridoli as opposed to Lower Devonian, Gedinne-Ems) without the existence of transitional forms indicates the desirability of recognizing them as separate taxa.

Distribution. To date *Amsdenina* has been recognized in Ludlow age beds in Tennessee and Oklahoma, but also occurs elsewhere in the mid-continent, and also in Pridoli age beds in several localities in the Northern Appalachians.

Ecology. Amsdenina is known with both the *Dicaelosia-Orthostrophella* Community and by itself as a gypidulid, single taxon community.

Suborder Chonetidina Muir-Wood, 1955
    Superfamily Chonetacea Bronn, 1862
        Family Chonetidae Bronn, 1862
            Subfamily Chonetinae Bronn, 1862
                Genus *Amosina* Boucot, new genus

PLATE III

Figures 1—16 *Llanoella stephensi* new genus and species. Pillar Bluff Limestone. Type locality of Pillar Bluff Limestone on Pillar Bluff Creek, 5 miles southwest of Lampasas and just south of the Lampasas County line in northern Burnet County, Texas. The Barnes locality number is TF-273 (see p. 129 of Barnes, Cloud and Warren, 1947).

1.   Exterior of pedicle valve ($\times 2$). Note the rounded sulcus cross-section USNM No. 188543.
2.   Exterior of pedicle valve ($\times 2$). Note the more angular margins of the sulcus. USNM No. 188548.
3—5.   Side view, top view and posterior view of brachial valve ($\times 2$). Note the angular margins of the fold, the trace of the median septum and the undulating crest of the fold. USNM No. 188544.
6-8.   Side view, top view and posterior view of brachial valve ($\times 2$). Note the angular sides of the fold and the grooved top of the fold. USNM No. 188545.
9—13.   Brachial view, anterior view, pedicle view, posterior view and side view of holotype. Note the median septum trace present in both valves, and the rounded lateral margins of fold and sulcus. USNM No. 188546.
14—16.   Top view, posterior view and side view of brachial valve (x2). Note the weak lateral plications and the plicate fold. USNM No. 188547.

Chonetes fuertensis Kayser, 1897, *Z. Dtsch. Geol. Gesell.*, 49, pp. 274—317.

Chonetes fuertensis Kayser, 1897, Thomas, 1905, *Z. Dtsch. Geol. Gesell.*, 57, p. 260.

Chonetes fuertensis Kayser, 1897, Castellaro, 1959, *Rev. Asoc. Geol. Argentina*, XIII (1) pp. 51—53, plate III, fig. 17—20.

"Chonetes" fuertensis Kayser, 1897, Castellaro, 1966, Guia Paleontol. Argentina, I: Paleozoica, Seccion III, Faunas Siluricas, pp. 31—32, fig. a—e.

Type species: Chonetes fuertensis Kayser, 1897, pp. 300—301, plate X, fig. 3.

Diagnosis. Castellaro (1959) has provided the best diagnosis and description of the type species of *Amosina. Amosina* is characterized by its subrectangular, laterally elongate outline and very sulcate pedicle valve, with corresponding fold in the brachial valve. In addition there is a very prominent raised median striation and small median costate region composed of the prominent raised median striation and the adjacent finer striae. The margins of the sulcus are relatively angular. It is a relatively small shell.

Comparison. There is no other Silurian-Devonian chonetid with the external form of *Amosina. Mesolobus* of the Permo-Carboniferous is somewhat similar externally in that it has a prominent sulcus on the pedicle valve, but the lateral margins of the sulcus are rounded rather than angular and the median plication is much larger and has a very rounded cross-section rather than an angular cross-section. The internal structures of *Mesolobus* are very different from those of *Amosina*.

Description. Castellaro (1959) gives an adequate description of *A. fuertensis*, the only species assigned to the genus.

Discussion. Amosina is essentially a small, subrectangular *Strophochonetes* having a highly sulcate pedicle valve with angular margins and a raised median costal region with a prominent median striation. Internally the median septum and short anderidia of the brachial valve ally it closely to *Strophochonetes. Protochonetes stappenbecki*, the only other Malvinokaffric Silurian chonetid, is of ordinary chonetid form.

Distribution. Amosina is a Malvinokaffric Silurian chonetid genus known only from the Silurian of the Precordillera de San Juan, Argentina (Castellaro, 1959, 1966) and from the Titicaca region of southern Peru (Laubacher's locality 540). The precise stratigraphic range of *Amosina* in Argentina and Peru is poorly known but does not appear to exceed a Late Llandovery-Ludlow range. There are no known post-Ludlow derivatives of *Amosina*.

Ecology. Amosina occurs separately and as a member of the *Harringtonina* Community in both the Precordillera de San Juan in Argentina and in southern Peru. The small size of *Amosina* relative to other Silurian chonetids might have been ascribed to environmental factors while it was known only from the Precordillera de San Juan, but now that it is also known from southern Peru, where it has the same small size, this is considered a very unlikely possibility. It is notable as well that in these same regions the Silurian leptocoelid *Harringtonina* is also small as compared with both northern hemisphere North Silurian Realm leptocoelids *(Eocoelia)* as well as with Devonian leptocoelids of both hemispheres. These observations are in keeping with the generalization that Silurian brachiopods are normally much smaller than their Devonian descendents.

Superfamily Atrypacea
    Family Leptocoelidae Boucot and Gill, 1956
        Pacificocoelia, new genus

Type species: Leptocoelia murphyi Johnson, 1970, pp. 165—168, plate 45, fig. 13—19; plate 46, fig. 1—21.

Diagnosis. Leptocoelids possessing a well developed cardinal process, lacking dental lamellae, deeply impressed brachial valve muscle field with bounding ridges, and having an evenly convex brachial valve in all but small specimens.

Comparison. Pacificocoelia has a gently convex brachial whereas *Leptocoelia* has a flat brachial valve; there is no overlap in the known species of either genus relative to this character. *Pacificocoelia* resembles the Malvinokaffric Silurian leptocoelid genus *Harringtonina* in having a convex brachial valve but the latter genus possesses well developed dental lamellae. *Harringtonina* is probably the direct precursor of *Pacificocoelia. Leptocoelia* has a lightly impressed brachial valve muscle field, whereas *Pacificocoelia* has a deeply impressed brachial valve muscle field.

Distribution. Pacificocoelia is the "*Leptocoelia*" reported by Boucot et al. (1969) and Johnson (1970a) from central Texas, the Perija of Venezuela, Colombia, the Precordillera de San Juan in Argentina, Oklahoma, Nevada, Kazakhstan, New South Wales in Australia, and the Central Appalachians. It is also known in Manchuria (Hamada, 1971) and Texas (Wilson and Majewske, 1960, plate VI, fig. 7—8). The widely separated occurrences of *Pacificocoelia* probably include species with differing origins in detail, although ultimately derived from *Harringtonina*, as was *Leptocoelina.*

Stratigraphic range. Pacificocoelia is known only in beds of Early Devonian age.

Ontogeny. All leptocoelids, including *Eocoelia*, have flat brachial valves in their early growth stages. The retention of a flat brachial valve in both *Leptocoelia* sensu strictu and *Australocoelia* and in *Eocoelia* of large size suggests its primitive condition in *Eocoelia* (and *Eocoelia's* possible pre-Llandovery age common ancestor with *Harringtonina*) and a neotonic character in *Leptocoelia* and *Australocoelia*.

Discussion. The species of *Pacificocoelia* may be divided into two groups. The first group is characterized by deeply impressed musculature in the pedicle valve and includes the following species: *P. infrequens, P. biconvexa, P. nunezi,* and *P. murphyi* (Johnson's 1970a genus *Leptocoelina*, based on *L. squamosa*, is very close morphologically to *P. murphyi*: studies of the ontogeny of both *L. squamosa* and *P. murphyi* would be predicted to show great similarities). The second group includes forms with a very weakly impressed pedicle valve muscle field and includes *P. texana* and *P. acutiplicata* (study of numerous specimens of *P. acutiplicata* make clear that it is morphologically very distinct from other pacificocoelids or from species belonging to the other leptocoelid genera). The pacificocoelids from both New South Wales and Oklahoma are internally not yet known well enough to assign to either group one or group two.

Superfamily Spiriferacea
    Family Cyrtiidae Fredericks, 1919 (1924)
        Subfamily Eospiriferinae Schuchert and LeVene, 1929
            Genus *Hedeina* Boucot, 1963
                Subgenus *Macropleura* Boucot, 1963

Discussion. Brunton et al. (1967) have discussed the taxonomic consequences resulting from their finding that Linnaeus's original type of *Anomia crispa* is an eospiriferid. They make clear that *Hedeina*, with Linnaeus's species *A. crispa* as type, is an eospiriferid of the type originally assigned to the genus *Macropleura* by Boucot (with *Delthyris macropleura* as type species). I accepted this change in 1967 (see Brunton et al., 1967). Reconsideration of the species assigned to *Hedeina* (see Boucot, 1963a, p. 691) now makes it clear that all of the Devonian (Gedinne through Eifel) species are relatively large as contrasted with the Silurian species. I propose that *Macropleura* be revived from synonymy as a subgenus of *Hedeina*.

Revised definition of (Macropleura): Hedeinid-type eospiriferids with about twice the size of *Hedeina (Hedeina)*.

Species assigned to *H. (Macropleura):*
Spirifer macropleuroides Clarke, 1907
Delthyris macropleura Conrad, 1840
Spirifer sibiricus Tschernychew, 1893
Eospirifer? bascucanicus Rzhonsnitskaya, 1952
Spirifer perlamellosus var. *densilineata* Chapman, 1908
Spirifer vetulus Eichwald, 1840

Stratigraphic significance. Together with many other Devonian derivatives of Silurian ancestors *H. (Macropleura)* is considerably larger than its Silurian precursor *H. (Hedeina).* This distinction has both stratigraphic utility and evolutionary significance (the increase in size with time is a not unusual feature in many groups).

Family Delthyrididae Waagen, 1883
    Subfamily Acrospiriferinae Termier and Termier, 1949
        Genus *Vandercammenina* new genus

Type species: Spirifer trigeri de Verneuil, 1850, Classification des terrains paléozoiques du Department de la Sarthe, avec une liste des fossiles dévoniens et carbonifères. *Bull. Soc. Géol. Fr., Sér. 2,* VII, p. 781.

Diagnosis. Acrospiriferinids possessing unbranched lateral costae separated by u-shaped or v-shaped interspaces and having similar costae covering the fold and sulcus.

Comparison. Vandercammenina lacks the laterally bifurcating costae of both *Struveina* and *Multispirifer (Multispirifer* also has more complex cardinalia than either *Struveina* or *Vandercammenina).* The absence of either crural plates or anterior grooves on the lateral costae removes *Vandercammenina* from the Fimbrispiriferinae.

Distribution. Vandercammenina is present, although never in great abundance, in the Rhenish Complex of Communities of the Rhenish-Bohemian Region of the Old World Realm. The following localities may be cited: Nova Scotia (Boucot et al., 1958b, plate 1, fig. 6—9), Germany (Scupin, 1900), Belgium (Vandercammen, 1963; Beclard, 1895), Turkey, (Verneuil, 1869), France (Oehlert, 1889), Spain (Comte, 1938).

Stratigraphic range. Siegen and Ems.

Species assigned:

Spirifer trigeri de Verneuil, 1850
Spirifer bischofi A. Roemer, 1858, in: Giebel, *Silur. Fauna Unterharz,* p. 29, plate 4, fig. 3.
Un-named form, A.J. Boucot, in: Boucot et al., 1958, plate 1, fig. 6—9.

Spirifer bischofi var. *paucicosta* Scupin, 1900, in: Die Spiriferen Deutsch-lands, Paläontol. Abh., N.F., IV, p. 278, plate VII, fig. 4a, b, 5.

Discussion. The relations of the Early Devonian spiriferids with ribbed fold and sulcus to those of the Middle and Upper Devonian are uncertain (except for *Fimbrispirifer* and *Paraspirifer*).

Genus *Struveina* new genus

Type species: Spirifer daleidensis Steininger, 1853, *Geogr. Beschr. Eifel,* p. 71.

Diagnosis. Acrospiriferinids possessing branched costae laterally and on the fold and sulcus. The costae are separated by either u-shaped or v-shaped interspaces.

Comparison. Vandercammenina lacks branching costae lateral on the fold and sulcus, whereas *Struveina* has them. *Multispirifer* has more complex cardinalia than *Struveina.* The paraspiriferinid genus *Fimbrispirifer* has crural plates whereas *Struveina* lacks them.

Distribution. Struveina is a relatively rare member of the Rhenish Complex of Communities of the Rhenish-Bohemian Region of the Old World Realm. It occurs in Germany (Scupin, 1900), Belgium (Vandercammen, 1963), France (Oehlert and Davoust, 1879), England (Scupin, 1900), Spain (Comte, 1938) and Kazakhstan (Kaplun, 1961).

Stratigraphic range. Siegen, Ems and Eifel.

Species assigned:
Spirifer daleidensis Steininger, 1853
Spirifer jouberti Oehlert and Davoust, 1879
Spirifer ferronenis Comte, 1938
Spirifer parcefurcatus Spriestersbach, 1915
Eospirifer (Multispirifer) bifurcatus Kaplun, 1961
Eospirifer (Multispirifer) bifurcatus var. *plana* Kaplun, 1961

Discussion. Kaplun's (1961) forms assigned to *Eospirifer (Multispirifer)* lack the finely striate ornamentation of the Cyrtiadae (the fine ornament illustrated by Kaplun appears far coarser than that of eospiriferids) and the cardinalia (plate XIV, fig. 4) illustrated are not eospiriferid in form (the crural plates are too short and set too far apart; there is doubt that the specimen as illustrated belongs to her species *E. (M.) bifurcatus* as it lacks a fold).

Subfamily Hysterolitinae Termier and Termier, 1949

Discussion. Pitrat (1965) failed to recognize the Subfamily Hysterolitinae, and placed the type-genus *Hysterolites* in the Acrospiriferinae. This pro-cedure combines a group of taxa that have relatively distant relations at the Subfamily level. The new diagnosis given below results in a more phylogenetic classification.

Diagnosis. Solle (1952) provides good illustrations of the brachial valve cardinalia of *Hysterolites* itself, and Vandercammen (1963) provides good illustrations of the brachial valve cardinalia of *"Spirifer" solitarius* (the type-species of *Multispirifer*, the second genus belonging to the Hysterolitinae). Their illustrations show that the cardinalia of the Hysterolitinae are placed on an elevated "shelf" that is more or less in the same plane as the interarea. This placement of the cardinalia contrasts completely with that of the Acrospiriferinae where the cardinal process is placed at the bottom of a relatively open notothyrial cavity lacking a "shelf". On steinkerns this small, but significant, distinction may be easily observed in that the cardinalia of Hysterolitinae are partly overhung by the filling of the notothyrial cavity (which lacks any impress of cardinalia), whereas in Acrospiriferinae the filling of the notothyrial cavity is posteriorly indented by the cardinal process and structures lateral to it (crural plates if present).

Relationships. The fine ornamentation and cardinalia of the Hysterolitinae (Vandercammen, 1963) indicate that the group belongs to the Delthyrididae. Presumably the Subfamily was derived in pre-Siegen time from one of the many forms of *Howellella* then available, but we have no information at this time that suggests which species might provide the most reasonable ancestor. Many species have been assigned to *Hysterolites* since 1820, but the bulk of them possess cardinalia indicating an assignment to members of the Acrospiriferinae.

Genus *Multispirifer* Kaplun, 1961

Type species. Spirifer solitarius Krantz, 1857, Über ein neues bei Menzenberg aufgeschlossenes Petrefakten-Lager in den devonischen Schichten: *Verhandl. Naturhist. Verein Preuss. Rheinl. Westphalen*, 3, pp. 150—152, plate IX, fig. 1a, b.

Discussion. Kaplun (1961) based the genus *Multispirifer* on *S. solitarius* Krantz, but employed Kazakhstan material as the basis for the description of her genus. Unfortunately the Kazakhstan material (forms belonging to *Struveina* rather than *Multispirifer)* is unrelated generically to the type species of *Multispirifer*, although Kaplun was quite correct in recognizing that no genus existed in 1961 to receive her Kazakhstan forms. Kaplun (1961) may have been unduly influenced in her decision to place her forms in *Eospirifer* (with *Multispirifer* as a subgenus) by Scupin's (1900) earlier erroneous assignment of *"S." solitarius* Krantz to *Eospirifer.* It is necessary, therefore, to rediagnose and compare *Multispirifer* sensu strictu.

PLATE IV

Diagnosis. Hysterolitinids with anteriorly bifurcating costae and a costate fold and sulcus.

Comparison. Hysterolites possesses unbranched costae and a fold and sulcus uncomplicated by costae.

Stratigraphic range and distribution. Multispirifer is a relatively rare shell known with certainty only from the Siegen of Belgium and Germany.

Subfamily Paraspiriferinae Pitrat, 1965, emended Boucot
     Genus *Concinnispirifer* new genus
(Plate IV, 7—17)

Type species: Spirifer concinna Hall, 1857, *10th Ann. Rep. N.Y. St. Cab. Nat. Hist.,* pp. 60, 61, fig. 1—3.

PLATE IV
Cumberlandina, Concinnispirifer and *Duryeella,* new genera.

1—6   *Cumberlandina cumberlandiae* (Hall, 1857). Ridgely Sandstone. Western Maryland Railroad cut, Collier Mountain, East of Cumberland, Maryland. USNM Acc. No. 187800.
1.   Interior of brachial valve (× 1). USNM No. 126098.
2.   Exterior of pedicle valve (× 1). USNM No. 126098.
3.   Exterior of brachial valve (× 1). USNM No. 126098.
4.   Interior of pedicle valve (× 1). USNM No. 126098.
5.   Interior of pedicle valve (× 3). USNM No. 126098.
6.   Interior of brachial valve (× 3). USNM No. 126098.
7—9, 11. *Concinnispirifer concinnus* (Hall, 1857). Beds of Helderberg age. Locality 524c, East of Covington, Virginia. USNM No. 97859.
7.   Exterior (posterior view, × 1).
8.   Exterior (side view × 1).
9.   Exterior of brachial valve (× 1).
11.   Exterior of pedicle valve (× 1).
10, 12—17. *Concinnispirifer concinnus* (Hall, 1857). Dalhousie Shale (bed 9). Locality 36F-bed F of Schuchert, Stewarts Cove, Dalhousie, New Brunswick. USNM Acc. No. 120092.
10.   Exterior (side view, × 1). USNM No. 126101A;
12.   Exterior of pedicle valve (× 2). USNM No. 126101B.
13.   Exterior of pedicle valve (× 2). USNM No. 126101A.
14.   Exterior of brachial valve (× 1). USNM No. 126101A.
15.   Exterior of pedicle valve (× 1). USNM No. 126101A.
16.   Exterior (posterior view, × 1). USNM No. 126101A.
17.   Impression of interior of brachial valve (× 2). USNM No. 126101B.
18—20.   *Duryeella macra* (Hall, 1857). Onondaga Limestone. Mt. Marion, Catskill Quadrangle, N.Y. USNM No. 126061.
18.   Impression of interior of brachial valve (× 2).
19.   Impression of interior of brachial valve (posterior view, × 2).
20.   Impression of interior of brachial valve (posterior view, × 2).

Diagnosis. Concinnispirifer is erected to receive Paraspiriferinids having a normal spiriferoid fold and sulcus.

Comparison. Paraspirifer has an inflated form and an acuminate fold and corresponding sulcus that are lacking in *Concinnispirifer. Fimbrispirifer* has a costate fold and sulcus as opposed to the smooth fold and sulcus present in *Concinnispirifer.*

Exterior. Unequally biconvex shells with the pedicle valve having a greater degree of convexity than the brachial valve. The outline of the shells is subcircular to transversely elongate. The hinge line is straight and the maximum width is usually at the hinge line. The posterior portion of the lateral margin is relatively straight, but anteriorly it is evenly rounded as is the anterior margin. The brachial valve bears a prominent fold which has an angular cross-section with the apex rounded. The fold is 2 to 3 times as wide as the first lateral plication. There are usually about ten to twelve plications on each flank. The lateral plications have a subrectangular or low-rounded cross-section and are separated by narrow, sharp interspaces. The middle of the antero-medial plications is usually grooved in well-preserved specimens. The interarea of the pedicle valve is gently apsacline to orthocline and is relatively long. The beak of the pedicle valve is usually incurved. The interarea of the brachial valve is apsacline and incurved. The delthyrial cavity is bordered by narrow deltidial plates inclined normally to the interarea. The deltidial plates may be apically conjunct and the resulting cover extends over about two-thirds of the delthyrium. The interarea is smooth except for growth lines which parallel the hinge line. The sides of the delthyrium include an angle of about sixty degrees. The anterior commissure is uniplicate and crenulate. The fine ornamentation consists of growth lamellae crossed by radial striae which terminate as a fringe of minute spines on the edge of each growth lamella.

Pedicle interior. Short dental lamellae border the delthyrial cavity and support the short, pointed teeth which are located at the median edges of the hinge line. The dental lamellae tend to become obsolete due to the deposition of secondary material in the umbonal cavities as well as in the delthyrial cavity. The muscle field is deeply impressed into the thick deposit of secondary material which floors the posterior of the valve. The muscle field is tripartite and consists of a narrow, elongate, median adductor track flanked by a pair of elliptical diductor impressions. The muscle field does not extend anteriorly beyond the midlength and is usually about one-third to one-quarter as wide as the maximum width of the valve. The peripheral portions of the interior are crenulated by the impress of the external ornamentation but the remainder is relatively smooth due to the deposition of secondary material.

Brachial interior. The cardinalia consist of a ctenophoridium in the posterior part of the notothyrial cavity and discrete hinge plates whose bottom sides extend below the base of the dental sockets as blade-like extensions. The hinge plates are inclined medially and are posteriorly supported by fulcral plates; the latter forming the bottom of the dental sockets. A thin myophragm bisects the muscle field which is usually not discernible.

Species assigned:

Spirifer concinna Hall, 1857.
Un-named shell from the Slite Beds, Slite Cement Company Quarry, Slite, Gotland (U.S.N.M. Locality No. 10, 033; 56-G-33).

Stratigraphic range. Beds of Gedinne (Helderberg) age in the Appohimchi Subprovince of North America. Beds of Late Wenlock age on Gotland.

Geographic range. Eastern North America and Gotland.

Subfamily Mucrospiriferinae Pitrat, 1965 (= Mucrospiriferinae Boucot, 1959).

Discussion. Pitrat (1965, H686-687) assigned the genera *Mucrospirifer*, *Amoenospirifer*, *Brevispirifer*, *Strophopleura*, and *Tylothyris* to the Mucrospiriferidae (here reduced to Subfamily rank). Carter (1972) included *Mucrospirifer*, *Brevispirifer*, *Amoenospirifer*, and *Apousiella* (= *Bouchardopsis* Maillieux, 1933) in the Mucrospiriferinae, and specifically excluded *Eleutherokomma* (because of its denticulate appearing hinge line), *Strophopleura*, and *Tylothyris*. Reconsideration of the Mucrospiriferinae here results in assigning *Mucrospirifer*, *Eleutherokomma*, *Brevispirifer*, *Mediospirifer*, *Apousiella*, *Duryeella* and *Cumberlandina* to the Subfamily but rejecting *Amoenospirifer*. The absence of lamellose mucrospiriferinid type ornamentation on *Amoenospirifer* is inconsistent with placement in the Mucrospiriferinae (the Delthyrinae may be a better receptacle). *Eleutherokomma* is so similar to *Mucrospirifer* that it would appear unreasonable to remove it from the Mucrospiriferinae despite the possible presence of a denticulate hinge line. (Many denticulate brachiopod taxa are included in familial units made up chiefly of non-denticulate taxa.)

Diagnosis. Members of the Delthyridae possessing closely spaced lamellose growth lamellae. Radial fimbriae or spines are rarely observed, but on well-preserved material are seen to be present. A convex, triangular deltidium of varying size is present rather than a pair of linear, ribbon-like deltidial plates.

Groups. The Mucrospiriferinae may, for convenience, be divided into two groups of taxa; those with a prominent groove down the fold (and a corresponding plication in the sulcus) and those lacking such a groove. The taxa with this median groove include *Mucrospirifer, Eleutherokomma, Brevispirifer,* and *Apousiella.* The earliest such mucrospiriferinid known to me is "*Spirifer*" *macra* pars Boucot, 1959, non Hall, 1857 (plate 93, fig. 5, 6, not 1—4, 7—10) from strata of Ems age, with an assignment to *Mucrospirifer* being most reasonable.

Genus *Duryeella,* new genus
(Plate IV, 18—20)

Type species: Spirifer macra Hall, 1857, *10th Rep. N.Y. St. Cab. Nat. Hist.,* p. 134.

Spirifera macra Hall, 1857, in: Hall, 1867, *Paleontol. N.Y.* IV (1), pp. 190, 191, plate 27, fig. 17—28.

Diagnosis. Mucrospiriferinids possessing a gently convex brachial valve and a pedicle valve having about twice the convexity of the brachial valve. The interarea of the pedicle valve is almost pyramidal, and is steeply apsacline. The fold is convex and undivided by a median groove. The cardinalia of the brachial valve consist of a simple ctenophoridium. The brachial valve muscle field is weakly impressed, and is divided medially by a low myophragm. This genus is named in recognition of Miss Frances Duryee's patient services in further unraveling the Early Paleozoic palynological record.

Comparison. Duryeella differs from the mucrospiriferinids having a medial groove down the brachial fold. *Duryeella* is very similar to *Mediospirifer,* and in fact may be thought of as ancestral to that genus, but the lack of a bifurcated, bulbous ctenophoridium removes it from that genus as does the lack of a well impressed posterior and anterior pair of adductor muscle scars in the brachial valve. *Cumberlandina* has an almost orthocline pedicle valve interarea, that of *Duryeella* being steeply apsacline, the valves of *Cumberlandina* are almost equally convex whereas those of *Duryeella* are very unequally biconvex.

Stratigraphic and geographic distribution. Duryeella is known only within beds of Schoharie (Ems) and Onondaga (Eifel) age in the Eastern Americas Realm — chiefly from New York. A few specimens from Brazil probably belong to this genus (derived from the Amazon Basin region Eastern Americas Realm faunas).

Description. An adequate description is given in Hall (1867).

Genus *Cumberlandina*, new genus
(Plate IV, 1—6)

Type species: Spirifer cumberlandiae Hall, 1857, *10th Rep. N.Y. St. Cab. Nat. Hist.*, p. 63.
Spirifer cumberlandiae Hall, 1857, in: Hall, 1859, *Paleontol. N.Y.*, III (1), pp. 421, 422, plate 96, Fig. 9a—h.

Diagnosis. Mucrospiriferinid possessing subequally convex valves, an un-grooved fold (and unplicate sulcus), with an orthocline pedicle valve inter-area.

Comparison. See under Comparison of *Duryeella.*

Stratigraphic and geographic distribution. Cumberlandina is recognized to date only from beds of Oriskany age (Siegen) in Maryland, a part of the Eastern Americas Realm.

Description. An adequate description is given in Hall (1859). Hall (1857, p. 63) points out the similarity of *Cumberlandina* to *Mucrospirifer* ("*S.*" *cumberlandiae* to "*S.*" *mucronata*). However, the concentric growth lamellae of *Cumberlandina* are not nearly as pronounced as those belong-ing to other genera of the Mucrospiriferinae, but this feature may be interpreted as merely a transitional feature relating the Mucrospiriferinae to older members of the Delthyrinae from which they were probably derived.

Superfamily Stringocephalacea King, 1850
    Family Centronellidae Hall and Clarke, 1895, emended Cloud, 1942

Discussion. Cloud (1942) emended the terebratuloid Family Centronel-lidae to include the Subfamilies Rensselaeriinae, Eurythyrinae, Centronel-linae and Amphigeniinae. At the same time he erected the new Family Rhipidothyridae to include the Subfamilies Globithyrinae and Rhipido-thyrinae. Within the Globithyrinae he included the genera *Prorensselaeria, Globithyris* and *Rhenorensselaeria.* He showed that the early ontogeny of *Globithyris* included a form closely allied morphologically to the adults of *Prorennselaeria* in all regards. The new material of *Prorensselaeria* figured by Boucot and Johnson (1967a) plus reconsideration of the previously figured specimens indicates that *Prorennselaeria* is also morphologically intermediate between *Nanothyris* and *Rensselaeria.* The stratigraphic posi-tion of *Prorensselaeria* is consistent with it, or a morphologically similar antecedent *Nanothyris* derivative having given rise to *Rensselaeria.* This new data strongly supports Cloud's earlier conclusion that *Prorensselaeria,* the oldest and presumably ancestral genus of the Rhipidothyridae, was derived from the Centronellidae.

Cloud also included the genus *Rhenorensselaeria* in his Subfamily Globithyrinae. *Rhenorensselaeria*, however, has a deeply impressed, complex pedicle valve muscle field (see Cloud, 1942; Boucot et al., 1967, for illustrations) unlike that of any other rhipidothyrid or centronellid. However, the pedicle valve muscle field of *Rhenorensselaeria* is deeply impressed in a manner similar to *Lievinella*, new genus. *Rhenorensselaeria* is here made the type genus of the new Family Rhenorensselaeridae that also contains *Lievinella*. The Rhenorensselaeridae may have had a common origin together with the Centronellidae, from which the Rhipidothyridae arose, but the former point us still unclear. *Lievinella* possesses crural plates, a feature common to the taxa of both the Rhipodothyridae and Centronellidae, but it is conceivable that this is a polyphyletic feature. For the moment, however, is is reasonable to assume that the Rhenorensselaeridae and Centronellidae had a common pre-Devonian ancestry (Fig. 21).

Family Rhenorensselaeridae, new family

The new Family is defined to include primitive terebratuloids having crural plates in the brachial valve or a median septum formed from conjunct crural plates, together with a deeply impressed pedicle valve muscle field. The genera *Lievinella* and *Rhenorensselaeria* are included.

Genus *Lievinella*, new genus

Type species: Rensselaerina primaeva Barrois, Pruvost, Dubois, 1920.

Diagnosis. Rhenorensselaerid brachiopods with well developed short crural plates in the brachial valve and a deeply impressed pedicle valve muscle field. Narrow sulci are present in most specimens of both valves, but these may be of specific importance only.

Comparison. Lievinella differs from the only other genus in the Family in lacking a median septum in the brachial valve. It is not known whether *Lievinella* has a small knob on each half of the hinge plate, features present in *Rhenorensselaeria*. *Lievinella* differs from all taxa of the Centronellidae and Rhipidothyridae in having a deeply impressed pedicle valve muscle field.

Description. The type species is well described and illustrated by Barrois et al. (1920). The specimen illustrated in their plate XIV, fig. 21 is here selected as the holotype of the species *L. primaeva*.

Discussion. Lievinella is interpreted (Fig. 21) to be the precursor of *Rhenorensselaeria* in a manner parallel to *Prorensselaeria* being a precursor of *Globithyris*. The deeply impressed pedicle valve muscle field in both genera of the Rhenorensselaeridae makes it reasonable to conclude that they evolved independently of the genera of the Centronellidae and Rhipidothyridae. It is concluded that the brachial valve median septum of *Rhenorensselaeria* is a feature developed parallel to that present in *Globithyris*. It is concluded that both of these septate genera had ancestors possessing discrete crural plates (during the ontogeny of *Globithyris* these plates fuse medially to form the septum; it is concluded that the ancestor of *Rhenorensselaeria* lacked a median septum and that the ontogeny of *Rhenorensselaeria* will be found to be similar to that of *Globithyris* as regards the formation of the median septum).

Subfamily Centronellinae Waagen, 1882
    Genus *Centronella* Billings, 1859
        "*Centronella*" *arcei* (Ulrich, 1892)

Ulrich (1892, in Steinmann, pp. 53—56) described the species ?*Centronella arcei* from the Devonian of Bolivia. Cloud (1942, p. 74) subsequently rejected ?*C. arcei* from the terebratuloids on the basis that Ulrich cited it as spire-bearing. However, Ulrich (1892, p. 54) stated "Weder an Schliffen, noch an angebrochenen Exemplaren habe ich Andeutungen einer Schleife oder einer Spirale entdecken können." Examination of shells from Bolivia indicates that ?*C. arcei* is indeed a centronellid. Similar material occurs in the Devonian of the Sierra de la Ventana (Castellaro, 1966, p. 115) where it has been cited as *Cryptonella* sp. (I have prepared an extensive collection that leaves no doubt on this point). Branisa (1965, p. 51) introduced a *nomina nudum Proboscidina abastoflorum* based on material which, although unfigured and undescribed, is similar if not specifically identical to ?*C. arcei*. Isaacson is in the process of studying abundant, well preserved material belonging to this taxon. There is no doubt, however, that ?*C. arcei* is a Malvinokaffric centronellid similar in many regards to both *Oriskania* and *Centronella* sensu strictu.

APPENDIX II

Supporting Comments for Stratigraphic Range Data and Taxonomic Relationship Information Shown in Fig. 27

In order to discuss the implications for rates of appearance and disappearance of taxa at the generic level, as well as of total number of taxa per

unit of time, I have prepared Fig. 27 on which are shown known stratigraphic ranges, and where available, what are considered to be reasonable
phylogenetic relationships between genera. Some of this information has
not been previously discussed or considered. Therefore, it is necessary to
present here some documentation and justification for certain points. This
information and discussion will be presented Superfamily by Superfamily,
following the order of presentation employed in the *Treatise on Invertebrate Paleontology* (Volume H).

Superfamily Dalmanellacea

The spelling of the generic term *Dicaelosia* has aroused discussion.
Many students of the dalmanellids still prefer the spelling *Dicoelosia*.
However, Cloud (1948) has reviewed the problem and given the basis for
preferring *Dicaelosia* as the correct spelling.

Superfamily Chonetaca

The poorly known genus *Australostrophia*, restricted to the Malvinokaffric Realm Devonian, is now known to be a member of the Chonostro-
phiidae (Boucot, 1974).

Superfamily Atrypacea

The Early Llandovery species *Plectatrypa henningsmoeni* Boucot and
Johnson, 1967b, is indicated to be ancestral to the Early Devonian genus
Karpinskia. This conclusion is based on the simple cardinalia present in
both taxa plus the relatively long dental lamellae in *Karpinskia* which
resemble similar structures in *Plectatrypa henningsmoeni*. The similar
structures in *P. henningsmoeni* are (see illustrations in Boucot and Johnson, 1967b) obviously ridges that bound the longitudinally elongate
muscle field (very similar in many regards to that of *Carinatina*). The
problem here is whether one may deduce that the long dental lamellae of
Karpinskia, in addition to having given rise to the structures in *Carinatina*
could also have been considered to have arisen from the similar structures
present in *P. henningsmoeni*. Until detailed shell structure studies are
made of all three taxa some uncertainty will remain with regard to this
question.

Superfamily Spiriferacea

(1) Pitrat's (1965) Subfamilies Paraspiriferinae and Fimbrispiriferinae
are synonymized. The inclusion by Pitrat (1965) of *Euryspirifer*, a natural

member of the Acrospiriferinae as it lacks crural plates (characteristic of the Paraspiriferinae, revised) and anteriorly bifurcating or anteriorly grooved lateral plications (characteristic of the Paraspiriferinae, revised) in the Paraspiriferinae is inconsistent with the morphology of *Euryspirifer*. All of the genera here assigned to the Paraspiriferinae, revised (see Fig. 27 for inferred phylogenetic relations) possess typical delthyridid fine ornamentation of concentric growth lamellae fringed with spines, short, delthyrid type crural plates in the brachial valve, and flattish lateral costae that bear either anterior grooves *(Paraspirifer, Concinnispirifer, Fimbrispirifer divaricatus)* or anteriorly bifurcating costae (formed from originally anterior grooves as in *F. venustus)* on the flattish lateral costae. The members of the Paraspiriferinae, revised are not to be confused with the Old World Realm Early and Middle Devonian spiriferids (except for species of *Paraspirifer* in Europe) that lack both crural plates in the brachial valve and flattish costae that are either grooved or bifurcate anteriorly (*Vandercammenina, Multispirifer, Struveina*) and belong to the Acrospiriferinae and Hysterolitinae.

(2) The ambocoelid genus *Pustalatia* (sic. *Vitulina*, sic. *Pustulina*) has been reported from beds of Eifel age in northwest Africa (Villemur and Drot, 1957) but the specimens lack the pustulose ornamentation characteristic of *Pustulatia* and are, therefore, placed in *Plicoplasia. Pustulatia* has also been reported from the Devonian of the Malvinokaffric Realm but has in all cases turned out to be *Plicoplasia.*

(3) Derivation of the Ambocoelidae. A morphologic basis is now available for ascertaining the relations of the Ambocoelidae. The earliest representatives occur in Early Wenlock age beds (zone of *M. riccartonensis)* of the Cape Phillips Formation on Baillie-Hamilton Island. These are shells of howellellid external form with a few broad lateral plications separated by similarly broad interspaces. Internally tooth tracks are present in the pedicle valve but dental lamellae are absent anterior to the tooth tracks. In the brachial valve the hinge plates are of delthyrid form, but are supplemented by a medial mound-like cardinal process. It is reasonable in view of this morphology to predict that the ancestral pre-Wenlock junction of the Ambocoelidae with a howellellid member of the Delthyridae will have a morphology very close to that known for Late Llandovery *Howellella.* The removal of dental lamellae and the introduction of a simple cardinal process are all that is required to make the transition. Therefore the Family Ambocoelidae is here assigned to the Spiriferacea.

Notes

[1] Worldwide swings in available nutrient supply will, following Valentine's (1971a) deductions, certainly have an effect on rate of evolution, with increases favoring a lower rate of evolution and decreases an increased rate. During time intervals of provincialism changes in nutrient supply affecting one biogeographic unit more than another will have corresponding effects on rates of evolution in that unit. Measurement of such changes in nutrient supply are difficult. One possibility may be to compare maximum number of taxa per community per small area in what are concluded to be similar environments. For example, a marked difference in number of taxa in a particular high-diversity community (the same Community Group) over a period of time might be ascribed to change in food supply. Naturally enough, very rapid changes in nutrient supply may manifest themselves as catastrophic events of one sort or another.

[2] The Hunsruck Schiefer of the German Lower Devonian contains the most fully described Siluro-Devonian Benthic Assemblage 6 fauna studied to date (see Kutscher, 1970, for a brief description and introduction to the appropriate literature). Sturmer and Bergstrom (1973) have provided additional data concerning the depth significance of the Hunsruck fauna and summarized opinions past and present.

[3] There are numerous examples from both the fossil record and the Recent which show that these environmental gradients are paralleled by morphologic gradients: for example, the shell size-salinity gradient so well known in the post-Pleistocene of the Baltic (Segerstrale, 1957) and Lake Champlain-St. Lawrence (Goldring, 1932) regions. There is a more limited body of evidence indicating that such phenotypic gradients are accompanied by genotypic gradients. Thus, it is possible that the phenomenon of geographic speciation is merely the resultant of interaction between a combined environmental gradient-genotypic gradient in time for all variables possible; natural selection reduced to its essence. The possibility that morphologic gradients will parallel environmental gradients opens up wide vistas for additional study. Such study may provide the paleoecologist with a powerful tool for unraveling somewhat more of the environmental complexities of the past with real data rather than mere speculation based on recent examples only.

[4] Rubel (1970) has provided a most ingenious approach to the problem of defining communities in terms of recurring taxic associations when faced with the problem of employing Silurian core material from Esthonian boreholes. His results are entirely consistent with those obtained from outcrop Silurian studied elsewhere both by himself and others.

[5] Nor should we feel any qualms at defining communities based on what is obviously only a small portion of the original biota. How many communities defined for living materials represent the entire biota?

[6] Futuyma (1973, p. 444) has pointed out that dependent or interdependent organisms arising under conditions of co-evolution should have abundances that vary together ("the abundance of one species should serve as a fairly accurate predictor of the abundance of each of several other species"). Given an adequate number of samples from a large enough geographic region (in order to minimize the possibility of chance associations) Futuyma's suggestion might be tested with samples of fossils.

[7] Possibly the methodology outlined by Cohen (1970) might aid in arriving at statistically meaningful conclusions about dependence relations among the taxa found in fossil collections (if so, the data currently on hand will have to be quantified!), unless the associated physical characteristics of the environment were always identical (a most unlikely possibility worldwide).

[8] "Quiet-water" and "rough-water" as used here refer to habitats in which current activity is either sufficient or insufficient to disarticulate bivalves and brachiopods, to sort or not to sort the shells of any one species into normally distributed populations lacking the distributions skewed towards an abundance of small individuals that have been predicted for some organisms with planktic larval stages and high infant mortalities, and the occurrence or non-occurrence of these situations with either fine-grained sediment or coarse-grained sediment. When all of the factors suggest little turbulence or transportation (even if shells are not actually observed in living position) then a quiet-water environment is inferred, and vice versa. The term "rough-water" is used here in reference to conditions capable of disarticulating bivalved invertebrates and sorting the disarticulated products. It is not necessarily meant to imply breaking waves or any particular depth. It is likely that a variety of biologic factors (burrowing organisms, predators, deposit feeders, scavengers) are capable of disarticulating shells that grew in a quiet-water environment, but associated physical and biologic evidence should aid in deciding that this is the case (petrography of the beds, sedimentary structures, evidences of burrows, evidences of predation and the like).

As used here the terms "rough-water" and "quiet-water" are relative in the sense that different bivalved taxa, either closely related or unrelated, have widely differing rates of disarticulation (different sizes of the same species may also have different disarticulation rates). However, associated sediment types and sedimentary structures do indicate that what I here term "rough-water" does not overlap very much with what I term "quiet-water". Rapid burial will obviously prevent disarticulation under rough-water conditions. Experience with Silurian-Devonian brachiopods does suggest that the community assignments here allotted to the rough-water and quiet-water categories are workable for Benthic Assemblage 3 and more offshore positions, all interpreted to be subtidal, but that difficulty is encountered in what is interpreted to be the intertidal region (Benthic Assemblages 1 and 2).

For example, most collections of *Atrypella* (*Atrypella* Community, quiet-water, low-diversity) occur as multitudes of articulated shells, commonly in a fine-grained matrix. A few collections of *Atrypella* are found in a disarticulated condition mixed with disarticulated shells belonging to other genera; such collections are interpreted as having been subjected to rough-water conditions after death. *Kirkidium* (*Kirkidium* Community, rough-water, low-diversity) on the other hand, is virtually never found in an articulated condition, and is commonly present in sand-size sediment; such collections are inferred to have lived in rough water, as well as being deposited after death in rough water. The rare articulated specimens of *Kirkidium* are almost always small to very small specimens that can easily be interpreted as having lived and died among large shells and been protected from disarticulating currents by a "blanket" of large shells above them.

Lewis (1964) makes clear the strong correlation on rocky shores between taxa present in the intertidal zone and rough water as contrasted to quiet-water conditions. Lewis's conclusions clearly apply to rocky intertidal regions of the past, and it is probably reasonable to extrapolate these conclusions to other substrates. Lewis's conclusions fit well with what is known of coral reef communities of the present in both the intertidal and upper subtidal regions, which suggests that reef environments of the past should be examined with Lewis's conclusions in mind.

An excellent example of what may be accomplished in relating the petrographic character of sedimentary rocks to environmental questions is provided by Anderson and Makurath's (1973) analysis indicating that gypidulinid community shells occur in a high-energy, rough-water environment.

[9] See Kohn, 1958, pp. 574—575; Warmke and Almodovar, 1963, for examples.

[10] Brenchley and Pickerill (1973) and Pickerill (written communication, 1973) report that Lower Caradoc benthic communities in Wales and the Welsh Borderland are distributed in a manner indicating strong correlation with substrate type among other things; a conclusion conforming very well, of course, with what is found for the rough- and quiet-water faunas dealt with here as well as with fossil faunas of diverse age and the benthos of the Recent.

[11] Paine's (1966,1971) studies of the correlation between number of carnivores and number of herbivores in the shallow-water marine environment may provide additional explanation for the differences in diversity between Benthic Assemblage 2 on the one hand and Benthic Assemblages 3—5 on the other. Presumably the intertidal regions of the past will have had a more reduced number of carnivores which will have permitted a lower number of herbivores than would have been predicted in the subtidal region with its higher number of carnivores.

Complicating any conclusions drawn about the concept of higher subtidal taxic diversity being related, at least in part, to the presence of a higher diversity of prey-specific carnivores that keep down the population size of individual prey taxa so as to permit a higher diversity of prey taxa, is the problem that there are differing degrees of prey specificity among carnivores and that some carnivores have the capability of "switching" from one prey taxon to another as abundance of the favored prey taxon goes down (Murdoch, 1969).

[12] Later discussion (see "Deep Sea") of the exceptionally low population densities characteristic of deep-sea benthos coupled with the almost total absence of bivalves in Cenozoic deep-sea cores and the possibility of a much higher calcium carbonate compensation depth may help to explain this anomaly.

[13] There are a number of simple statistical techniques for resolving a large number of faunal lists (fewer incorporate abundance data as well) into categories of recurrent taxa that may be considered as communities (see Valentine's, 1973, ch. 7, for a discussion of some of the possibilities), but until one deals with more than a few hundred similar faunas the simple, graphical technique described here is entirely adequate to do the job, particularly as it capitalizes on the impressions gained by the paleontologists while preparing and sorting the collections (the memory capacity of the interested,

bipedal computer should not be underestimated when trying to cope with small numbers of samples, although for large numbers of samples problems arise in trying to analyse for events that occur at lower levels of significance).

[14] Cummins' (1969) estimate of 450 ft. as the lower limit of Benthic Assemblage 5, based on an interpretation of Ziegler's (1965) interbedded lava flow-animal community data, is happily consistent with my own impression. Cummins' assignment of 150 ft. as the lower limit of Benthic Assemblage 2, is, however, an extrapolation that is not forced by the Ziegler data, as the beds above the flow at the critical Tortworth locality are of C_6 age, those below the flow of C_5 age and the community assignments immediately below and above are not as well supported by published data as might be wished, plus the fact that the complicating possibilities of both eustatic rise in sea level and vertical movements of the area must be considered when dealing with this fine a vertical scale over what may be a lengthy time interval of many thousands or even more years. Cummins' (1969) assignment of 50 ft. to Benthic Assemblage 1 and 100 ft. to all the other Benthic Assemblages (2 through 5) although consistent with the very limited Ziegler (1965) data is hardly to be considered as demonstrated. The important thing about the interbedded lava flow-animal community data, as interpreted by Cummins, is that it provides results of the same order of magnitude as those obtained independently from the lower limit of the photic zone, lower limit of active reef growth, and lower limit of mean low tide assigned to Benthic Assemblage 2 from taxic-diversity and shell-size deductions.

[15] Sheehan's (1973b) Benthic Assemblage 6 Ashgill fauna from southern Sweden is characterized by small shells ("A striking feature of the brachiopods is their uniformly small size") which further serves to emphasize this general relation.
    Elles' (1939, fig. 1) placement of "large brachiopods" in the Benthic Assemblage 2 position for the Ordovician and Silurian is worthy of note!

[16] The potential relation of food supply to shell size is also suggested by Vermeij and Porter (1971) who found that molluscs in two areas of low productivity in the intertidal region were characteristically smaller than those in areas of high productivity, while also indicating that the generalization for the Silurian-Devonian regarding larger size shells in the intertidal than in the subtidal region is a generalization and not a law.

[17] The upper limit of Benthic Assemblage 1 is taken as mean high-water level (Fig.15) and the lower limit of Benthic Assemblage 2 is taken as mean low-water level (Fig.15).

[18] Information about Silurian conodonts, provided by C. Rexroad, arrived too late for compilation in Table II. However, Rexroad's (written communication, 1974) estimates (normalized by me with the Wenlock taken as unity) are as follows: Pridoli 0.4, Ludlow 1.3, Wenlock 1.0, Llandovery 1.7. Note how well Rexroad's estimates fit those provided by everyone else; it is difficult to avoid concluding that the general agreement reached by the paleontologists reflects some underlying "average" rate of evolution that has affected all animal groups in about the same manner.

[19] Ayala et al. (1973) make very clear that a highly specialized bivalve (*Tridacna maximus*) living in a reef environment of the type one suspects would be subject to easy extinction as contrasted with level-bottom bivalves, still has a very high degree of genetic variability. The experiment of Ayala et al. casts serious doubt on the conclu-

sions of Bretsky and Lorenz (1970) that genetic variability may be correlated with various adaptive and evolutionary paths followed by organisms although the number of shallow-water marine organisms whose genetic characters have been studied to date is admittedly something less than an adequate sample on which to base a reliable conclusion. Jones (1973), however, concludes after an intensive study of some European land snails: "Selection by the ecological environment and the genetic environment may therefore both be important in controlling the genetic structure of snail populations" and this suggests that the currently very limited available data on marine molluscan genetic constitution is probably inadequate for arriving at the conclusions desired by the various workers cited above.

[20] Vermeij's (1971) observation that "high intertidal neritids and limpets able to live on a wide range of substrata tend to be species of limited geographic distribution" is a direct contradiction of the Bretsky and Lorenz (1970) surmise that taxa adapted to a wide range of environments have broader geographic ranges than more specialized forms.

Dayton and Hessler (1972) using the same deep-sea data as employed by Sanders (1969) to set up the "Time-Stability Hypothesis" reject the hypothesis and suggest that continued biological disturbance by predators (actually their "croppers" who utilize both varied live and dead food sources) in a very low population density environment act to give rise to the observed high density by permitting a variety of organisms to develop that actually employ essentially the same food sources. Dayton and Hessler point out that possibilities for the varied niches called for in Sanders' hypothesis probably do not exist, and cannot be employed to explain the varied diversity actually observed.

[21] Vermeij (1972) presents convincing evidence that high intertidal molluscs show a far higher degree of endemism than do subtidal molluscs in direct contradiction of the Bretsky and Lorenz (1970) surmise.

[22] Thayer's (1973) recent paper, although employing far less data than used to compile Fig.36, makes the same point that there is no consistent trend departing from shoreline in terms of average time duration of taxa (Fig.36 points out that the percentage of cosmopolitan, long-lived taxa stays about the same as does the percentage of endemic, short-lived taxa). Thus, Thayer too develops the data in a manner that opposes the conclusions arrived at by Bretsky and Lorenz (1970).

[23] Pianka (1966) has skillfully summed up most aspects (except for Valentine's more recent comments about trophic-resource stability as a control for diversity) of the taxic diversity question but indicates that we are far from arriving at an answer that fits all cases. The overall question of taxic diversity is undoubtedly an instance of a number of interacting controls that vary in importance from locality to locality and from time to time (on both short- and long-time scales with organic evolution being significant in the latter case). It is very doubtful that there will be found to be any simple, one factor control over the bulk of the taxic diversity problems.

[24] Boss (1971) has summarized the data on numbers of marine, fresh-water and terrestrial molluscs which indicate that there are far more marine than fresh-water molluscs (note particularly the relative numbers of bivalve and gastropod species in the two environments!) which supports the conclusion that there are far fewer fresh-water than

marine species (the numbers of estuarine and brackish forms are presumably intermediate). Boss's (1971) summary of the number of terrestrial gastropods (unusually
high) might appear to be unusual until the many isolating mechanisms on land are
considered as well as the potentialities for small population size.

[25] Hecht and Agan (1972) point out a similar relation for Miocene to Recent, shallow-
water bivalves occurring along the eastern coast of North America, for which southern
diversity is higher than northern and in which the southern bivalves belong to newer
taxa than the northern taxa. It is likely that Hecht and Agan's northern taxa belong to
groups that cover a far larger region than do their southern taxa enabling the correlation of rapid evolution with small population size to be made.

[26] Going back to the Lower Carboniferous, Mamet (1973) has pointed out that the
distribution, taxonomic diversity and degree of endemism of Foraminifera may be
broken down into three latitudinal groupings: (1) a Tethyan Realm with a rich, highly
diversified fauna high in endemics; (2) a Taimyr-Alaska Realm with a poor, little
diversified fauna low in endemics; and (3) a Kuznetz-North American Realm intermediate not only geographically but also faunally between the first two. The similarity
of this Lower Carboniferous story developed for the Foraminifera is strikingly similar
to that presented for the Mesozoic and Cenozoic by Stehli et al. (1969). Similar
interpretations in regard to rate of evolution and size of populations may be applied to
the Lower Carboniferous story as with the post-Paleozoic story.

[27] If we were able to extrapolate Vermeij's (1972) conclusion that intertidal, and
particularly high intertidal, molluscs show a far higher degree of endemism than subtidal molluscs, to the fossil record a case could be made for more rapid evolution in
the far smaller intertidal area than in the far larger subtidal region. Present paleontologic data has not been synthesized in a manner permitting this possibility to be
tested. In any event, Vermeij's conclusion is consistent with the thesis presented in this
book.

[28] This situation might be taken to suggest that the long-term genetic pool of marine
invertebrates is located for the most part in the lower-diversity level-bottom regions
rather than in the higher-diversity reef environments. Should such a deduction be
correct it might even be extended by analogy to the rain forest. Possibly one should
consider that the rain forest with its high taxic diversity is not necessarily the prime
genetic pool from which temperate and other taxa of plants and animals are in largest
part derived.

[29] Much of the shallow-water, tropical marine diversity is accounted for by the reef
environment. The sharp contrast in total diversity between polar and tropical shallow-
water benthos is thus due in largest part to the presence or absence of the reef
environment. The undersaturated condition of polar, cold surface waters relative to
calcium carbonate contrasts with the saturated condition found in tropical regions.
Thus, the largest part of the diversity contrast may be ascribed to the impossibility of
building, and maintaining for thousands of years, biogenic, calcium-carbonate-based
reef structures far from the tropical regions. However, there is still a residual contrast,
community by community (if one pays close attention to analogous or homologous
communities one at a time) from the polar regions to the tropical regions in total
diversity.

[30] Taylor (1971) and Brander et al. (1971) summarize data on the taxic diversity of a number of Indian Ocean invertebrates on both the East African shore and on oceanic islands. Both papers make clear that the number of taxa on the East African shore, in similar niches, is higher than on the corresponding oceanic islands, which raises the possibility that MacArthur's (1972, for summary) thesis regarding the lower taxic diversity to be found on offshore islands for a variety of reasons as shown by birds may also apply to shallow-water benthos.

[31] For example, Walmsley and Boucot's (1974) treatment of isorthid species distribution provides more biogeographic resolution than does mere consideration of the genera and subgenera. They note that distribution of isorthid species in the North Silurian Realm during the Llandovery heralds developments observed later at the subgeneric and generic level.

[32] Phyletic rate of evolution is a pure synonym of phyletic rate of extinction in those cases where lineages extend through a time interval, but if a lineage terminates at the end of an interval then the appearance of the taxon in question will be by phyletic evolution whereas its end will be counted as terminal extinction, not as phyletic extinction.

[33] Valentine (1969, pp. 693—694) points out the presence of extinction highs in the Late Ordovician, Late Devonian, Permo-Triassic and Late Cretaceous. He comments that the Late Ordovician and Late Devonian extinction highs are not accompanied by corresponding diversity lows, although such is the case for the Permo-Triassic. Actually the Late Ordovician and Late Devonian (see Fig. 22) extinction highs are accompanied by diversity lows, *but* if one does not employ original sources, relying instead on secondary *Treatise*-type sources that provide inadequate time subdivision, the marked diversity lows accompanying the extinction highs are submerged by earlier or later diversity highs (Ashgill diversity high, Lower-Middle Llandovery diversity low, Late Llandovery diversity high with the "Lower Silurian" diversity averaging out the entire Llandovery to provide a high; Frasne low diversity added to Famenne low diversity averaging out to moderate diversity).

Valentine is not to be blamed for being unaware of this defect in his source material, as employing original sources of both taxonomic and correlation information across the critical time intervals would have been a truly monumental chore, but this problem does point out the care with which secondary sources with low time precision, and also low geographic precision, must be employed in compilations that are time-sensitive or geographic-sensitive. Many of the features we are examining occur during a relatively short time interval and in relatively small geographic regions.

[34] Even the planktic acritarchs (A. Loeblich, written communication, 1973)!

[35] The Bretsky and Lorenz (1970) surmise that worldwide extinction events may be in large part genetically controlled, and result from the accumulation of taxa having a very low degree of genetic variability, is unsupported by the recently available genetic data of Valentine et al. (1973). Valentine et al. find that a bivalve, one would conclude on the Bretsky and Lorenz criteria (stable environment, high-diversity community, highly specialized, etc.) to have a low degree of genetic variability, has, in fact, one of the highest degrees of genetic variability yet discovered in an animal.

[36] The important extinction event (high terminal extinction rate) at the end of the Triassic effected many groups (see Wiedmann, 1973 for an account of its effects on ammonoids) but occurs during a time interval of relatively high cosmopolitanism in both the preceding Late Triassic and succeeding Early Jurassic (although it does coincide, as pointed out by Wiedmann, 1973, p. 183, with an important regressive event).

[37] Wiedmann (1973, p. 186) has recognized this relation in connection with various regressive events associated with Mesozoic extinction events (high terminal extinction rate) followed by subsequent transgressive events associated with taxic diversification (high rate of cladogenetic evolution).

[38] After having concluded here that a high level of provincialism is conducive to increased rates of evolution (because of the smaller size of interbreeding populations) it is comforting to read (Kauffman, 1973, p. 358) that Cretaceous bivalvia show the same effect: "Biogeographically restricted taxa evolved faster than more widely distributed forms...".

[39] Stanley (1973a) has provided an ecologic explanation for the evolution of the Precambrian biota that may answer many of the objections to earlier ideas.

[40] The "time intervals" referred to here are of slightly differing relative length (see p. 62, section entitled "Absolute time and relative time" for a discussion of the relative lengths of these time intervals). They have been employed here as "units" because they represent the current, fossil-based time subdivisions practical for recognition purposes on a worldwide basis. In some regions of the world each of these units may be further subdivided for purposes of local correlation, but not on a worldwide basis at present (although future advances in our knowledge of fossils will certainly make such subdivision on a worldwide basis practical). As these "time intervals" are of somewhat differing relative length one *must* keep this in mind when considering the discussion of changing rates that follows. The rate changes discussed here are large enough that the departures from unity of the "time intervals" do modify the conclusions in some cases. See following section for comments.

[41] This interpretation of the Pridoli decline in numbers assumes that the Pridoli is equivalent in time duration to the preceding Ludlow and following Gedinne, which is not the case. See p. 68 for the preferred interpretation which takes the shorter time duration of the Pridoli into account. The difference in length of the Pridoli, if permitted to pass unnoticed, would have given rise to a very unwarranted, fallacious interpretation!

[42] These changes, based on information provided by the brachiopods, are paralleled by those of the tetracorals (Oliver, 1973, p. A125, finds the percent of Eastern Americas Realm endemics among the tetracorals to be as follows: Pridoli 15%, Gedinne 57%, Siegen (no data), Ems 90%, Eifel, 62%, Givet 45%, Frasne 0%).

[43] Bramlette (1965) has emphasized the possible effect on reduction in available nutrient supply in the Late Cretaceous being an important, if not the most important factor, in the marked Late Cretaceous extinction event. At the same time Valentine (1971a) has emphasized the potential importance of a reduced available nutrient supply in increasing rate of cladogenetic evolution. It is critical to understand that there is no

conflict in having the co-occurrence of a high terminal extinction rate and a high rate of cladogenetic evolution both being related to a single variable, reduced available nutrient supply in this instance. Following this logic it is reasonable (following Bramlette's lead) to have the increased available nutrient supply of the Early Tertiary correlated with a markedly decreased rate of terminal extinction and also a decreased rate of cladogenetic evolution insofar as the nutrient supply factor is involved. Increased cladogenetic rate of evolution in the Early Tertiary for the stocks crossing the Cretaceous-Tertiary boundary may be conceived of as due to factors of increased provincialism that more than account for any decrease due to increased available nutrient supply. Needless to say, reduced nutrient supply should also act to increase rate of phyletic evolution due to its governing the overall population size.

[44] Finally, it is difficult to believe that the apparent absence of a deep-sea and bathyal benthos was due to a lack of available nutrients. The rich and varied life present on the Lower Paleozoic continental platforms and shelves indicates a rich supply of nutrients. It is difficult to conceive that the surface waters of the contemporary seas would not also have been provided with dissolved nutrient materials as this would require a global surface water circulation pattern capable of restricting the continental-platform water masses to those platforms with no possibility for extensive exchange with the oceanic surface waters. Once it is admitted that surface water nutrients would have been present in the oceans it is very difficult to imagine them staying in an azoic condition for very long. If the unlikelihood of an azoic surface water condition is admitted then it follows that a steady "rain" of planktic debris to the bottom would have occurred as well as an oceanic circulation that would have provided both particulate and dissolved nutrients to the deep-sea floor. Once such a nutrient distribution pattern is admitted for the ocean floor it is difficult to see how it could have persisted for long in an azoic condition. The study of life, both past and present, suggests that available niches provided with nutrient materials do not remain vacant for very long, unless the physical conditions are so extreme as to preclude the existence of protoplasm which certainly is not the case for the bottoms of the deep oceans. One might appeal to the presence of widespread colder waters capable of raising the calcium carbonate compensation depth to a position near the margin of the continental shelf in order to dissolve any traces of a shelly deep-sea or bathyal fauna (although it would be interesting to have it investigated by meteorologists and oceanographers with a speculative bent).

Another view of the apparent absence of an Early Paleozoic shelly benthos is obtained by speaking with marine geologists. Bivalves of the present deep sea have very, very low population densities (the Sanders and Hessler Anchor Dredge must be dragged for at least several hours across the ocean floor in order to come up with its "rich haul"). Bivalves are virtually unknown in deep-sea Cenozoic cores probably due to the initially very low population densities and the postdepositional effects of solution. Consideration of this information for the present may permit one to conclude that if the Early Paleozoic deep seas did indeed have a rich (by present-day standards) bivalve fauna the chances for preservation are very low.

Menzies et al. (1973) have concluded that the great bulk of living deep-sea benthos had its origins in the post-Paleozoic shelf fauna. However, this view still affords little insight into whether or not there was or was not a prolific and diverse Paleozoic deep-sea benthos. The major extinction event of the Permo-Triassic boundary interval might have left a Paleozoic deep-sea benthic fauna relatively unaffected (because of the different parameters involved), but if Menzies et al. are correct in concluding that the present deep-sea benthos is derived ultimately from Cenozoic and Mesozoic shelf ancestors it is reasonable to conclude that the Paleozoic deep-sea benthos, if present, might have been slowly superseded by Mesozoic and Cenozoic replacements.

[45] However, Mamet's (1973) description of three foraminiferal Realms for the Northern Hemisphere Lower Carboniferous that have boundaries parallel to present latitudes is similar to the well documented post-Paleozoic story developed in Hallam's *Atlas*.

[46] "Normal" was used in the sense of the most abundant type of fauna found in the region. It had no environmental connotation.

[47] This surmise has been recently confirmed by Niebuhr (1973) who found a minimum of five brachiopod-based communities during the course of a partial traverse (probably Benthic Assemblage 3 plus overlap into 4) across the platform margin in Eureka County, Nevada, while studying beds of *pinyonensis* zone, Ems age.

[48] The quiet-water level-bottom communities characterized by abundant plicate spiriferids are commonly preserved as disarticulated shells. The disarticulation is thought to result from the activity of both bioturbating organisms and of weak currents. The quiet-water reef and reef-related communities (Bohemian and Uralian Complexes of Communities, for example) are taxonomically very different than the quiet-water level-bottom communities characterized by abundant plicate spiriferids. Clearly it is not the quiet-water environment that spells the taxonomic difference. The differences have been related to the reef as opposed to non-reef environment that may possibly include such variables as type and abundance of food supply or others not yet considered. It is reasonable to conclude that the topographic, small-scale complexities of the reef and reef-related environment might, at equivalent depths, be subject to far weaker post-mortem current activities than similar depth level-bottom communities as an explanation for the differences in disarticulation. The alternate explanation that reef and reef-related shells disarticulate with greater difficulty appears far less likely because of the varied taxa present in both environments.

[49] The concept of evolution affecting the ecosystem as a unit by means of natural selection as discussed by Dunbar (1960, 1968, 1972) is somewhat supported by the community evolution picture developed here. The conclusion reached here that Community Groups (it is unclear just how the "Community Groups" of this treatment compare with the "Ecosystem" of Dunbar, and others) contain taxa that tend to evolve together rather than "passing in the night" does suggest that many, if not the overwhelming majority of taxa tend to evolve as an ecological unit. However, the question of whether the evolution of one taxon in an ecological unit depends on that of others in the same ecological unit is not as yet answered by the paleontologic data. It may well be, if the ecologic unit is as much a biologic expression of the appropriate conjunction of physical variables as of anything else that the ecologic unit does not consist largely of dependent and interdependent taxa, at least as far as level-bottom units are concerned.

[50] Alberstadt and Walker (1973) have supplied additional information from older reef complexes on the problem of community succession (their "ecological succession") that complements the work of Lecompte, as has Nicol (1962).

[51] Richards (1972) cites similar Richmond, Late Ordovician examples.

[52] The shaly faunas may contain similar taxa to the hard substrate faunas, but the former must possess adaptive features enabling them to colonize and thrive on a muddy bottom (see, for example, Schumann's, 1969, account of different pedicle types in brachiopods that adapt them for bottoms with varying sediment type.).

[53] Physically controlled community successions are exemplified by the community sequences encountered in the Permo-Carboniferous cyclothems and by the faunas occurring above and below lava flows; in both cases rapid changes in depth as well as marine to non-marine transitions are involved.

Dayton (1971) describes an additional type of community succession in which competition for space results in a hierarchical succession of taxa. This hierarchical succession may be interrupted by physical events of a destructive nature that permit the succession to begin all over again, or changed by different physical conditions which permit or exclude certain taxa. Recognition of such situations in the fossil record might be very difficult. The occurrence of varying taxa and abundances of the same taxa in what would appear to be the same position could be explained by such situations. The physical variables involved in Dayton's examples are battering by drift logs, wave exposure, and desiccation.

[54] The high-energy, turbulent environment of the gypidulinid Communities, as well as the role of the massive deposits of secondary shell material in the beak region (present in many other rough-water genera additional to the gypidulinids as pointed out elsewhere in this book) in maintaining the anterior margin upwards, is well documented by Anderson and Makurath (1973).

[55] It must be emphasized that both linguloid and orbiculoid brachiopods occur in associations well removed from the Orbiculoid-Linguloid Community! The shells of these two groups are widely distributed. It is the overall association and geographic position as well as taxic content that helps one to recognize the Orbiculoid-Linguloid Community. Pickerill (1973) has emphasized these points in his description of some linguloid specimens occurring well removed from any possibility of belonging to the Orbiculoid-Linguloid Community, although in living position, from the Ordovician of Wales.

[56] Osgood and Szmug (1972) notwithstanding: the presence of a few linguloids in growth position together with some bits of fossil wood is inadequate evidence for stretching the range of *Zoophycos* into a Benthic Assemblage 2 position; a subtidal marine tongue is just as likely.

[57] Von Arx (1962, p. 191) points out how surface water salinity in the oceans today is largely controlled by the interplay between evaporation rate and annual rainfall.

[58] The presence of cold climate in the Malvinokaffric, Gondwana region from the Ordovician through the Permian (with the exception of Australia for the Ordovician through Devonian) is supported by the presence of widespread tillites and associated glacial phenomena in the Late Ordovician and Permo-Carboniferous. In addition the presence of the highly provincial Silurian and Devonian, low-diversity Malvinokaffric faunas in regions lacking any reef limestones, or limestones (except for Australia) of any type (except for scattered shells) is consistent with the presence of cooler waters. The Ordovician invertebrate faunas of North Africa, southern and central Europe

appear to be low-diversity, provincial units that may be similarly interpreted as cold-water types, with limestones notably absent. This situation does not appear to hold for the Cambrian. Therefore, one may conclude that the Malvinokaffric, Gondwana region from the Ordovician through the Permian was one characterized by high provincialism, due in large part to a colder climate.

[59] Independent evidence favoring restricted circulation on the Silurian platforms has been provided by J. Gray (in Gray and Boucot, 1973). Gray finds that abundant Silurian spores of land-plant type (although not necessarily derived from land plants) occur only in the very near-shore (Benthic Assemblage 1) or brackish-water positions; almost never seaward. She interprets this in comparison with distribution of pollen and spores in modern oceans, where riverine and ocean currents are effective in widespread distribution of the plant microfossils, as evidence favoring the absence of either strong currents crossing the platforms of the Silurian, or of rivers powerful enough to transport large amounts of spore-bearing water a great distance away from shore. This Silurian circulation pattern is markedly different from that encountered in the Devonian (D.C. McGregor, oral communication, 1972) or in the post-Devonian (Gray and Boucot, 1972).

[60] Sonnenfeld's (1964) work indicating the need for organic material, and possibly bacteria, to make secondary dolomitization possible is consistent with the occurrence of biotic dolomitization subtidally below the sediment—water interface as well as in the non-marine to intertidal environment posited by Sonnenfeld.

[61] Wiedmann (1973) finds that Mesozoic ammonoids show a pattern of regression correlated with extinction (high rate of terminal extinction) followed by evolutionary diversification (high rate of cladogenetic evolution) accompanying the subsequent transgression.

[62] See section on "Orogeny and provincialism" for definition of orogeny employed here.

[63] Unless the disconformity between the Lower Caballos and Lower Arkansas Novaculite with the Upper Caballos Formation and Upper Arkansas Novaculite respectively should reflect a Middle Devonian event to the south in the Llanorian region (Wilson and Majewske, 1960).

[64] First noted in North Africa by Sougy and Lecorche (1963).

[65] Ziegler and McKerrow (1974) have reviewed the occurrence of marine red beds in the Early Paleozoic, chiefly the Silurian, of Europe and the Central Appalachians. They conclude that the red coloring material, hematite, was derived chiefly from deeply weathered terrains that I would conclude to have been very warm in nature as well as intermittently moist, and then transported to the site of deposition without being subject to reduction. Similar red marine sediments are also well known in northern Maine and adjacent New Brunswick (the "Manganese" district of Aroostook County). These marine red beds provide additional circumstantial evidence for the presence of warm climates in the North Silurian Realm region.

[66] Woodrow et al. (1973) have ably summarized the data concerning widespread occurrences of carbonate nodules, tubes and the like in Devonian non-marine rocks of the Old Red Sandstone and its correlatives of the circum-North Atlantic region where intimate association with dipnoan fish as well all indicate the presence of a tropical wet and dry, monsoonal type of climate.

[67] In an earlier paper (Boucot and Johnson, 1973) what is here termed the North Silurian Realm was during the Llandovery referred to as the "Cosmopolitan Llandovery Realm", whereas during the Late Silurian its two subdividions (the Uralian-Cordilleran and North Atlantic Regions) were not assigned formally to a Realm. I have decided to employ the all-embracing term 'North Silurian Realm" in order to further emphasize the contrast between northern and southern (Malvinokaffric Realm) faunas.

[68] The emphasis here on the overall cosmopolitan nature of the North Silurian Realm during the Lower Silurian should not obscure the fact that a certain low level of endemism may be recognized from place to place. This low level of endemism may be of great significance when trying to understand rates of evolution in particular groups or organisms. For example, among the stricklandid brachiopods, it is notable that the bulk of the North American Platform was occupied by the genus *Microcardinalia*, with *Stricklandia* sensu strictu known only in marginal regions of the Appalachians and British Columbia. Needless to say, in terms of the known relatively rapid phyletic evolution of *Stricklandia* this is very convenient in reducing the area occupied and the presumed size of the population. This example taken from the stricklandids should make clear how critical it is in all such problems to have as high quality taxonomic data as possible; it will not do when attempting paleoecologic and biogeographic analysis with an evolutionary end in mind to settle for anything less than the most meticulous taxonomy! Also noteworthy in the late Upper Llandovery is the restriction of the atrypacean genus *Pentlandella* to Britain and a limited portion of the Baltic region. Generalizations about the overall provincial or cosmopolitan fauna of a region or of a continent or more should not permit one to overlook the possibilities of high endemism affecting particular taxa nor should the high endemism characteristic of particular taxa blind one to the overall cosmopolitanism of the largest part of a fauna.

[69] New unpublished collections kindly made available to me by the Greenland Geological Survey, from northwestern Greenland contain a brachiopod fauna (including subrianinids) that is assigned to the Uralian-Cordilleran Region.

[70] The marked, worldwide increase in brachiopod size that occurs near the beginning of the Devonian has a bearing on the precise age of the Malvinokaffric Realm Devonian as that Realm's brachiopods are typically large. Boucot (1973) and Stanley (1973b) have provided discussions of the overall tendency observed in many groups of organisms for increase in size during evolution (Copes Rule). Llandovery through Lower Gedinne brachiopods are relatively small as contrasted with later Devonian shells (the Pentamerinae are the most conspicuous Silurian exceptions in that they are very large shells, although with no Devonian descendents for comparison). Late Gedinne shells begin to show a marked increase in size over Lower Gedinne forms. Siegen and later Devonian brachiopods are typically large. Thus it is reasonable to conclude that all of the Malvinokaffric Realm Devonian shells must be of post-Gedinne age from considerations of size alone.

[71] Previous discussions of Devonian biogeography considering brachiopods (Boucot et al., 1969, for example) have employed the term "Appalachian Province" for what is here entitled "Eastern Americas Realm", and similar discussions of tetracoral distributions (Oliver, 1968, for example) have employed the term "Eastern North America Province" for what is here entitled "Eastern Americas Realm". Oliver and I have discussed the desirability of employing a uniform nomenclature for units on which we agree with the result that we have both adopted Eastern Americas Realm with the following subdivisions: Amazon-Colombia Subprovince, Nevada Subprovince and Appohimchi Subprovince (for the area between Gaspe and Hudson Bay on the northeast and Chihuahua to the southwest). At the current level of our understanding of Devonian biogeography this nomenclature should prove adequate for the "Appalachian type" faunas of both North and South America.

[72] Scrutton (1973) has most recently raised this difficult question in connection with the age of some Devonian tetracoral faunas from the Perija in Venezuela. Scrutton compares the Perija corals with similar taxa in the upper, Eifel-age portions of the Onondaga in the Appohimchi Subprovince and logically deduces an Eifel age for the Perija material. However, the brachiopods of this region include a number of taxa that have not previously been definitely recorded from strata as young as the Eifel (although Scrutton tends to discount this evidence) including (Bowen, 1972) such things as *Pacificocoelia* (Bowen's *Leptocoelia flabellites*), large *Eodevonaria* (small forms of *Eodevonaria* are known from the Appohimchi Subprovince Onondaga), *Dalejina* (Bowen's *Rhipidomella*), and so forth. The critical point is not to argue about the known ranges of a number of miscellaneous taxa, tetracoral, brachiopod or microfossil, but to realize that these Amazon-Colombian Subprovince faunas yield a variety of taxa that in the Appohimchi Subprovince are not associated with each other in the same bed. Therefore we are faced with the ever difficult correlation problem of deciding whether one group of taxa represents precursors of a group appearing later in the Appohimchi Subprovince (the bulk of the Onondaga-type tetracorals) or another group represents holdovers that disappeared earlier in the Appohimchi Subprovince (the bulk of the "typical" Ems-age brachiopods). Questions of this sort are difficult to answer definitively at the moment; one should try to keep an open mind until the evidence (still lacking) becomes overwhelming.

[73] About the only taxon that is known to be endemic to the Amazon-Colombian Subprovince is the poorly known chonetid *"Chonetes" freitasi* Rathbun. The unique, coarse ornament of this chonetid indicates that it probably is a new genus, but the lack of adequate specimens showing clearly the characters of the brachial interior prevents certainly regarding the precise nature and taxonomic affinity of the species. *Dictyostrophia* of the Colombian Devonian, although known only from this Subprovince, is probably too rare to have much significance at this time.

[74] Notable is the absence of *Leptocoelia, Costellirostra* and *Eatonia* (many other typical Appohimchi Subprovince taxa such as *Hipparionyx* and *Beachia* are also absent but their absence is probably due in largest part to the absence in the Nevadan Subprovince due to post-Lower Devonian erosion to the east or dolomitization, of Benthic Assemblage 2 communities such as the *Hipparionyx* Community).

[75] Lenz and Pedder (1972) list a rich Siegen-age fauna from the Yukon. This fauna, the *Gypidula* sp.1-*Davidsoniatrypa* fauna, consists largely of a *Vagrania-Skenidioides* Community, Benthic Assemblage 4—5, Old World Realm group of taxa that is closely

allied with Cordilleran Region Ems-age faunas in Nevada (Benthic Assemblage 4—5 *pinyonensis* zone particularly) and also with *Quadrithyris* zone, later Gedinne faunas in Nevada that herald the Cordilleran Region in a biogeographic sense. Fig. 25 does not separate out a Cordilleran Region for the Gedinne-Siegen, although enough information is available to see that significant differences are present that contrast the area with co-occurring Tasman Region and Rhenish-Bohemian Region faunas of the same age. It may be inferred that to the west of the area occupied by the Eastern Americas Realm, Benthic Assemblage 3, Nevadan Subprovince Siegen-age faunas in central Nevada we should find deeper-water, Benthic Assemblage 4—5 faunas of Cordilleran Region, Old World Realm type that correspond to the *Gypidula* sp. *1-Davidsoniatrypa*, *Vagrania-Skenidioides* Community. Whether or not these predicted *Vagrania Skenidioides* Community faunas would occur westerly in Nevada or adjacent northern California to the Eastern Americas Realm, Benthic Assemblage 3 faunas of the bulk of the Nevada *Trematospira* zone fauna described to date or to the west of deeper-water Eastern Americas Realm faunas of *Dicaelosia-Hedeina* Community type remains to be determined (it is important to keep in mind here that the known Lower Devonian faunas of northern California belong to the Old World Realm, although too little is known about them to enable a placement in a Region to be made, but the work of Alfred Potter should settle this question in the future).

[76] McKerrow and Ziegler (1972) have questioned the position of the Ouachita Belt during the Early Paleozoic. They place it in entirety (including a large segment of the Piedmont as well as the Florida Platform) on the west side of their Gondwanaland in a position well removed from North America. No evidence is given for this decision. The facies relations between Ouachita Belt rocks and those occurring to the northwest on the North American Platform are so well known for the Late Cambrian through Devonian interval as to require little comment (for example, see Wilson, 1954; McBride et al., 1969). The well-known blocks of platform dolomite and limestone, similar to those occurring to the northwest on the edge of the North American Platform in the Cambro-Ordovician, occurring in turbidite deposits in the Ordovician of the Marathon Region Ouachita Facies rocks have no reasonable source on the western side of the Old World. Many other such examples might be cited. The very uniform Cambro-Devonian stratigraphies of the Mediterranean region are grossly unlike the stratigraphies of either the Florida Platform or the Ouachita region (where, for example, are the Old World equivalents of the Devonian chert represented by the Arkansas Novaculite and Caballos Chert, the Silurian clastics of the Blaylock and Missouri Mountain Slate, as well as the Ordovician and Late Cambrian units). The Platform Carbonate Dagger Flat and Marathon units of the Marathon region have no western Old World lithologic equivalents. The absence of the widespread glacio-marine and glacio-fluviatile horizon present in the Mediterranean region from the Ouachita region is noteworthy in this same regard as is the completely different litho- and biostratigraphy of the Cambro-Ordovician. All of these things can be "explained away" by appeals to one or another type of evidence, but one is left with the impression that the Ouachita Belt has little to do with anything known from the Mediterranean region. Finally, it is of more than passing interest that Keller and Cebull (1973) in a paper entitled "Plate tectonics and the Ouachita System in Texas, Oklahoma, and Arkansas" do not even consider the possibility that the Ouachita Facies Belt was ever far removed from the North American continent!

Additionally, the occurrence on the platform to the north of the Ouachita Line, in Pecos County, Texas, of bedded chert similar to the Caballos certainly ties in the Ouachita Facies to the platform in such a manner that the McKerrow-Ziegler hypothesis may be discounted (Jones, 1944, p. 1043: "On the outcrop in the Marathon region of the Big Bend, the Devonian Caballos Formation is 200—600 ft. thick, almost entirely novaculite and chert. The Plymouth Oil Company's Levy No. 1, Pecos County, penetrated 418 ft. of Devonian, about 90 per cent chert. The rig time and rock bits necessary to drill through this chert make drilling to the Silurian, Simpson, and Ellenburger reservoirs enormously expensive.").

[77] W.A. Oliver's (written communication, 1972) identification of the Schoharie (Ems) tetracoral *Acinophyllum davisi* Stumm from Carrington's Jemison Chert locality 152 (Chilton County, Alabama) makes clear that the Jemison (and also, therefore, the underlying Butting Ram) is of Early Devonian age. Oliver also makes clear that this coral is a typical Eastern Americas Realm taxon; this makes it safe to assign the Jemison to the Eastern Americas Realm, a conclusion not very surprising in view of its proximity to typical Appohimchi Subprovince localities to the northwest (in the Valley and Ridge Province of the Birmingham region), and also to the presence of abundant *Meristella* (an Eastern Americas taxon). Carrington (1972b) reports spores of Ems age from the Jemison.

[78] Dubatolov (1972) has recently considered the biogeographic distribution of Devonian tabulate corals. I have not yet had an opportunity to carefully consider this publication, but cursory examination indicates that many of the biogeographic units (particularly those in the Soviet Union) correspond to those employed here for brachiopod-based units, although different names have been employed in many instances. The actual sampling is, perhaps unavoidably because of the lack of study of tabulates outside of the Soviet Union, heavily weighted in favor of the biogeographic units in the Soviet Union. Previous biogeographic studies, chiefly those dealing with brachiopods, have been heavily relied on as a guide to the units present outside of the Soviet Union. No attempt appears to have been made, particularly for data acquired within the Soviet Union, to consider the problem of whether some of the regional differences in tabulate faunas have automatically been concluded to reflect biogeographic rather than ecologic control. Regardless, Dubatolov's efforts indicate the possibility of using the tabulates, which are very widespread in the Silurian-Devonian, extensively in the future for biogeographic purposes.

[79] Barnes et al. (1973) repeat and expand a bit on the earlier conclusion of Nicoll and Rexroad, but I still feel that until more work has been done on the possibilities of explaining the known distribution patterns for Lower Silurian conodonts in terms of planktic depth zonation as opposed to provincialism we should not make hard and fast conclusions. Barnes et al.'s conclusion that Late Silurian conodonts are *less* provincial than are Early Silurian forms, flying in the face of the evidence afforded by other groups of megafossils, is still more reason for caution.

[80] Westermann's (1973) conclusions about the widely varying lower depth limit of cephalopods, as evidenced by their widely differing resistance to septal failure by implosion, suggests an additional morphologic criterion to be examined in terms of depth zonation for all groups of septate cephalopods.

Wiedmann (1973, pp. 167,190) has suggested that ammonoids may have been depth-zoned and that smooth forms may have occurred deeper than ornamented types.

[81] It is logical to conclude that, if all other factors are held equal, organisms arriving quickly at reproductive maturity will evolve more rapidly than those which arrive very slowly at reproductive maturity. However, the fossil record indicates that the effect of widely differing population size far outweighs the effects of widely differing period needed to reach reproductive maturity. For example, the time needed for reaching sexual maturity in proboscidians is far greater than in rodents, oysters, or scallops but the small populations of proboscidians (highly provincial as well as requiring large areas for each individual as compared with the others) have evolved far more rapidly than the others. One is forced to conclude, therefore, that the effect of differing time interval needed to arrive at reproductive maturity is far lower than is size of inter-breeding population in determining rate of evolution.

[82] For anyone retaining a degree of skepticism about the very high positive correlation between population size and rate of evolution the obligate cavernicoles, the troglobites are most instructive! Barr (1968, *Evol. Biol.*, 2: 36—102) has provided an excellent summary of the organic diversity modified for a completely troglobitic existence, and also provides an extensive introduction to the literature on the subject. He summarizes the nature of the profound morphologic modifications found in many groups of troglobites including blindness, loss of pigment in the outer layers, loss of wings, development of euryphagous habits, etc. Barr also makes clear that most of the cave systems in which these troglobites occur, and in which they presumably evolved from "outside" ancestors, are of no more than Pleistocene age, or possibly Plio-Pleistocene age in some instances. Troglobite populations are very small compared to those present in related species "outside" and population densities tend to be very low as well. The relatively minute troglobite populations contrast wildly with those of related species known from "outside" so that the correlation between very small population and the very rapid morphologic changes "inside" as compared to the virtual absence of such rank changes "outside" is hard to contest. The fact that the morphologic changes accompanying troglobitic existence may be interpreted as rapid loss of structures and functions having no selective value "inside" is of interest but aside from the point about the correlation about rate of change and population size. Loss of vision, loss of pigmentation and the like in salamanders, fish and other taxa during a portion of the Pleistocene in these small populations are certainly examples of most rapid evolution and can be compared with profit to the most rapid changes known in the cichlid fishes of African lakes and changes in proboscideans occurring in comparably short intervals where population sizes are also deduced to have been very small.

[83] Other things remaining equal (environmental heterogeneity, etc.), an area with a large food supply should support larger populations of each species than should one with a small food supply. Therefore, it is reasonable to deduce that the forms occurring in the area of small food supply should evolve more rapidly. Our inability to measure nutrients independently in the geologic record is, therefore, regrettable.

[84] A lower depth limit for the Lower Paleozoic marine shelly fauna is provided by Denton's (1973, *Proc. R. Soc. Lond.*, B, 185: 272—299) summary of lower depth limit data at which *Nautilus* has been netted and also the maximum hydrostatic pressure the *Nautilus* shell is capable of withstanding before imploding (about 500 m and 600 m, respectively). In other words it is unreasonable to expect that nautiloids will be found unbroken below about this depth today or that fossil nautiloids and ammonoids in an unbroken condition will reflect depths far lower than those estimated for living *Nautilus*.

An additional lower depth limit for the Lower Paleozoic marine shelly fauna is provided by Lewis's (1965, *Can. J. Zool.*, 43: 1049—1074) data for percentages of filter-feeding and deposit-feeding benthos occurring at various depths seaward from Barbados. Lewis found that there was a sharp change from predominantly filter-feeding shelly benthos on the continental shelf depth to deposit-feeding predominance once the continental margin depth and the upper limits of the bathyal zone had been passed. Abyssal benthos, of course, are almost exclusively deposit feeders. The presence of abundant filter feeding brachiopods out to the limit of Benthic Assemblage 5 is consistent with a continental margin/upper bathyal zone position in view of Lewis's data.

The rarity of bedded, geosynclinal-type cherts (lydites) associated with Benthic Assemblage 5 and nearer-shore deposits is also consistent with Benthic Assemblage 5 having a lower limit near the continental margin.

[85] Even a casual inspection of the 25 odd volumes of the *Treatise on Invertebrate Paleontology* published to date indicates that most of the stratigraphically long-ranging taxa are widespread geographically whereas most of the taxa with short stratigraphic ranges are geographically restricted. This generalization cuts across phyletic and class boundaries. The case with vertebrate distributions is similar, although it must be noted that most vertebrates tend to be more geographically restricted than most marine invertebrates and the former also tend to have shorter stratigraphic ranges than the latter which conforms very well to the generalization. Stanley's conclusion, finally, is refuted by the fact that geographically widespread, more cosmopolitan rudistids and hermatypic corals, in common with most other groups of animals, were more long-ranging in time than were geographically more restricted, provincial taxa of the same groups. It is hard to conceive of any support for widespread rudistids and hermatypic corals having been less competitive than were geographically restricted taxa of the same groups!

References

Adams, J.E. and Rhodes, M.L., 1960. Dolomitization by seepage refluxion. *Bull.Am. Assoc. Petroleum Geologists*, 44: 1912—1920.

Adey, W.H. and Macintyre, I.G., 1973. Crustose coralline algae: a re-evaluation in the geological sciences. *Bull. Geol. Soc. Am.*, 84: 883—904.

Aldridge, R.J., 1972. Llandovery conodonts from the Welsh Borderland. *Bull. Brit. Mus. (Nat. Hist.)*, 22 (2): 127—231.

Amos, A. and Boucot, A.J., 1963. A revision of the brachiopod family Leptocoeliidae. *Palaeontology*, 6: 440—457.

Amsden, T.W., 1949. Stratigraphy and paleontology of the Brownsport Formation (Silurian) of western Tennessee. *Yale Univ. Peabody Mus. Nat. Hist., Bull.*, 5: 1—138.

Amsden, T.W., 1951. Brachiopods of the Henryhouse Formation (Silurian) of Oklahoma. *J. Paleontol.*, 25: 69—96.

Amsden, T.W., 1973. Brachiopods of the Edgewood Formation (in preparation).

Anderson, E.J., 1971. Environmental models for Paleozoic communities. *Lethaia*, 4: 287-302.

Arkell, W.J., 1956. *Jurassic Geology of the World.* Hafner, New York, N.Y., 806 pp.

Ayala, F.J., 1972. Competition between species. *Am. Scientist*, 60: 348-357.

Barnes, V.E., Cloud, P.E., Jr. and Warren, L.E., 1947. Devonian rocks of central Texas. *Bull. Geol. Soc. Am.*, 58: 125-140.

Barrois, C., Pruvost, P. and Dubois, G., 1920. Faune Siluro-Devonienne de Lievin. *Mém. Soc. Géol. Nord*, 2 (2): 225 pp.

Beclard, F., 1895. Les Spirifères du Coblenzien Belge: *Mém. Soc. Belge Géol.*, 9: 240 pp.

Behrens, E.W. and Watson, R.L., 1969. Differential sorting of pelecypod valves in the swash zone. *J. Sed. Petrol.*, 39: 159—165.

Berdan, J.M., Berry, W.B.N., Boucot, A.J., Cooper, G.A., Jackson, D.E., Johnson, J.G., Klapper, G., Lenz, A.C., Martinsson, A., Oliver, W.A., Rickard, L.V. and Thorsteinsson, R., 1969. Siluro-Devonian boundary in North America. *Geol. Soc. Am. Bull.*, 80: 2165—2174.

Berry, W.B.N., 1972. Early Ordovician bathyurid province lithofacies, biofacies, and correlations—their relationship to a proto-Atlantic Ocean. *Lethaia* 5: 2165—2174.

Berry, W.B.N. and Boucot, A.J. (Editors) 1970. Correlation of the North American Silurian rocks. *Geol. Soc. Am. Spec. Pap.*, 102: 289 pp.

Berry, W.B.N. and Boucot, A.J., 1971. Depth distribution of Silurian graptolites. *Geol. Soc. Am., Abstr. Ann. Meet.*, p.505.

Berry, W.B.N. and Boucot, A.J. (Editors), 1972a. Correlation of the South American Silurian rocks. *Geol. Soc. Am., Spec. Pap.*, 133: 59 pp.

Berry, W.B.N. and Boucot, A.J., 1972b. Silurian graptolite depth zonation. *Proc. 24th Int. Geol. Congr. Montreal, Sect. 7, Paleontol.*, 59—65.

Berry, W.B.N. and Boucot, A.J. (Editors), 1972c. Correlation of the Near-Eastern and Southeast Asian Silurian rocks. *Geol. Soc. Am., Spec. Pap.*, 137: 65 pp.

Berry, W.B.N. and Boucot, A.J., 1973a. Glacio-eustatic control of Late Ordovician-Early Silurian platform sedimentation and faunal changes. Geol. Soc. Am. Bull., 84: 275—284.

Berry, W.B.N. and Boucot, A.J. (Editors), 1973b. Correlation of the British Silurian rocks. Geol. Soc. Am., Spec. Pap., (in press).

Berry, W.B.N. and Boucot, A.J. (Editors), 1973c. Correlation of the African Silurian rocks. Geol. Soc. Am., Spec. Pap., 147: 83 pp.

Berry, W.B.N. and Boucot, A.J. (Editors), 1973d. Correlation of the Australian Silurian rocks. Geol. Soc. Am., Spec. Pap., (in press).

Berry, W.B.N. and Boucot, A.J. (Editors), 1973e. Correlation of the East Asian Silurian rocks. Geol. Soc. Am., Spec. Pap. (in preparation).

Berry, W.B.N. and Boucot, A.J. (Editors), 1973f. Correlation of the Asiatic Russian Silurian rocks. Geol. Soc. Am., Spec. Pap. (in preparation).

Berry, W.B.N. and Boucot, A.J. (Editors), 1973g. Correlation of the European Silurian rocks. Geol. Soc. Am., Spec. Pap. (in preparation).

Bofinger, V.M., Compston, W. and Gulson, B.L., 1970. A Rb-Sr study of the Lower Silurian State Circle Shale, Canberra, Australia. Geochim. Cosmochim. Acta, 34: 433—445.

Bottino, M.L. and Fullagar, P.D., 1966. Whole-rock rubidium-strontium age of the Silurian-Devonian boundary in northeastern North America. Bull. Geol. Soc. Am., 77: 1167—1176.

Boucek, B., 1968. Significance of dacryoconarid tentaculites and graptolites. Int. Symp. Devon. System, Alberta Soc. Petroleum Geologists, II: 1275-1281.

Boucot, A.J., 1959. Brachiopods of the Lower Devonian rocks at Highland Mills, New York. J. Paleontol., 33: 727—769.

Boucot, A.J., 1963a. The Eospiriferidae. Palaeontology, 5: 682—711.

Boucot, A.J., 1963b. The globithyrid facies of the Lower Devonian. Senckenb. Lethaia, 44: 79—84.

Boucot, A.J., 1968a. Origins of the Silurian fauna. Geol. Soc. Am., Abstr. Ann. Meet., 33—34.

Boucot, A.J., 1968b. Silurian and Devonian of the northern Appalachians. In: Studies of Appalachian Geology, Wiley—Interscience, New York, N.Y., pp. 83—94.

Boucot, A.J., 1969. The Soviet Silurian: recent impressions. Bull. Geol. Soc. Am., 80: 1155—1162.

Boucot, A.J., 1970. Practical taxonomy, zoogeography, paleocology, paleogeography and stratigraphy for Silurian and Devonian brachiopods. Proc. N. Am. Paleontol. Conv., F: 566—611.

Boucot, A.J., 1971, Malvinokaffric Devonian marine community distribution and implications for Gondwana. Anal. Acad. Brasil Cienc., 43 (suplemento): 23--49.

Boucot, A.J. and Harper, C.W., 1968. Silurian to lower Middle Devonian Chonetacea. J. Paleontol., 42: 143—176.

Boucot, A.J. and Heath, E.W., 1969. Geology of the Moose River and Roach River synclinoria, northwestern Maine. Maine Geol. Surv. Bull., 21: 117 pp.

Boucot, A.J. and Johnson, J.G., 1967a. Paleogeography and correlation of Appalachian Province Lower Devonian sedimentary rocks. Tulsa Geol. Soc. Digest, 3, (Symp. Vol.): 53 pp.

Boucot, A.J. and Johnson, J.G., 1967b. Silurian and Upper Ordovician atrypids of the genera Plectatrypa and Spirigerina. Norsk Geol. Tidsskr., 47: 79—101.

Boucot, A.J. and Johnson, J.G., 1972. Callicalyptella, a new genus of notanopliid brachiopod from the Devonian of Nevada: J. Paleontol., 46: 299—302.

Boucot, A.J. and Johnson, J.G., 1973. Silurian brachiopod zoogeography. In: A. Hallam (Editor), *Atlas of Palaeobiogeography*. Elsevier, Amsterdam, pp. 59—66.

Boucot, A.J. and Yochelson, E., 1966. Paleozoic gastropods from the Moose River synclinorium, northern Maine. *U.S. Geol. Surv., Prof. Pap.*, 503A: 20 pp.

Boucot, A.J., Brace, W. and DeMar, R., 1958a. Distribution of brachiopod and pelecypod shells by currents. *J. Sed. Petrol.*, 28: 321—332.

Boucot, A.J., MacDonald, G.J.F., Milton, C. and Thompson, J.B., Jr., 1958b. Metamorphosed Middle Paleozoic fossils from central Massachusetts, eastern Vermont, and western New Hampshire. *Bull. Geol. Soc. Am.*, 69: 855—870.

Boucot, A.J., Harper, C.W. and Rhea, K., 1959. Geology of the Beck Pond area. *Maine Geol. Surv., Spec. Geol. Stud. Ser. 1*: 33 pp.

Boucot, A.J., Harper, C.W. and Rhea, K., 1966. New Scotland depositional history of the Beck Pond region, Somerset County, Maine. *Maine Geol. Surv., Contrib. Geol. Maine, Bull.*, 18: 23 pp.

Boucot, A.J., Cumming, L.M. and Jaeger, J., 1967. Contributions to the age of the Gaspe Sandstone and Gaspe Limestone. *Geol. Surv. Can., Pap.*, 67-25: 27 pp.

Boucot, A.J., Johnson, J.G. and Talent, J.A., 1969. Early Devonian brachiopod zoogeography. *Geol. Soc. Am., Spec. Pap.*, 119: 106 pp.

Boucot, A.J., Gauri, K.L. and Southard, J., 1970. Silurian and Lower Devonian brachiopods, structure and stratigraphy of the Green Pond Outlier in south eastern New York. *Palaeontographica*, 135 (A): 59 pp.

Boucot, A.J., Brookins, D. Forbes, W. and Guidotti, C.V., 1972. Staurolite Zone Caradoc (Middle—Late Ordovician) Age, Old World Province brachiopods from Penobscot Bay, Maine. *Bull. Geol. Soc. Am.*, 83: 1953—1960.

Boucot, A.J., Dewey, J.F., Dineley, D.L., Fyson, W.K., Hickox, C.F., McKerrow, W.S. and Ziegler, A.M., 1973. Geology of the Arisaig area: *Geol. Soc. Am., Spec. Pap.*, 139.

Boucot, A.J., Gray, J., Lee, M.H., Rohr, D.M. and Smith, R.E., 1974. Determining the lower depth limits of planktic fossils. (In prep.)

Bourque, P.A., 1972. Commentaires et illustration de la faune, 89 pp. (unpublished).

Bowen, P.Z., 1967. Brachiopoda of the Keyser Limestone (Silurian-Devonian) of Maryland and adjacent areas. *Geol. Soc. Am., Mem.*, 102: 103 pp.

Branisa, L., 1965. *Los Fosiles Guias de Bolivia*. Serv. Geol. Bolivia, 282 pp.

Bretsky, P.W., Jr., 1969. Evolution of Paleozoic benthic marine invertebrate communities. *Palaeogeogr. Palaeclimatol. Palaeoecol.*, 6: 45—59.

Bretsky, P.W., 1970. Upper Ordovician ecology of the central Appalachians. *Peabody Mus. Nat. Hist., Yale Univ., Bull.*, 34: 150 pp.

Bretsky, P.W., 1973. Evolutionary patterns in the Paleozoic bivalvia: documentation and some theoretical considerations. *Bull. Geol. Soc. Am.*, 83: 1—11.

Bretsky, P.W. and Lorenz, D.M., 1970. Adaptive response to environmental stability: a unifying concept in paleoecology. *Proc. N. Am. Paleontol. Conv.*, F: 522—550.

Brunton, C.H.C., Cocks, L.R.M. and Dance, S.P., 1967. Brachiopods in the Linnaean Collection. *Proc. Linn. Soc. Lond.*, 178: 161—183.

Buzas, M.A., 1972. Patterns of species diversity and their explanations. *Taxon*, 21: 275—286.

Calef, C.F., 1972. Diversity and density of brachiopods across the Shelf in Upper Silurian (Ludlow) times. *Geol. Soc. Am., Ann., Meet., Abstr.*, p.465.

Calver, M.A., 1968a. Distribution of Westphalian marine faunas in northern England and adjoining areas. *Proc. Yorksh. Geol. Soc.*, 37: 1—72.

Calver, M.A., 1968b. Coal Measures invertebrate faunas. In: D. Murchison and T.S. Westoll (Editors), *Coal and Coal-Bearing Strata*. Elsevier, Amsterdam, pp.147—177.

Carls, P., 1969. Die Conodonten des tieferen Unter-Devons der Guadarrama (Mittel-Spanien) und die Stellung des Grenzbereiches Lochkovium/Pragium nach der rheinischen Gliederung. *Senckenb. Lethaea*, 50 (4): 303—355.

Carls, P., 1973. *Baturria* n.g. Carls: Proschizophoriinae (Brachiopods) aus hohem Silur und tiefem Devon S.Aragons (Spanien). *Senckenb. Lethaea* (in press).

Carrington, T.J., 1972a. Meta-Paleozoic rocks, Chilton County, Alabama (Guide to Alabama geology). *Geol. Soc. Am., Ann. Meet., 21st, Guidebook Field Trips*, 1-1 — 1-29.

Carrington, T.J., 1972b. Occurrence of fossil spores in metasedimentary rocks of Early Devonian age, Chilton County, Alabama. *Geoscience Man*, 4: 128—129.

Carter, J.L., 1972. Two new genera of lamellose spiriferacean brachiopods. *J. Paleontol.*, 46: 729—734.

Caster, K.E. and Kjellesvig-Waering, E.N., 1964. Upper Ordovician eurypterids of Ohio. *Paleontogr. Am.*, IV: 301—358.

Clarke, J.M., 1900. The Devonian Mollusca of the State of Para. In: *The Paleozoic Faunas of Para, Brazil— Arch. Museu Nacl. Rio de Janeiro*, 10: 23—100.

Clarke, J.M., 1913. Fossiles Devonianos do Parana. *Serv. Geol. Mineral. Brasil Mon., 1:* 353 pp.

Cloud, P.E., 1942. *Terebratuloid Brachiopoda of the Silurian and Devonian. Geol. Soc. Am., Spec. Pap.*, 38: 182 pp.

Cloud, P.E., Jr., 1948. *Dicaelosia* versus *Bilobites. J. Paleontol.*, 22: 373—374.

Cloud, P.E., Jr., 1962. Environment of calcium carbonate deposition west of Andros Island, Bahamas. *U.S. Geol. Surv., Prof. Pap.*, 350: 138 pp.

Cloud, P.E., Jr. and Barnes, V.E., 1948. *The Ellenburger Group of central Texas.* Univ. Texas, Austin, Texas, 473 pp.

Cocks, L.R.M., 1972. The origin of the Silurian *Clarkeia* fauna of South America, and its extension to West Africa. *Palaeontology*, 15: 623—630.

Comte, P., 1938, Brachiopodes devoniens des gisements de Ferrones (Asturies) et de Sabero (Leon). *Ann. Paleontol.*, 27: 6—88.

Cooper, G.A. and Phelan, T., 1966. *Stringocephalus* in the Devonian of Indiana. *Smithsonian Misc. Coll.*, 151:1—20.

Copper, P., 1967. Adaptions and life habits of Devonian atrypid brachiopods. *Paleogeogr. Palaeoclimatol. Palaeoecol.*, 3: 363—379.

Cordoba, D.A., 1964. *Map of the Geology of Apizolaya Quadrangle (east half), Northern Zacatecas, Mexico.* Thesis, Univ. of Texas, Austin (unpublished).

Cowen, R., Gertman, R. and Wiggett, G., 1973. Camouflage patterns in *Nautilus*, and their implications for cephalopod paleobiology. *Lethaia*, 6: 201—213.

Craig, G.Y., 1967. Size frequency distributions of living and dead populations of pelecypods from Bimini, Bahamas, B.W.I. *J. Geol.*, 75: 34—35.

Deffeyes, K.S., Lucia, F.J. and Weyl, P.K., 1965. Dolomitization of Recent and Plio-Pleistocene sediments by marine evaporite waters on Bonaire, Netherlands Antilles. In: *Dolomitization and Limestone Diagenesis — Soc. Econ. Petrol. Mineral., Spec. Pap.*, 13: 71-88.

Dobzhansky, T., 1970. Genetics and the Evolutionary Process. Columbia University Press, 505 p.

Edwards, D., 1973. Devonian floras. In: A Hallam (Editor), *Atlas of Palaeobiogeography.* Elsevier, Amsterdam, pp.105—116.

Ehlers, G.M. and Kesling, R.V., 1970. Devonian strata of Alpena and Presque Isle counties, Michigan. *Guidebook Field Trips, Geol. Soc. Am., North-Central Sect.*, 130 pp.

Ekdale, A.A., 1973. Relation of invertebrate death assemblages to living benthic communities in Recent carbonate sediments along eastern Yucatan coast. *Bull. Am. Assoc. Petroleum Geologists*, 57 (4): 777.

Eldredge, N. and Gould, S.J., 1972. Punctuated equilibria: an alternative to phyletic gradualism. In: *Models in Paleontology*. Freeman, Cooper, New York, N.Y., pp.82—115.

Fischer, A.G., 1960. Latitudinal variations in organic diversity. *Evolution*, 14: 50-73.

Fisher, R.A., Corbett, A.S. and Williams, C.B., 1943. The relationship between the number of species and the number of individuals in a random sample of an animal population. *J. Animal Ecol.*, 12: 42—58.

Fisher, W.L. and Rodda, P.U., 1967. Stratigraphy and genesis of dolomite, Edwards Formation (Lower Cretaceous) of Texas. *Proc. 3rd Forum Geol. Ind. Minerals, Spec. Distr. Publ., 34, State Geol. Surv. Kansas*, 52—75.

Font-Altaba, M., and Closas, J., 1960. A bauxite deposit in the Paleozoic of Leon Spain. *Econ. Geol.*, 55: 1285—1290.

Forney, G. and Boucot, A.J., 1973. Silurian-Early Devonian gastropod biogeography (in preparation).

Friedman, G.M. and Sanders, J.E., 1967. Origin and occurrence of dolostones. In: G.V. Chilingar, H.J. Bissell and R.W. Fairbridge (Editors), *Carbonate Rocks (Developments in Sedimentology, 9A* Elsevier, Amsterdam, pp.267—348.

Friend, P.F. and House, M.R., 1964. The Devonian Period. In: W.B. Harland (Editor), *The Phanerozoic Time Scale — Q.J. Geol. Soc. Lond.*, 120S: 233—236.

Fuchs, G., 1971. Faunengemeinschaften und Fazieszonen im Unterdevon der Osteifel als Schlüssel zur Paläogeographie. *Notizbl. Hess. Landesamt Bodenforsch.*, 99: 78—105.

Fullagar, P.D. and Bottino, M.L., 1970. Rb-Sr whole-rock ages of Silurian-Devonian volcanics from eastern Maine. *Maine Geol. Surv. Bull.*, 23: 49—52.

Gabrielse, H., 1963. McDame map-area, Cassiar District, British Columbia. *Geol. Surv. Can., Mem.*, 319: 138 pp.

Garrels, R.M. and Mackenzie, R.T., 1971. *Evolution of Sedimentary Rocks*. Norton, 397 pp.

Gibson, T.G. and Buzas, M.A., 1973. Species diversity: patterns in modern and Miocene Foraminifera of the eastern margin of North America. *Bull. Geol. Soc. Am.*, 84: 217—238.

Gill, E.D., 1973. Application of recent hypotheses to changes of sea level in Bass Strait, Australia. *R. Soc. Vict. Proc. Bass Strait Symp.*, 85: 117—124.

Gooch, J.L. and Schopf, T.J.M., 1973. Variability in the deep sea: relation to environmental variability. *Evolution*, 26: 545—552.

Gray, J. and Boucot, A.J., 1971. Inverse acritarch-chitinozoan-scolecodont and trilete spore abundance relationships: a shoreline finding guide for the Silurian. *Geol. Soc. Am., Abstr. Annual Meet., Cordilleran Sect.*, 127—128.

Gray, J. and Boucot, A.J., 1972. Palynological evidence bearing on the Ordovician Silurian paraconformity in Ohio. *Geol. Soc. Am., Bull.*, 83: 1299—1314.

Gray, J. and Boucot, A.J., 1973. Stratigraphy and paleoecology of pre-Devonian "land" and "land vascular" plant remains (in preparation).

Greiner, H., 1973. Ordovician-Silurian stratigraphy and contact relations in northern New Brunswick. *Geol. Soc. Am., Abstr. Program NE Sect. 8th Meet.*, 170.

Hallam, A., 1972. Diversity and density characteristics of Pliensbachian-Toarcian molluscan and brachiopod faunas of the North Atlantic margins. *Lethaia*, 5: 389—412.

Hallam, A. (Editor), 1973. *Atlas of Palaeobiogeography*. Elsevier, Amsterdam, 531 pp.

Hamada, T., 1971. Early Devonian brachiopods from the Lesser Khingan District of northeast China. *Paleontol. Soc. Japan, Spec. Pap.*, 15:98 pp.

Harder, E.C., 1949. Stratigraphy and origin of bauxite deposits. *Bull. Geol. Soc. Am.*, 60: 887—908.

Harland, W.B. (Editor), 1964. The Phanerozoic time scale. *Q. J. Geol. Soc. Lond.*, 120S:458 pp.

Harper, C.W., Boucot, A.J. and Johnson, J.G., 1973. Evolution of the Stropheodontidae (in preparation).

Harris, L.D., 1973. Dolomitization model for Upper Cambrian and Lower Ordovician carbonate rocks in the eastern United States. *J. Res. U.S. Geol. Surv.*, 1: 63—78.

Hedgpeth, J.W., 1957. Classification of marine environments. *Geol. Soc. Am., Mem.*, 67 (1): 17—27.

Hessler, R.R. and Sanders, H.L., 1967. Faunal diversity in the deep sea. *Deep-Sea Res.*, 14: 65—78.

Holland, C.H., 1971. Silurian faunal provinces? In: F.A. Middlemiss, P.F. Rawson and G. Newall (Editors), *Faunal Provinces in Space and Time—Geol. J., Spec. Issue*, 4: 61—76.

House, M.R., 1971. Devonian faunal distributions. In: F.A. Middlemiss, P.F. Rawson and G. Newall (Editors), *Faunal Provinces in Space and Time—Geol. J., Spec. Issue*, 4: 77—94.

Hsu, K.J. and Siegenthaler, C., 1969. Preliminary experiments on hydrodynamic movement induced by evaporation and their bearing on the dolomite problem. *Sedimentology*, 12: 11—25.

Hume, G.S., 1954. The Lower Mackenzie River area, Northwest Territories and Yukon. *Geol. Surv. Can., Mem.*, 273:118 pp.

Hutt, J.E., Berry, W.B.N. and Rickards, R.B., 1972. Some major elements in the evolution of Silurian and Devonian graptoloids. *Proc. 24th Int. Geol. Congr., Sect. 7, Paleontol.*, 163—173.

Ingels, J.C., 1963. Geometry, paleontology, and petrography of Thornton Reef Complex, Silurian of northeastern Illinois. *Bull. Am. Assoc. Petroleum Geologists*, 47: 405—440.

Isaacson, P.E., 1974. First South American occurrence of *Globithyris:* its ecological and age significance in the Malvinokaffric realm. *Paleo*, 48: 778—784.

Isaacson, P.E., 1973. A Devonian marine and non-marine clastic wedge in the Central Andes: implications for a western land source (in preparation).

Jackson, J.B.C., 1972. The ecology of the molluscs of *Thalassia* communities, Jamaica, West Indies, 2. Molluscan population variability along an environmental stress gradient. *Mar. Biol.*, 14: 304—337.

Johnson, J.G., 1970a. Great Basin Lower Devonian brachiopods. *Geol. Soc. Am. Mem.*, 121:421 pp.

Johnson, J.G., 1970b. Taghanic onlap and the end of North American Devonian provinciality. *Bull. Geol. Soc. Am.*, 81: 2077—2106.

Johnson, J.G., 1970c. Early Middle Devonian Brachiopods from Central Nevada. *J. Paleontol.*, 44: 252—264.

Johnson, J.G., 1971a. A quantitative approach to faunal province analysis. *Am. J. Sci.*, 270: 257—280.

Johnson, J.G., 1971b. Lower Givetian brachiopods from Central Nevada. *J. Paleontol.*, 45: 301—326.

Johnson, J.G. and Boucot, A.J., 1973. Devonian brachiopod zoogeography. In: A. Hallam (Editor), *Atlas of Palaeobiogeography*. Elsevier, Amsterdam, pp. 89—96.

Johnson, J.G. and Flory, R.A., 1972. A Rasenriff fauna from the Middle Devonian of Nevada. *J. Paleontol.*, 46: 892—899.

Johnson, J.G., Murphy, M.A. and Boucot, A.J., 1968. Lower Devonian faunal succession in central Nevada. *Int. Symp. Devon. System, Alberta Soc. Petrologists Geologists*, 2: 679—691.

Kaljo, D., 1972. Facies control of the faunal distribution in the Silurian of the Eastern Baltic Region. *Int. Geol. Congr., 24th, Sect. 7, Paleontol.*, 544—548.

Kaplun, L.I., 1961. Brakhiopodi Nizhnego Devona Severnogo Pribalkhashaya. In: *Materiali po Geologii i Poleznim Iskopaemim Kazakhstana, I (26), Stratigr. Paleontol.*, pp. 64—114.

Kauffman, E.G., 1970. Population systematics, radiometrics and zonation—a new biostratigraphy. *Proc. North Am. Paleontol. Conv.*, F: 612—666.

Kauffman, E.G., 1972. Evolutionary rates and patterns of North American Cretaceous mollusca. *Proc. 24th Int. Geol. Congr. Sect. 7, Paleontol.*, 174—189.

Kauffman, E.G., 1973. Cretaceous bivalvia. In: A. Hallam (Editor), *Atlas of Paleobiogeography*. Elsevier, Amsterdam, pp. 353—384.

Keen, M.C., 1972. Evolutionary patterns of Tertiary ostracods and their use in defining stage and epoch boundaries in western Europe. *Int. Geol. Congr., 24th, Sect. 7, Paleontol.*, 190—197.

Kegel, W., 1953. Contribuicao para o estudo do Devoniano da Bacia do Parnaiba. *Div. Geol., Mineral., Serv. Graf. Inst. Brasil. Geogr. Estatistica, Bol.* 141:48 pp.

Keller, G.R. and Cebull, S.E., 1973. Plate tectonics and the Ouachita System in Texas, Oklahoma, and Arkansas. *Bull. Geol. Soc. Am.*, 83: 1659—1666.

Khudoley, K.M. and Meyerhoff, A.A., 1972. Paleogeography and geologic history of Greater Antilles. *Geol. Soc. Am., Mem.*, 129:199 pp.

Kjellesvig-Waering, E.N., 1958. The genera, species and subspecies of the Family Eurypteridae, Burmeister, 1845. *J. Paleontol.*, 32: 1107—1148.

Kjellesvig-Waering, E.N., 1961. The Silurian Eurypterida of the Welsh borderland. *J. Paleontol.*, 35 (4): 789—835.

Kjellesvig-Waering, E.N., 1973. A new Silurian *Slimonia* (Eurypterida) from Bolivia. *J. Paleontol.*, 47: 549—550.

Klein, G. de V., 1972. Sedimentary model for determining paleotidal range: reply. *Bull. Geol. Soc. Am.*, 83: 539—546.

Kohn, A.J., 1959. The ecology of *Conus* in Hawaii. *Ecol. Monogr.*, 29: 47—90.

Kummel, B., 1973. Lower Triassic (Scythian) molluscs. In: A. Hallam (Editor), *Atlas of Palaeobiogeography*. Elsevier, Amsterdam, pp. 225—234.

Ladd, H.S. Tracey, J.I., Jr. and Gross, G.M., 1970. Deep drilling on Midway atoll. *U.S. Geol. Surv., Prof. Pap.*, 680-A: A1—A22.

Lanphere, M.A., 1967. Age of primary metamorphism of the Abrams mica schist, Klamath Mountains, California. *Geol. Soc. Am., Spec. Pap.*, 101:118.

Laporte, L.F., 1963. Algae of the Manlius Limestone. In: *Stratigraphy, Facies Changes and Paleoecology of the Lower Devonian Helderberg Limestone and the Middle Devonian Onondaga Limestones—Geol. Soc. Am., Guidebook, Field Trip*, 1:9.

Lecompte, M., 1958. Les récifs paléoziques en Belgique. *Geol. Rundsch.*, 47: 384—401.

Lecompte, M., 1961. Facies marine et stratigraphie dans le Dévonien de la Belgique. *Ann. Soc. Géol. Belg.*, 85: B17—B57.

Lecompte, M., 1968. Le Devonien de la Belgique et le nord de la France. *Int. Symp. Devon. Syst., Alberta Soc. Petrol. Geol.*, 1: 15—52.

Lesperance, P.J. and Bourque, P.A., 1971. The Synphoriinae: an evolutionary pattern of Lower and Middle Devonian trilobites. *J. Paleontol.*, 45: 182—208.

Lever, J., 1958. Quantitative beach research, 1. The left—right phenomenon: sorting of lamellibranch valves on sandy beaches. *Basteria*, 22: 21—51.

Levinton, J.S., 1972. Genetic polymorphism and environmental heterogeneity: some paleontological possibilities. *Abstr. Annual Meet. Geol. Soc. Am.*, 577—578.

Levinton, J., 1973. Genetic variation in a gradient of environmental variability: marine bivalvia (Mollusca): *Science*, 180 (4081): 75—76.

Lochman-Balk, C., 1971. The Cambrian of the craton of the United States. In: *Cambrian of the World, 1.* Wiley-Interscience, New York, N.Y., pp. 79—167.

Lowenstam, H.A., 1950. Niagaran reefs of the Great Lakes area. *J. Geol.*, 58: 430—487.

Ludvigsen, R., 1972. Late Early Devonian dacryoconarid tentaculites, northern Yukon Territory. *Can. J. Earth Sci.*, 9: 297—318.

MacArthur, R.H. and Wilson, E.O., 1963. An equilibrium theory of insular zoogeography. *Evolution*, 17: 373—387.

Maksimova, Z.A., Modzalevskaya, E.A., Kaplun, L.I. and Senkevitch, M.A., 1972. Nizhnii Devon tikhookeanskoi paleobiogeografischeskoi oblasti na Territorii S.S.R.R. *Sov. Geol.*, 3: 27—43.

Mamet, B.L., 1971. De l'emploi des Foraminifères pour une zonation du Carbonifère inférieur en Europe Occidentale. *Int. Kongr. Stratigr. Geol. Carbon, 7., Krefeld, Zusammenfass. Vorträge Veröffentl.*, p. 108.

Manten, A.A., 1971. *Silurian Reefs of Gotland (Developments in Sedimentology, 13).* Elsevier, Amsterdam, 539 pp.

Margalef, R., 1963. On certain unifying principles in ecology. *Am. Naturalist*, 97: 357—374.

Martin-Kaye, P., 1951. Sorting of lamellibranch valves on beaches in Trinidad. *Geol. Mag.*, 88: 432—434

Massa, D. and Jaeger, H., 1971. Données stratigraphiques sur le Silurien de l'ouest de la Libye. *Proc. Brest Ordovician-Silurian Symp.*, pp. 313—321.

Mayr, E., 1963. *Animal Species and Evolution.* Harvard Univ. Press, Cambridge, Mass., 797 pp.

Mayr, E., 1965. Numerical phenetics and taxonomic theory. *Syst. Zool.* 14:73—97.

Mayr, E., 1970. *Populations, Species, and Evolution.* Harvard University Press, Cambridge, Mass., 453 pp.

McBride, E.F., (and co-workers), 1969. *A Guidebook to the Stratigraphy, Sedimentary Structures and Origin of the Flysch and Pre-Flysch Rocks of the Marathon Basin, Texas. (Guidebook Ann. Meet. A.A.P.G. and S.E.P.M.).* Dallas Geol. Soc., Dallas, Texas, 104 pp.

McDougall, I., Compston, W. and Bofinger, V.M., 1966. Isotopic age determinations on Upper Devonian rocks from Victoria, Australia: a revised estimate for the age of the Devonian-Carboniferous boundary. *Bull. Geol. Soc. Am.*, 77:1075—1088.

McGregor, D.C., Sanford, B.V. and Norris, A.W., 1970. Palynology and correlation of Devonian formations in the Moose River basin, northern Ontario. *Geol. Assoc. Can., Proc.*, 22:45—54.

McKerrow, W.S. and Ziegler, A.M., 1972. Paleozoic oceans. *Nature Phys. Sci.*, 240:92—94.

McLaren, D.J., 1970. Presidential address: time, life and boundaries. *J. Paleontol.*, 44:801—815.

McLaughlin, R.E., 1970. Palynology of core samples of Paleozoic sediments from beneath the coastal plain of Early County, Georgia. *Geol. Surv. Georgia, Inf. Circ.*, 40:27 pp.

Meyerhoff, A.A., 1970a. Continental drift: implications of paleomagnetic studies, meteorology, physical oceanography, and climatology. *J. Geol.*, 78:1—51.

Meyerhoff, A.A., 1970b. Continental drift, 2: high-latitude evaporite deposits and geologic history of Arctic and North Atlantic Oceans. *J. Geol.*, 78:406—444.

Modzalevskaya, E.A., 1968. Biostratigraphic subdivision of the Devonian in the Far East and Transbaikal region, U.S.S.R.. *Proc. Int. Symp. Devon. System, Calgary*, 1967, 1:543—556

Modzalevskaya, E.A., 1969. *Polevoy Atlas Siluriiskoy, Devoniskoy, i Rannekamenu-golnoy fauni Dal'nego Vostoka.* Geol. S.S.S.R., VSEGEI, Dal'nevostochnoye Geol. Upravleniye, 327 pp.

Morales, P.A., 1965. A contribution to the knowledge of the Devonian faunas of Colombia. *Bol. Geol., Univ. Ind. Santander*, 19:51—111.

Murray, J.W., 1968. Living foraminifers of lagoons and estuaries. *Micropaleontology*, 14:435—455.

Neuman, R.B., 1972. Brachiopods of Early Ordovician volcanic islands: *Proc. 24th Int. Geol. Congr., Sect. 7, Paleontol.*, pp. 297—302.

Newell, N.D., 1956. Catastrophism and the fossil record. *Evolution*, 10:97—101.

Newell, N.D., 1967. Revolutions in the history of life. In: C.C. Albritton, Jr. (Editor), *Uniformity and Simplicity — A Symposium on the Principle of the Uniformity of Nature. Geol. Soc. Am., Spec. Pap.*, 89: 63—91.

Newell, N.D., 1971. An outline history of tropical organic reefs. *Am. Mus. Novitates*, 2465:37 pp.

Nicoll, R.S. and Rexroad, C.B., 1968. Stratigraphy and conodont paleontology of the Salamonie Dolomite and Lee Creek Member of the Brassfield Limestone (Silurian) in southeastern Indiana and adjacent Kentucky. *Bull. Indiana Geol. Surv.*, 40:1—73.

Nikiforova, I.I. and Sapelnikov, V.P., 1971. Noviye rannesiluriiskiye Virgianidae (brachiopods): *Pal. Zh.*, 1971:47—56.

Noble, J.P.A. and Ferguson, R.D., 1971. Facies and faunal relations at edge of early mid-Devonian carbonate shelf, south Nahanni River area, N.W.T. *Bull. Can. Petrol. Geol.*, 19:570—588.

Norris, A.W. and Uyeno, T.T., 1971. Stratigraphy and Conodont faunas of Devonian Outcrop Belts, Manitoba. *Geol. Assoc. Can., Spec. Pap.*, 9:209—223.

Oehlert, D.P., 1889. Sur le Devonien des environs d'Angers. *Bull. Soc. Géol. Fr., Sér.*, 3, 17:742—791.

Oehlert, D.P. and Davoust, A., 1879. Sur le Devonien du department de la Sarthe. *Bull. Soc. Géol. Fr., Ser. 3*, 7:697—717.

Oliver, W.A., Jr., 1968. Succession of rugose coral faunas in the Lower and Middle Devonian of eastern North America. *Int. Symp. Devon. System, Calgary, Alta. Soc. Petrol. Geol.*, 2:733—744.

Oliver, W.A., Jr., 1973a. *Devonian Coral Endemism in Eastern North America and its Bearing on Paleogeography (Special Papers in Palaeontology, 12)* Paleontol. Assoc., pp. 318—319.

Oliver, W.A., Jr., 1973b. Endemism and evolution of Late Silurian to Middle Devonian rugose corals in eastern North America (In press).

Ormiston, A.R., 1968. Lower Devonian trilobites of Hercynian type from the Turkey Creek Inlier, Marshall County, south-central Oklahoma. *J. Paleontol.*, 42:1186—1199.

Ormiston, A.R., 1972. Lower and Middle Devonian trilobite zoogeography in northern North America. *Int. Geol. Congr., 24th, Sect. 7, Paleontol.*, 594—604.

Paine, R.T., 1971. A short-term experimental investigation of resource partitioning in a New Zealand rock-intertidal habitat. *Ecology*, 52:1096—1106.

Palmer, A.R., 1965. Biomere—a new kind of stratigraphic unit. *J. Paleontol.*, 39:149—152.

Palmer, A.R., 1972. Problems of Cambrian biogeography. *Int. Geol. Congr., 24th, Sect. 7, Paleontol.*, 310—315.

Palmer, A.R., 1973. Cambrian trilobites. In: A. Hallam (Editor), *Atlas of Paleobiogeography*. Elsevier, Amsterdam, pp. 3—12.

Pantoja-Alor, J., and Robison, R.A., 1967. Paleozoic sedimentary rocks in Oaxaca, Mexico. *Science*, 57:1033—1035.

Pitrat, C., 1965. Spiriferidina. In: R.C. Moore (Editor), *Treatise on Invertebrate Paleontology, Part H*, pp. H667—H728.

Plumstead, E.P., 1969. Three-thousand million years of plant life in Africa. *Geol. Soc. S. Afr.* (Annexure to V. LXXII):72 pp.

Poole, F.G. and Hayes, P.T., 1971. Depositional framework of some Paleozoic strata in northwest Mexico and southwestern United States. *Ann. Meet. Geol. Soc. Am., 67th, Cordill. Sect., Abstr.*, 3(2):179.

Potter, A.W. and Boucot, A.J., 1971. Ashgillian, Late Ordovician branchiopods from the eastern Klamath Mountains, California. *Geol. Soc. Am., Abstr. Progr.* 3/2:180—181.

Raup, D.M. and Stanley, S.M., 1971. *Principles of Paleontology*. Freeman, New York, N.Y., 388 pp.

Rickard, L.V., 1963. The Helderberg Group. In: *Stratigraphy, Facies Changes and Paleoecology of the Lower Devonian Helderberg Limestones and the Middle Devonian Onondaga Limestones — Geol. Soc. Am., Guidebook, Field Trip.*, 1:1—8.

Robison, R.A., 1972. Mode of life of agnostid trilobites. *Int. Geol. Congr., 24th, Sect. 7, Paleontol.*, 33—40.

Rowe, G.T., 1971. Benthic biomass and surface productivity. In: J.D. Costlow, Jr. (Editor), *Fertility of the Sea*. Gordon and Breach, pp. 441—454.

Rubel, M., 1970. On the distribution of brachiopods in the lowermost Llandovery of Estonia. *Eesti NVS Teaduste Akad. Toimetised, 19 (Keemia, Geol.)* 69—79.

Sanders, H.L., 1969. Benthic marine diversity and the stability-time hypothesis. In: *Diversity and Stability in Ecological Systems*. Biol. Dept., Brookhaven Natl. Lab., Brookhaven, pp. 71—81.

Sanders, H.L. and Hessler, R.R., 1969. Ecology of the deep-sea benthos. *Science*, 163:1419—1424.

Savage, N.M., 1968. *Australirhynchia*, a new rhynchonellid brachiopod from the Lower Devonian of New South Wales. *Palaeontology*, 11:731—735.

Savage, N.M., 1969. New spiriferid brachiopods from the Lower Devonian of New South Wales. *Palaeontology*, 12:472—487.

Savage, N.M., 1970. New atrypid brachiopods from the Lower Devonian of New South Wales. *J. Paleontol.*, 44(4):655—668.

Savage, N.M., 1971. Brachiopods from the Lower Devonian Mandagery Park Formation, New South Wales. *Palaeontology*, 14:387—427.

Scott, G., 1940. Paleoecological factors controlling the distribution and mode of life of Cretaceous ammonoids in the Texas area. *J. Paleontol.*, 14:299—323.

Scupin, H., 1900. Die Spiriferen Deutschlands. *Paläontol. Abh., N.F.*, 4:1—140.

Seddon, G. and Sweet, W.C., 1971. An ecologic model for conodonts. *J. Paleontol.*, 45:869—880.

Seilacher, A., 1967. Bathymetry of trace fossils. *Mar. Geol.* 5:413—428.

Shaw, A.B., 1964. *Time in Stratigraphy*. McGraw-Hill, New York, N.Y., 364 pp.

Sheehan, P.M., 1973a. The relation of Late Ordovician glaciation to the Ordovician-Silurian changeover in North American brachiopod faunas. *Lethaia*, 6:147—154.

Sheehan, P.M., 1973b. Brachiopods from the Jerrestad Mudstone (Early Ashgillian, Ordovician) from a boring in southern Sweden. *Geol. Palaeontol.*, 7: 59—76.

Simpson, G.G., 1953. *The Major Features of Evolution*, Columbia Univ. Press, 434 pp.

Smith, D.W.G., Baadsgaard, H., Folinsbee, R.E. and Lipson, J., 1961. K/Ar age of Lower Devonian bentonites of Gaspe, Quebec. *Bull. Geol. Soc. Am.*, 72:171—174.

Smith, Roy, 1972. *Taxonomy and Ecology of Atrypella spp. and the Atrypella Community*. M.S. thesis, Oregon State Univ. (unpublished).

Solle, G., 1952. Neue Ergebnisse paläontologischer Arbeitstechnik. *Paläontol. Z.*, 26:255—264.

Sougy, J. and Lecorche, J.P., 1963. Sur le nature glacière de la base de la série de Garat el Hammoueid (Zemmour, Mauritanie septentrionale): *Compt. Rend. Acad. Sci.*, 256:4471—4474.

Speden, I.G., 1966. Paleoecology and the study of fossil benthic assemblages and communities. *N.Z.J. Geol. Geophys.*, 9:408—423.

Stanton, R.J. and Evans, I., 1972. Community structure and sampling requirements in paleoecology. *J. Paleontol.*, 46:845—858.

Stehli, F.G. and Wells, J.W., 1971. Diversity and age patterns in hermatypic corals. *Syst. Zool.*, 20:115—126.

Stehli, F.G., Douglas, R.G. and Newell, N.D., 1969. Generation and maintenance of gradients in taxonomic diversity. *Science*, 164:947—949.

Struve, W., 1963. Das Korallen-Meer der Eifel vor 300 Millionen Jahren — Funde, Deutungen, Probleme. *Natur Museum*, 93:237—276.

Summerson, C.H., 1959. Evidence of weathering at the Silurian-Devonian contact in central Ohio. *J. Sediment. Petrol.*, 29:425—429.

Talent, J.A., Campbell, K.S.W., Davoren, P.J., Pickett, J.W. and Telford, P.G., 1972. Provincialism and Australian Early Devonian faunas. *J. Geol. Soc. Austr.* 19:81—97.

Thorsteinsson, R., 1968. Geology of the arctic Archipelago. In: R.J.W. Douglas (Editor), *Geology and Economic Minerals of Canada* — *Geol. Surv. Can., Econ. Geol. Rep.*, 1:547—590.

Trettin, H., 1965. Middle Ordovician to Middle Silurian carbonate cycle, Brodeur Peninsula, northwestern Baffin Island. *Can. Petrol. Geol. Bull.*, 13:155—180.

Trettin, H.P., 1971. Preliminary notes on Lower Paleozoic geology, Foxe Basin, northeastern Melville Peninsula, and parts of northern and central Baffin Island. *Geol. Surv. Can.* (Open File Rep.), 151 pp.

Urey, H.C., 1973. Cometary collisions and Geological Periods. *Nature*, 242 (March 2):32—33.

Valentine, J.W., 1967. The influence of climatic fluctuations on species diversity within the Tethyan Provincial System. *Syst. Assoc. Publ.*, 7:153—166.

Valentine, J.W., 1968. Climatic regulation of species diversification and extinction. *Bull. Geol. Soc. Am.*, 79:273—276.

Valentine, J.W., 1969. Patterns of taxonomic and ecological structure of the shelf benthos during Phanerozoic time. *Palaeontology*, 12:684—709.

Valentine, J.W., 1971a. Resource supply and species diversity patterns. *Lethaea*, 4:51—61.

Valentine, J.W., 1971b. Plate tectonics and shallow-marine diversity and endemism, an actualistic model. *Syst. Zool.*, 20:253—264.

Valentine, J.W., Ayala, F.J., Zumwalt, G.S. and Hedgecock, D., 1973. Mass extinctions and genetic polymorphism in *Tridacna. Abstr. Programs 1973, Cordilleran Sect., Geol. Soc. Am.*, 116—117.

Vandercammen, A., 1963. Spiriferidae du Devonien de la Belgique. *Inst. R. Sci. Nat. Belg., Mém.*, 150:1—177.

Verneuil, P.E.P. de, 1869. Appendix à la faune dévonienne du Bosphore. In: *Tchihatcheff, Asie Mineure — Paléontologie 1866*.

Villemur, J.R. and Drot, J., 1957. Contribution à la faune devonienne du Bassin de Taoudeni. *Bull. Soc. Géol. Fr., Sér. 6*, 7:1077—1082.

Von Arx, W.S., 1962. *An Introduction to Physical Oceanography*. Addison-Wesley, New York, N.Y., 422 pp.

Wallace, P., 1972. Populations and paleo-environments in the Devonian of the Cantabrian Cordillera, north Spain. *Int. Geol. Congr., 24th, Sect. 7, Paleontol.*, 121—129.

Warme, J.E., 1969. Live and dead molluscs in a coastal lagoon: *J. Paleontol.*, 43:141—150.

Watkins, R., 1973. Carboniferous faunal associations and stratigraphy, Shasta County, northern California. *Bull. Am. Assoc. Petroleum Geologists*, 57:1743—1764.

Watkins, R., and Boucot, A.J., 1973. Evolution of Silurian brachiopod communities along the southeastern coast of Acadia (in preparation).

Watkins, R., Berry, W.B.N. and Boucot, A.J., 1973. Why "Communities"? *Geology* 1:55—58.

Weisbord, N.E., 1926. Venezuelan Devonian fossils. *Bull. Am. Paleontol.*, 11:34 pp.

Wetherill, G.W., 1965. Geochronology of North America. *Nat. Acad. Sci. Nat. Res. Council, Publ.* 1276:315 pp.

Whittaker, R.H., 1972. Evolution and measurement of species diversity. *Taxon*, 21(2/3):213—251.

Williams, A., 1957. Evolutionary rates of brachiopods. *Geol. Mag.*, 94:201—211.

Wilson, J.L., 1954. Late Cambrian and Early Ordovician trilobites from the Marathon Uplift, Texas. *J. Paleontol.*, 28:249—285.

Winter, J., 1971. Brachiopoden-Morphologie und Biotop—ein Vergleich quantitativer Brachiopoden-Spektren aus Ahrdorf-Schichten (Eifelium) der Eifel. *N. Jb. Geol. Paläontol., Monatsh.*, 2:102—132.

Winterer, E.L. and Murphy, M.A., 1960 Silurian reef complex and associated facies, central Nevada. *J. Geol.*, 68:117—139.

Wolfart, R. and Voges, A., 1968. Beiträge zur Kenntnis des Devons von Bolivien. *Beih. Geol. Jb.*, 74:241 pp.

Zenger, D.H., 1972. Significance of supratidal dolomitization in the Geologic Record. *Bull. Geol. Soc. Am.*, 83:1—12.

Ziegler, A.M., 1965. Silurian marine communities and their environmental significance. *Nature*, 207:270—272.

Ziegler, A.M., Boucot, A.J. and Sheldon, R.P., 1966. Silurian pentameroid brachiopods preserved in position of growth. *J. Paleontol.*, 40:1032—1036.

Ziegler, A.M., Cocks, L.R.M. and Bambach, R.K., 1968. The composition and structure of Lower Silurian marine communities. *Lethaia*, 1:1—27.

REFERENCES ADDED IN PROOF

Alberstadt, L.P. and Walker, K.R., 1973. Stages of ecological succession in Lower Paleozoic reefs of North America. *Geol. Soc. Am., Abstr. Programs*, 5 (7): 530—532.

Anderson, E.J. and Makurath, J.H., 1973. Paleoecology of Appalachian gypidulid brachiopods. *Palaeontology*, 16: 381—390.

Arkhangelsky, A.D., 1937. Bauxite deposits confined to the Paleozoic. *Trans. All-Union Sci. Res. Inst. Econ. Mineral.*, 112, (2): 115 pp.

Arnold, S.J., 1972. Species densities of predators and their prey. *Am. Naturalist*, 106: 220—236.

Ayala, F.J., Hedgecock, D., Zumwalt, G.S. and Valentine, J.W., 1973. Genetic variation in *Tridacna maximus*, an ecological analog of some unsuccessful evolutionary
lineages. *Evolution*, 27: 177—191.

Bader, R.G., 1954. The role of organic matter in determining the distribution of
pelecypods in marine sediments. *J. Mar. Res.*, 13: 32—47.

Barnes, C.R., Rexroad, C.B. and Miller, J.F., 1973. Lower Palcozoic conodont provincialism. *Geol. Soc. Am., Spec. Pap.*, 141: 157—190.

Boss, K.J., 1971. Critical estimate of the number of Recent Mollusca. *Occas. Pap.
Mollusks*, 3: 81—135.

Boucot, A.J., 1973. *Glypterina*, new genus, the Ordovician Ptychopleurellid; two new
occurrences. *J. Paleontol.*, 47: 136—137.

Boucot, A.J., 1974. Redefinition of *Australostrophia* (Brachiopoda, Devonian). *J.
Paleontol.*, in press

Bowen, J.M., 1972. Estratigrafia del pre-Cretaceo en la parte norte de la sierra de
Perija. *Bol. Geol., Min. Minas Hidrocarburos, Dir. Geol., Publ. Espec., 5, Mem.
Cuarto Congreso Geol. Venez.*, 2: 729—761.

Bradley, J., 1973. *Zoophycos* and *Umbellula* (Pennatulacea): their synthesis and identity. *Palaeogeogr., Palaeoclimatol. Palaeoecol.*, 13: 103—128.

Bramlette, M.N., 1965. Massive extinctions in biota at the end of Mesozoic time.
Science, 148: 1696—1699.

Brander, K.M., McLeod, A.A.Q.R. and Humphreys, W.F., 1971. Comparison of species
diversity and ecology of reef-living invertebrates on Aldabra atoll and at Watamu,
Kenya. In: D.R. Stoddart and Sir M. Yonge (Editors), *Regional Variation in Indian
Ocean Coral Reefs. Symp. Zool. Soc. Lond.*, 28: 397—432.

Brenchley, P.J. and Pickerill, R.K., 1973. Recognition of Caradoc Communities. *Paleontol., Assoc. Circ.*, 73a: 1.

Bretsky, P.W., 1973. A reflection on genetics, extinction and the "Killer Clam".
Geology, 1: 157.

Brodkorb, P., 1971. Origin and evolution of birds. In: D.S. Farner, J.R. King and K.C.
Parkes (Editors), *Avian Biology*. Academic Press, London, pp. 19—55.

Coates, A.G., 1973. Cretaceous Tethyan coral-rudist biogeography related to the evolution of the Atlantic ocean. In: *Organisms and Continents Through Time — Special
Papers in Palaeontology, 12*. Paleontol. Assoc., pp. 169—174.

Cohen, J.E., 1970. A Markov contingency-table model for replicated Lotka-Volterra
systems near equilibrium. *Am. Naturalist*, 104: 547—560.

Connell, J.H., 1963. Territorial behavior and dispersion in some marine invertebrates.
Res. Popul. Ecol., 5: 87—101.

Cummins, W.A., 1969. Patterns of sedimentation in the Silurian rocks of Wales. In: A.
Wood (Editor), *The Precambrian and Lower Palaezoic rocks of Wales*. University of
Wales Press, pp. 219—238.

Dayton, P.K., 1971. Competition, disturbance, and community organization: the pro-
vision and subsequent utilization of space in a rocky intertidal community. *Ecol.
Monogr.*, 41: 351—389.

Dayton, P.K. and Hessler, R.R., 1972. Role of biological disturbance in maintaining
diversity in the deep sea: *Deep-Sea Res.*, 19: 199—208.

Dubatolov, V.N., 1972. Zoogeografiya Devonskikh Morei Evrazii. *Tr. Inst. Geol.
Geofiz., Sibirsk. Otdel, Akad. Nauk S.S.S.R.*, 157: 128 pp.

Dunbar, M.J., 1960. The evolution of stability in marine environments, natural selection at the level of the ecosystem. *Am. Naturalist*, 94: 129—136.

Dunbar, M.J., 1968. *Ecological Development in Polar Regions — a Study in Evolution*.
Prentice-Hall, Englewood Cliffs, N.J., 119 pp.

Dunbar, M.J., 1972. The ecosystem as unit of natural selection. In: E.S. Deevey, (Editor), *Growth by Intussusception—Trans. Conn. Acad, Arts Sci.*, 44: 113—130.

Elles, G.L., 1939. Factors controlling graptolite succession and assemblages. *Geol. Mag.*, 76: 181—187.

Freulon, J.M., 1964. Etude géologique des séries primaires du Sahara Central. *Publ. Centre Rech. Zones Arides, C.N.R.S., Sér. Géol.*, 3: 198 pp.

Futuyma, D.J., 1973. Community structure and stability in constant environments. *Am. Naturalist*, 107: 443—446.

Gibbs, R.J., 1967. Geochemistry of the Amazon River system. *Bull. Geol. Soc. Am.*, 78: 1203—1231.

Goldring, W., 1932. The Champlain Sea. *N.Y. State Museum Bull.*, 239/240: 153—194.

Hecht, A.D. and Agan, B., 1972. Diversity and age relationships in Recent and Miocene bivalves. *Syst. Zool.* 21 : 308—312.

Hessler, R.R., 1974. The structure of deep benthic communities from central oceanic waters: In: *Biology of the Oceanic Pacific*. Oregon State Univ. Press.

Hughes, N.F. (Editor), 1973. *Organisms and Continents Through Time—Special Papers in Palaeontology, 12*, Paleontol. Assoc., 334 pp.

Hutchinson, G.E., 1957. Concluding remarks. In: *Population Studies: Animal Ecology and Demography, Cold Spring Harbor Symposia on Quantitative Biology*, pp. 415—427.

Jago, J.B., 1973. Cambrian agnostid communities in Tasmania. *Lethaia*, 6: 405—422.

Jones, J.S., 1973. Ecological genetics and natural selection in molluscs. *Science*, 182: 546—552.

Jones, T.S., 1944. Dolomite porosity in Devonian of West Texas Permian basin. *Bull. Am. Assoc. Petrol. Geol.*, 28: 1043—1044.

Klein, G. de V., 1972b. Determination of paleotidal range in clastic sedimentary rocks. *Proc. Int. Geol. Congr., 24th, Sect.* 6: 397—405.

Kohn, A.J., 1958. Problems of speciation in marine invertebrates. In: A.A. Buzzati-Traverso (Editor), *Perspectives in Marine Biology*. Univ. Calif. Press, pp. 571—588.

Kutscher, F., 1970. Die Versteinerungen des Hunsruckschiefers. *Aufschluss*, 19: 87—100.

Lenz, A.C. and Pedder, A.E.H., 1972. Lower and Middle Paleozoic sediments and paleontology of Royal Creek and Peel River, Yukon, and Powell Creek, N.W.T. *Excursion A-14, 24th Int. Geol. Congr.*, 43 p.

Levinton, J.S., 1973. Genetic extinction hypothesis and its critics. *Geology*, 1: 157—158.

Lewis, J.R., 1964. *The Ecology of Rocky Shores*. English Univ. Press, 323 pp.

Livingstone, D.A., 1963. Chemical composition of rivers and lakes. *U.S. Geol. Surv., Prof. Pap.*, 440-G: 64 pp.

Lowe-McConnell, R.H., 1969. Speciation in tropical fresh-water fishes. *Biol. J. Linn. Soc.*, 1: 51—75.

Ludvigsen, R., 1970, Age and fauna of the Michelle Formation, Northern Yukon Territory. *Bull. Can. Petrol. Geol.*, 18: 407—429.

MacArthur, R.H., 1965. Patterns of species diversity. *Biol. Rev.* 40: 510—533.

MacArthur, R.H., 1969. Patterns of communities in the tropics. *Biol. J. Linn. Soc.*, 1: 19—30.

MacArthur, R.H., 1972. *Geographical Ecology*. Harper and Row, London, 269 pp.

Mamet, B., 1973. Foraminiferal zonation of the Lower Carboniferous: stratigraphic implications. *Geol. Soc. Am. Abstr. Programs*, 5 (7): 725.

Menzies, R.J., George, R.Y. and Rowe, G.T., 1973. *Abyssal Environment and Ecology of the World Oceans.* Wiley, New York, N.Y., 488 pp.

Middlemiss, F.A., Rawson, P.F. and Newall, G. (Editors), 1971. Faunal provinces in space and time. *Geol. J.,* Spec. Issue 4: 236 pp.

Miller, A.K., 1949. The last surge of the nautiloid cephalopods. *Evolution,* 3: 231—238.

Mohanti, M., 1972. The Portilla Formation (Middle Devonian) of the Alba Syncline, Cantabrian Mountains Prov. Leon, Northwestern Spain: Carbonate Facies and Rhynchonellid Palaeontology: *Leidse Geol. Meded.,* D. 48, Afl. 2: 135—205.

Murdoch, W.W., 1969. Switching in general predators: experiments on predator specificity and stability of prey populations. *Ecol. Monogr.,* 39: 335—354.

Nicol, D., 1962. The biotic development of some Niagaran reefs — an example of an ecological succession or sere. *J. Paleontol.,* 36: 172—176.

Niebuhr, W.W., II, 1973. Paleoecology of the *Eurekaspirifer pinyonensis* Zone, Eureka county, Nevada. Thesis, Oregon State University, Corvallis, Ore., (unpublished).

Ogden, J.C., Brown, R.A. and Salesky, N., 1973. Grazing by the echinoid *Diadema antillarum. Science,* 182: 715—717.

Oliver, W.A., 1973. Devonian paleobiogeography. *U.S. Geol. Surv., Prof. Pap.,* 800A: A125.

Osgood, R.G., Jr. and Szmug, E.J., 1972. The trace fossil Zoophycos as an indicator of water depth. *Bull. Am. Paleontol.,* 62 (271): 21 pp.

Paine, R.T., 1966. Food web complexity and species diversity. *Am. Naturalist,* 100: 65—76.

Pianka, E.R., 1966. Latitudinal gradients in species diversity: A review of concepts. *Am. Naturalist,* 100: 33—46.

Pickerill, R.K., 1973. *Lingulasma tenuigranulata—*palaeoecology of a large Ordovician linguloid that lived within a Strophomenid-Trilobite Community. *Palaeogeogr. Palaeoclimatol. Palaeoecol.,* 13: 143—156.

Porter, J.W., 1972. Predation by *Acanthaster* and its effect on coral species diversity. *Am. Naturalist,* 106: 487—492.

Purdy, E.G., 1964. Sediments as substrates. In: J. Imbrie and N. Newell (Editors), *Approaches to Paleoecology,* Wiley, New York, N.Y., pp. 238—271.

Raup, D.M., 1972. Taxonomic diversity during the Phanerozoic. *Science,* 177: 1065—1071.

Rensch, B., 1960. *Evolution Above the Species Level.* Columbia Univ. Press, 419 pp.

Rhodes, F.H.T., 1966. The course of evolution. *Proc. Geol. Assoc.,* 77: 1—54.

Richards, R.P., 1972. Autecology of Richmondian brachiopods (Late Ordovician of Indiana and Ohio). *J. Paleontol.,* 46: 386—405.

Ronov, A.B., 1973. Evolution of rock composition and geochemical processes in the sedimentary shell of the earth. *Sedimentology,* 19: 157—172.

Schumann, D., 1969. "Byssus"-artige Stielmuskel-Konvergenzen bei artikulaten Brachiopoden. *Neues Jhrb. Geol. Paläontol., Abh.,* 133: 199—210.

Scrutton, C.T., 1973. Palaezoic coral faunas from Venezuela, 2. Devonian and Carboniferous corals from the sierra de Perija. *Bull. Brit. Museum (Nat. Hist.), Geol.,* 23: 223—282.

Segerstrale, S.G., 1957. Baltic Sea. In: J.W. Hedgpeth (Editor), *Treatise on Marine Ecology and Paleoecology. Geol. Soc. Am., Mem.,* 67: 751—800.

Simpson, G.G., 1944. *Tempo and Mode in Evolution.* Columbia University Press, 237 pp.

Sonnenfeld, P., 1964. Dolomites and dolomitization: a review: *Bull. Can. Petrol. Geol.,* 12: 101—132.

Sougy, J., 1964. *Les Formations paleozoiques du Zemmour Noir.* Gap, Hautes-Alpes, 695 pp.

Sturmer, W. and Bergstrom, J., 1973. New discoveries on trilobites by X-rays. *Paläontol. Z.*, 47: 104—141.

Stanley, S.M., 1937a. An ecological theory for the sudden origin of multicellular life in the Late Precambrian. *Proc. Natl. Acad. Sci.*, 70: 1486—1489.

Stanley, S.M., 1973b. An explanation for Cope's rule. *Evolution*, 27: 1—26.

Stanley, S.M., 1973c. Effects of competition on rates of evolution, with special reference to bivalve mollusks and mammals. *Syst. Zool.*, 22: 486—506.

Stehli, F.G., 1971. Tethyan and Boreal Permian faunas and their significance. *Smithsonian Contrib. Paleobiol.*, 3: 337—345.

Stehli, F.G., McAlester, A.L. and Helsley, C.E., 1967. Taxonomic diversity of Recent bivalves and some implications for geology. *Bull. Geol. Soc. Am.*, 78: 455—466.

Surlyk, F., 1973. Distribution and zonation of the epifauna in a Cretaceous rocky shore community. *Palaeontol. Assoc. Circ.*, 73a: 5—6.

Taylor, J.D., 1971. Reef-associated molluscan assemblages in the western Indian Ocean. In: D.R. Stoddart and Sir M. Yonge (Editors), *Regional Variation in Indian Ocean Coral Reefs*, Academic Press, London, pp. 501—534.

Thayer, C.W., 1973. Taxonomic and environmental stability in the Paleozoic. *Science*, 182: 1242—1243.

Ulrich, A., 1892. Palaeozoische Versteinerungen aus Bolivien. In: G. Steinmann, *Beiträge zur Geologie und Palaeontologie von Südamerika*, 1, pp. 1—116.

Valentine, J.W., 1973. *Evolutionary Paleoecology of the Marine Biosphere.* Prentice-Hall, Englewood Cliffs, N.J., 511 pp.

Valentine, J.W., Hedgecock, D., Zumwalt, G.S., Ayala, F.J., 1973. Mass extinctions and genetic polymorphism in the "Killer Clam," *Tridacna. Bull. Geol. Soc. Am.*, 84: 3411—3414.

Van Valen, L., 1973. A new evolutionary law. *Evolution Theory*, 1: 1—30.

Vermeij, G.J., 1971. Substratum relationships of some tropical Pacific intertidal gastropods. *Mar. Biol.*, 10: 315—320.

Vermeij, G.J., 1972. Endemism and environment: some shore molluscs of the tropical Atlantic. *Am. Naturalist*, 106: 89—101.

Vermeij, G.J. and Porter, J.W., 1971. Some characteristics of the dominant intertidal molluscs from rocky shores in Pernambuco, Brazil. *Bull. Mar. Sci.*, 21: 440—454.

Walmsley, V.G., and Boucot, A.J., 1974. The phylogeny, taxonomy and biogeography of Silurian and Early to Mid Devonian Isorthinea (Brachiopoda). *Palaeontographica* (in press).

Warmke, G.L. and Almodovar, L.R., 1963. Some associations of marine mollusks and algae in Puerto Rico. *Malacologia*, 1: 163—177.

Westermann, G.E.G., 1973. Strength of concave septa and depth limits of fossil cephalopods. *Lethaia*, 6: 373—404.

Wiedmann, J., 1973. Evolution or revolution of ammonoids at Mesozoic system boundaries. *Biol. Rev.*, 48: 159—194.

Wilson, J.L. and Majewske, O.P., 1960. Conjectured Middle Paleozoic history of central and west Texas. In: *Aspects of the Geology of Texas: a Symposium—Bur. Econ. Geol., Univ. Texas, Publ.*, 6017: 65—86.

Woodrow, D.L., Fletcher, F.W. and Ahrnsbrak, W.F., 1973. Paleogeography and paleoclimate at the deposition sites of the Devonian Catskill and Old Red facies. *Bull. Geol. Soc. Am.*, 84: 3051—3064.

Ziegler, A.M., and McKerrow, W.S., 1974. Silurian marine Red Beds. *J. Geol.*, (in press.)

INDEX

Abrams Mica Schist, 296
absence of intermediate forms, 332
absolute depth, 49—53
— time, 62, 63, 66, 151, 152
— — and relative time, 62—69
abundance, 27,
abundant taxa characteristic of Siluro-
    Devonian level-bottom communities,
    241—263
Acadian orogeny, 64, 296, 326
acritarchs, 50, 80, 151, 189, 191, 209,
    213, 254, 323
Acrospirifer, 176, 216, 245, 246, 248,
    251, 253, 258, 259
Aegiria, 176
Aesopomum, 176, 246, 262
agnostids, 255, 336, 338, 339
Algae, 48—50, 195, 207, 237, 241, 270
Aliconchidium, 261
Amazon-Colombian Subprovince, 140,
    210, 305, 318, 319, 321, 328—331,
    392
— — of the Eastern Americas Realm,
    316—319
Ambocoelia, 159, 176, 244—246, 262
Ambocoelid Community, 16, 18, 225,
    244, 245
Amoenospirifer, 371
ammonoid environments, 335, 336
ammonoids, 66, 80, 105, 106, 113,
    114, 187, 190, 217, 254, 335, 336,
    345,
Amosina, 245, 306,
— Community, 16, 210, 245, 306
Amphigenia, 145, 245, 304, 318
— Community, 16, 211, 214, 232, 245,
    263
Amphistrophia, 176, 249, 260, 324
Amsdenina, 249, 357, 359
Anabaia, 213

analogous communities, 54, 139, 227,
    228, 230, 236, 237, 244
Anastrophia, 248, 354
Anathyris, 245, 250, 259
—-*Pradoia* Community, 17, 245
Ancillotoechia, 176, 260
Ancylostrophia, 357
Anderson's Model, 43—47
Anoplia, 245, 317
Anopliidae, 58
Anoplotheca, 259
Anoptambonites, 205
Antirhynchonella, 176
Antler orogeny, 296
Apousiella, 371, 372
Appalachian Province, 392
— "unit", 334
appearance, 192
— rate, 3, 90, 119, 120, 146—148, 293

Appohimchi Subprovince, 16, 140,
    155, 159, 160, 183, 210, 211, 213,
    214, 222—226, 231, 250, 259, 281,
    304, 305, 311, 313, 317—319,
    321—323, 326, 328—330, 335,
    392, 394
aragonite, 288, 290
area, 8, 11, 54, 57, 61, 69, 86, 87,
    110—112, 117, 126, 131, 137, 145,
    187, 198, 343, 344
Arisaig Group, 256
— Volcanics, 64, 65, 298
Arkansas Novaculite, 276, 323
    393
Arthrophycus, 245, 248
— Community, 13, 245
articulated shells, 206, 217
Ashern Formation, 325
Ashgill, 126—128, 200, 298, 300
associations, 27

Я НЕ Я И ПОШАД НЕ МЕНЯ